SUBSTANCE
AND FUNCTION
and
EINSTEIN'S THEORY
OF RELATIVITY

Both Books Bound as One

by
ERNST CASSIRER

Authorized Translation by
WILLIAM CURTIS SWABEY, Ph.D.
and
MARIE COLLINS SWABEY, Ph.D.

DOVER PUBLICATIONS, INC.
NEW YORK

1931 Moritz Schlick founder of Vienna Circle

International Standard Book Number: 0-486-20050-7
Library of Congress Catalog Card Number: A54-10117

Manufactured in the United States of America
Dover Publications, Inc.
180 Varick Street
New York, N. Y. 10014

PREFACE

The investigations contained in this volume were first prompted by studies in the philosophy of mathematics. In the course of an attempt to comprehend the fundamental conceptions of mathematics from the point of view of logic, it became necessary to analyse more closely the function of the concept itself and to trace it back to its presuppositions. Here, however, a peculiar difficulty arose: the traditional logic of the concept, in its well-known features, proved inadequate even to characterize the problems to which the theory of the principles of mathematics led. It became increasingly evident that exact science had here reached questions for which there existed no precise correlate in the traditional language of formal logic. The content of mathematical knowledge pointed back to a fundamental form of the concept not clearly defined and recognized within logic itself. In particular, investigations concerning the concepts of the series and of the limit, the special results of which, however, could not be included in the general exposition of this book, confirmed this view and led to a renewed analysis of the principles of the construction of concepts in general.

The problem thus defined gained more general meaning when it became clear that it was in no way limited to the field of mathematics, but extended over the whole field of exact science. The systematic structure of the exact sciences assumes different forms according as it is regarded in different logical perspectives. Thus an attempt had to be made to advance from this general point of view to the forms of conceptual construction of the special disciplines,— of arithmetic, geometry, physics and chemistry. It did not accord with the general purpose of the enquiry to collect special examples from the particular sciences for the support of the logical theory, but it was necessary to make an attempt to trace their systematic structures as wholes, in order that the fundamental unitary relation by which these structures are held together might be revealed more distinctly. I did not conceal from myself the difficulty of carrying out such a plan; I finally resolved to make the attempt only because the value and significance of the preliminary work already accomplished within the special sciences became increasingly evident to me.

Particularly in the exact sciences, the investigator has turned from the special problems to the philosophical foundations with ever clearer consciousness and energy. Whatever one may judge in detail of the results of these researches, there can be no doubt that the logical problem has thereby been greatly and directly advanced. I have, therefore, sought to base the following exposition upon the historical development of science itself and upon the systematic presentation of its content by the great scientists. Although we cannot consider all the problems that arise here, nevertheless, the special logical point of view which they represent must be carried through and verified in detail. What the concept is and means in its general function can only be shown by tracing this function through the most important fields of scientific investigation and representing it in general outline.

The problem receives new meaning when we advance from purely logical considerations to the conception of *knowledge of reality*. The original opposition of thought and being breaks up into a number of different problems, which are, nevertheless, connected and held in intellectual unity by their common point of departure. Whenever, in the history of philosophy, the question as to the relation of thought and being, of knowledge and reality, has been raised, it has been dominated from the first by certain *logical* presuppositions, by certain views about the nature of the concept and judgment. Every change in this fundamental view indirectly produces a complete change in the way in which the general question is stated. The *system* of knowledge tolerates no isolated "formal" determination without consequences in all the problems and solutions of knowledge. The conception, therefore, that is formed of the fundamental nature of the concept is directly significant in judging the questions of fact which are generally considered under "Criticism of Knowledge," ("*Erkenntniskritik*") or "Metaphysics." The transformation which these questions undergo when regarded from the general point of view that is gained by criticism of the exact sciences and the new direction which their solution takes, Part II of the book attempts to show. Both parts, though seemingly separate in content, are united, nevertheless, in a philosophical point of view; both attempt to represent a single problem which has expanded from a fixed center, drawing ever wider and more concrete fields into its circle.

ERNST CASSIRER.

TRANSLATORS' PREFACE

It was thought that there was need for some comprehensive work in the English language on the philosophy of the exact sciences which would do full justice to the newer developments in mathematical and physical speculation while showing at the same time the historical connections of these tendencies. It seemed that the two works of Professor Ernst Cassirer herewith presented fulfilled these requirements best of all. The reader will find here a constructive and systematic survey of the whole field of the principles of the exact sciences from the standpoint of a logical idealism, which is historically derived from Kant, but which lacks the fatal rigidity of the latter's system. As Professor Cassirer develops his logical or critical idealism it becomes a doctrine of creative intelligence. His doctrine is neither idealism, pragmatism nor realism as these terms are understood in our English-speaking philosophy; it is rather a positivistic and non-static rationalism, which seeks to preserve the *spirit* which unites Plato, Descartes, Leibniz and Kant and to show how this spirit reaches its fulfillment in the modern development of mathematical and physical theory. *I DOUBT THIS. CASSIRER IS BOLDER THAN THIS CHARACTERIZATION NON WOULD ALLOW!*

The first part of the present book, *Substanzbegriff und Funktionsbegriff* was published in 1910, while the second part, which we have called the Supplement, *Zur Einstein'schen Relativitätstheorie*, appeared in 1921. The intervening period was, of course, one of immense importance for the philosophy of physics, since it marked the development of the new and revolutionary theory of relativity. In accordance with the fundamental maxim of his critical method Professor Cassirer based his analysis in 1910 on the historical state of science, which was still dominated by the Newtonian conceptions of space and time. On the ground of the same maxim, he has since taken account of the new theory of relativity and has, with good logical justification, seen in the latter the relative completion and realization of the historical tendency which he had described in his earlier works. Professor Cassirer's philosophy may be regarded as a fundamental epistemological "theory of relativity" which sets forth a general philosophical standpoint from which Einstein's theory is seen to be only the latest and most radical fulfillment of the motives

v

which are inherent in mathematical and physical science as such. While Professor Cassirer has had his fundamental principles confirmed rather than disproved by this recent development, his discussion in Chapter IV, Section VI, of *Substance and Function* must be taken in connection with his later statements.

With regard to the translation, the translators are aware that a good deal of the vigor and savor of the original has escaped in the process of substituting correct but colorless terms for the more vivid language of the original. Accuracy and clarity have been their chief aim. They alone are responsible for the italicized paragraph headings, which were inserted because it was thought that the book might be used as a text or reference work in connection with an advanced course in the Theory of Knowledge and that perhaps these guide-posts might help the student in finding his way through the difficult material.

Professor Cassirer himself kindly read the entire work in manuscript and, in a friendly letter, states that "*nach der Gesamteindruck* . . . *besteht für mich kein Zweifel dass der Sinn des Ganzen richtig getroffen und wiedergegeben ist.*" We wish herewith to express our hearty thanks to Professor Cassirer for permitting us to translate his works as well as for his trouble in reading the manuscript of the translation and for his courtesy in the whole transaction.

WILLIAM CURTIS SWABEY,
MARIE COLLINS SWABEY.

CONTENTS

PART I

THE CONCEPT OF THING AND THE CONCEPT OF RELATION

CHAPTER I

ON THE THEORY OF THE FORMATION OF CONCEPTS

CHAPTER II

THE CONCEPT OF NUMBER

vii

Chapter III

THE CONCEPT OF SPACE AND GEOMETRY

Chapter IV

THE CONCEPTS OF NATURAL SCIENCE

PART II

THE SYSTEM OF RELATIONAL CONCEPTS AND THE PROBLEM OF REALITY

CHAPTER V

ON THE PROBLEM OF INDUCTION

CHAPTER VI

THE CONCEPT OF REALITY

Chapter VII

SUBJECTIVITY AND OBJECTIVITY OF THE RELATIONAL CONCEPTS

Chapter VIII

ON THE PSYCHOLOGY OF RELATIONS

CONTENTS

SUPPLEMENT

EINSTEIN'S THEORY OF RELATIVITY CONSIDERED FROM THE EPISTEMOLOGICAL STANDPOINT

PART I

THE CONCEPT OF THING AND THE CONCEPT
OF RELATION

CHAPTER I

On the Theory of the Formation of Concepts

I

New developments in logic. The new view that is developing in contemporary philosophy regarding the foundations of theoretical knowledge is manifested perhaps nowhere as clearly as in the transformation of the chief doctrines of formal logic. In logic alone, philosophical thought seemed to have gained a firm foundation; in it a field seemed to be marked off that was assured against all the doubts aroused by the various *epistemological* standpoints and hypotheses. The judgment of Kant appeared verified and confirmed that here the steady and secure way of science had finally been reached. The further consideration that, as logic since the time of Aristotle had not had to retrace a single step, so also it had not advanced a single step, seemed from this point of view a confirmation of its peculiar certainty. Undisturbed by the continuous transformation of all material knowledge, it alone remained constant and without variation.

If we follow the evolution of science within the last decades more closely, however, a different picture of formal logic appears. Everywhere it is occupied with new questions and dominated by new tendencies of thought. The work of centuries in the formulation of fundamental doctrines seems more and more to crumble away; while on the other hand, great new groups of problems, resulting from the general mathematical theory of the manifold, now press to the foreground. This theory appears increasingly as the common goal toward which the various logical problems, that were formerly investigated separately, tend and through which they receive their ideal unity. Logic is thereby freed from its isolation and again led to concrete tasks and achievements. For the scope of the modern theory of the manifold is not limited to purely mathematical problems, but involves a general view which influences even the special methods of the natural sciences and is therein verified. But the systematic connection into which logic is thus drawn compels renewed criticism of its presuppositions. The appearance of unconditional certainty disappears; criticism now begins to be applied to those very doctrines

3

that have persisted unchanged historically in the face of profound changes in the ideal of knowledge.

The concept in Aristotelian logic. The Aristotelian logic, in its general principles, is a true expression and mirror of the Aristotelian metaphysics. Only in connection with the belief upon which the latter rests, can it be understood in its peculiar motives. The conception of the nature and divisions of being predetermines the conception of the fundamental forms of thought. In the further development of logic, however, its connections with the Aristotelian ontology in its special form begin to loosen; still its connection with the basic doctrine of the latter persists, and clearly reappears at definite turning points of historical evolution. Indeed, the basic significance, which is ascribed to the theory of the *concept* in the structure of logic, points to this connection. Modern attempts to reform logic have sought in this regard to reverse the traditional order of problems by placing the theory of the judgment before the theory of the concept. Fruitful as this point of view has proved to be, it has, nevertheless, not been maintained in its full purity against the systematic tendency which dominated the old arrangement. The intellectual tendency still shaping these new attempts revealed itself in that features crept into the theory of judgment itself, which could only be understood and justified by the traditional theory of the generic concept (*Gattungsbegriff*). The primacy of the concept, which they sought to lay aside, was once more implicitly acknowledged. The actual center of gravity of the system had not been changed but merely the external arrangement of its elements. Every attempt to transform logic must concentrate above all upon this one point: all criticism of formal logic is comprised in criticism of the general doctrine of the construction of concepts (*Begriffsbildung*).

Purpose and nature of the generic concept. The chief features of this doctrine are well-known and do not need detailed exposition. Its presuppositions are simple and clear; and they agree so largely with the fundamental conceptions, which the ordinary view of the world consistently uses and applies, that they seem to offer no foothold for criticism. Nothing is presupposed save the existence of things in their inexhaustible multiplicity, and the power of the mind to select from this wealth of particular existences those features that are *common* to several of them. When we thus collect objects characterized by possession of some common property into classes, and when we repeat this process upon higher levels, there gradually arises an ever firmer order and division of being, according to the series of factual similarities running through the particular things. The essential functions of thought, in this connection, are merely those of comparing and differentiating a sensuously given manifold. *Reflection*, which passes hither and thither among the particular objects in order to determine the essential features in which they agree, leads of itself to *abstraction*. Abstraction lays hold upon and raises to clear consciousness these related features,—pure, by themselves, freed from all admixture of dissimilar elements. Thus the peculiar merit of this interpretation seems to be that it never destroys or imperils the *unity* of the ordinary view of the world. The concept does not appear as something foreign to sensuous reality, but forms a *part* of this reality; it is a selection from what is immediately contained in it. In this respect, the concepts of the exact mathematical sciences stand upon the same plane as the concepts of the descriptive sciences, which are merely concerned with a superficial ordering and classification of what is given. Just as we form the concept of a tree by selecting from the totality of oaks, beeches and birch trees, the group of common properties, so, in exactly the same way, we form the concept of a plane rectangular figure by isolating the common properties which are found in the square, the right angle, the rhomboid, the rhombus, the symmetrical and asymmetrical trapezium and trapezoid, and which can be immediately seen and pointed out.[1] The well-known guiding principles of the concept follow of themselves from these foundations. Every series of comparable objects has a supreme generic concept, which comprehends within itself all the determinations in which these objects agree, while on the other hand, within this supreme genus, the sub-species at various levels are defined by properties belonging only to a part of the elements. In the same way that we ascend from the species to the higher genus by abandoning a certain characteristic, thereby drawing a larger range of objects into the circle, so by a reverse process, the specification of the genus takes place through the progressive addition of new elements of content. Hence, if we call the number of properties of a concept the magnitude of its *content*,[2] this magnitude increases as we descend from

[1] *Cf. e.g.*, Drobisch, *Neue Darstellung der Logik*, Ed. 4, Leipzig, 1875, §16 ff.; Überweg, *System der Logik*, Bonn, 1857, §51 ff.

[2] *Cf.* intension. (Tr.)

the higher concepts to the lower, and thus diminishes the number of species subordinate to the concept; while, when we ascend to the higher genus, this content will diminish as the number of species is increased. This increasing extension of the concept corresponds to a progressive diminution of the content; so that finally, the most general concepts we can reach no longer possess any definite content. The conceptual pyramid, which we form in this way, reaches its summit in the abstract representation of "something" under the all-inclusive being of which every possible intellectual content falls, but which at the same time is totally devoid of specific meaning.

The problem of abstraction. At this point in the traditional logical theory of the concept arise the first doubts concerning its universal validity and applicability. If the final goal of this method of forming concepts is entirely empty, the whole process leading to it must arouse suspicion. Such an outcome is unintelligible if the individual steps have fulfilled the requirements, which we are accustomed to make of every fruitful and concrete process of construction of scientific concepts. What we demand and expect of a scientific concept, first of all, is this: that, in the place of original indefiniteness and ambiguity of ideas, it shall institute a sharp and unambiguous determination; while, in this case, on the contrary, the sharp lines of distinction seem the more effaced, the further we pursue the logical process. And in fact, from the standpoint of formal logic itself, a new problem arises here. If all construction of concepts consists in selecting from a plurality of objects before us only the similar properties, while we neglect the rest, it is clear that through this sort of reduction what is merely a *part* has taken the place of the original sensuous *whole*. This part, however, claims to characterize and explain the whole. The concept would lose all value if it meant merely the neglect of the particular cases from which it starts, and the annihilation of their peculiarity. The act of negation, on the contrary, is meant to be the expression of a thoroughly positive process; what remains is not to be merely an arbitrarily chosen part but an "essential" moment by which the whole is *determined*. The higher concept is to make the lower intelligible by setting forth in abstraction the *ground* of its special form. The traditional rule, however, for the formation of the generic concept contains in itself no guarantee that this end will be actually achieved. In fact, there is nothing to assure us that the common properties, which we select from any arbitrary collection of objects, include the truly typical features, which characterize and determine the total structures of the members of the collection. We may borrow a drastic example from Lotze: If we group cherries and meat together under the attributes red, juicy and edible, we do not thereby attain a valid logical concept but a meaningless combination of words, quite useless for the comprehension of the particular cases. Thus it becomes clear that the general formal rule in itself does not suffice; that on the contrary, there is always tacit reference to another intellectual criterion to supplement it.

The metaphysical presuppositions of Aristotelian logic. In the system of Aristotle, this criterion is plainly evident; the gaps that are left in logic are filled in and made good by the Aristotelian metaphysics. The doctrine of the concept is the special link that binds the two fields together. For Aristotle, at least, the concept is no mere subjective schema in which we collect the common elements of an arbitrary group of things. The selection of what is common remains an empty play of ideas if it is not assumed that what is thus gained is, at the same time, the real *Form* which guarantees the causal and teleological connection of particular things. The real and ultimate similarities of things are also the creative forces from which they spring and according to which they are formed. The process of comparing things and of grouping them together according to similar properties, as it is expressed first of all in language, does not lead to what is indefinite, but if rightly conducted, ends in the discovery of the real essences of things. Thought only isolates the specific type; this latter is contained as an active factor in the individual concrete reality and gives the general pattern to the manifold special forms. The biological species signifies both the end toward which the living individual strives and the immanent force by which its evolution is guided. The logical doctrine of the construction of the concept and of definition can only be built up with reference to these fundamental relations of the real. The determination of the concept according to its next higher genus and its specific difference reproduces the process by which the real substance successively unfolds itself in its special forms of being. Thus it is this basic conception of *substance* to which the purely logical theories of Aristotle constantly have reference. The complete system of scientific defini-

tions would also be a complete expression of the substantial forces
which control reality.[3]

The concept of substance in logic and metaphysics. An understanding of Aristotle's logic is thus conditioned by an understanding of his
conception of being. Aristotle himself clearly distinguishes the
various sorts and meanings of being from each other; and it is the
essential problem of his theory of the categories to trace through
and make clear this division of being into its various subspecies.
Thus he also expressly distinguishes the existence, which is indicated
by mere relations in judgment, from existence after the fashion of a
thing; the being of a conceptual synthesis, from that of a concrete
subject. In all these quests for a sharper division, however, the
logical primacy of the concept of substance is not questioned. Only
in given, existing substances are the various determinations of being
thinkable. Only in a fixed thing-like substratum, which must first
be given, can the logical and grammatical varieties of being in general
find their ground and real application. Quantity and quality,
space and time determinations, do not exist in and for themselves,
but merely as properties of absolute realities which exist by themselves. The category of relation especially is forced into a dependent
and subordinate position by this fundamental metaphysical doctrine
of Aristotle. Relation is not independent of the concept of real being;
it can only add supplementary and external modifications to the
latter, such as do not affect its real "nature." In this way the
Aristotelian doctrine of the formation of the concept came to have a
characteristic feature, which has remained in spite of all the manifold transformations it has undergone. The fundamental categorical
relation of the thing to its properties remains henceforth the guiding
point of view; while relational determinations are only considered in
so far as they can be transformed, by some sort of mediation, into
properties of a subject or of a plurality of subjects. This view is in
evidence in the text-books of formal logic in that relations or connections, as a rule, are considered among the "non-essential" properties of a concept, and thus as capable of being left out of its definition without fallacy. Here a methodological distinction of great

[3] On the metaphysical presuppositions of the Aristotelian logic, *cf.* especially, Prantl, *Geschichte der Logik im Abendlande*, I; Trendelenburg, *Geschichte
der Kategorienlehre;* H. Maier, *Die Syllogistik des Aristoteles*, II, 2,
Tübingen, 1900, pp. 183 ff.

significance appears. The two chief forms of logic, which are especially opposed to each other in the modern scientific development, are distinguished—as will become clear—by the different value which is placed upon *thing-concepts* and *relation-concepts*.

II

The psychological criticism of the concept (*Berkeley*). If we accept this general criterion, we recognize further that the essential presupposition, upon which Aristotle founded his logic, has survived the special doctrines of the peripatetic metaphysics. In fact the whole struggle against the Aristotelian "concept realism" has been without effect upon this decisive point. The conflict between nominalism and realism concerned only the question of the metaphysical reality of concepts, while the question as to their valid logical definition was not considered. The reality of "universals" was in question. But what was beyond all doubt, as if by tacit agreement of the conflicting parties, was just this: that the concept was to be conceived as a universal genus, as the common element in a series of similar or resembling particular things. Without this mutual assumption, all conflict as to whether the common element possessed a separate factual existence or could only be pointed out as a sensuous moment in the individuals, would be essentially unintelligible. Moreover, the psychological criticism of the "abstract" concept, radical as it seems at first sight, introduces no real change here. In the case of Berkeley, we can follow in detail how his skepticism as to the worth and fruitfulness of the abstract concept implied, at the same time, a dogmatic belief in the ordinary definition of the concept. That the true scientific concept, that in particular the concepts of mathematics and physics, might have another purpose to fulfill than is ascribed to them in this scholastic definition,—this thought was not comprehended.[4] In fact, in the psychological deduction of the concept, the traditional schema is not so much changed as carried over to another field. While formerly it had been outer things that were compared and out of which a common element was selected, here the same process is merely transferred to *presentations* as psychical correlates of things. The process is only, as it were, removed to another dimension, in that it is taken out of the field of the physical

---

[4] For greater detail *cf.* my *Das Erkenntnisproblem in der Philosophie und Wissenschaft der neuern Zeit*, Vol. II, Berlin, 1907, pp. 219 ff.

into that of the psychical, while its general course and structure remain the same. When several composite presentations have a part of their content in common, there arises from them, according to the well-known psychological laws of simultaneous stimulation and fusion of the similar, a content in which merely the agreeing determinations are retained and all the others suppressed.[5] In this way no new, independent and special structure is produced, but only a certain division of presentations already given, a division in which certain moments are emphasized by a one-sided direction of attention, and in this way raised more sharply out of their surroundings. To the "substantial forms," which, according to Aristotle, represent the final goal of this comparing activity, there correspond certain fundamental elements, which run through the whole field of perception. And it is now asserted still more emphatically that these absolute elements alone, existing for themselves, constitute the real kernel of what is given and "real." Again the rôle of *relation* is limited as much as possible. Hamilton, with all his recognition of the Berkeleian theory, showed the characteristic function of relating thought. Against him, J. Stuart Mill emphasizes that the true positive being of every relation lies only in the individual members which are bound together by it, and that hence, since these members can only be given as individuals, there can be no talk of a general meaning of relation.[6] The concept does not *exist* save as a part of a concrete presentation and burdened with all the attributes of presentation. What gives it the appearance of independent value and underived psychological character is merely the circumstance that our attention, being limited in its powers, is never able to illumine the whole of the presentation and must of necessity be narrowed to a mere selection of parts. The consciousness of the concept is resolved for psychological analysis into consciousness of a presentation or part of a presentation, which is associatively connected with some word or other sensuous sign.

The psychology of abstraction. The "psychology of abstraction," according to this view, furnishes the real key to the logical meaning of every form of concept. This meaning is derived from the simple capacity of reproducing any given content of presentation. Abstract objects arise in every perceiving being in whom like

---

[5] *Cf., e.g.,* Überweg, *op. cit.,* §51.

[6] *Cf.* Mill, *An Examination of Sir William Hamilton's Philosophy,* London, 1865, p. 319.

determinations of the perceived have been given in repeated pre-sentations.[7] For these determinations are not confined to the particular moment of perception, but leave behind them certain traces of their existence in the psycho-physical subject. Since these traces, which must be thought of as unconscious during the time between the real perception and the recall, are again aroused by newly occurring stimuli of a similar sort, a firm connection is gradually formed between the similar elements of successive percep-tions. That which differentiates them tends more and more to disappear; it finally forms only a shadowy background on which the constant features stand out the more clearly. The pro-gressive solidification of these features that agree, their fusion into a unitary, indivisible whole, constitutes the psychological nature of the concept, which is consequently in origin as in function merely a totality of memory-residues, which have been left in us by percep-tions of real things and processes. The reality of these residues is shown in that they exert a special and independent influence upon the act of perception itself, in so far as every newly occurring content is apprehended and transformed according to them. Thus we stand here,—and this is sometimes emphasized by the advocates of this view themselves,—at a point of view closely akin to that of medieval "conceptualism;" real and verbal *abstracta* can only be taken from the content of perception because they are already contained in it as common elements. The difference between the ontological and psychological views is merely that the "things" of scholasticism were the beings copied in thought, while here the objects are meant to be nothing more than the contents of perception.

Weighty as this distinction may appear from the standpoint of *metaphysics*, the meaning and content of the *logical* problem are, nevertheless, not affected by it. If we remain in the sphere of this latter problem, we find here a common, fundamental belief regarding the concept, which has remained apparently unassailable throughout all the changes of the question. Yet precisely where there seems no conflict of opinion, the real methodological difficulty begins. Is the theory of the concept, as here developed, an adequate and faithful picture of the procedure of the concrete sciences? Does it include and characterize all the special features of this procedure; and is it able to represent them in all their mutual connections and specific

---

[7] *Cf.* especially B. Erdmann, *Logik*, Ed. 2, pp. 65 ff., 88 ff.

characters? With regard to the Aristotelian theory, at least, this question must be answered negatively. The concepts, which are Aristotle's special object and interest, are the generic concepts of the descriptive and classifying natural sciences. The "form" of the olive-tree, the horse, the lion, is to be ascertained and established. Wherever he leaves the field of *biological* thought, his theory of the concept at once ceases to develop naturally and freely. From the beginning, the concepts of geometry, especially, resist reduction to the customary schema. The concept of the point, or of the line, or of the surface cannot be pointed out as an immediate *part* of physically present bodies and separated from them by simple "abstraction." Even in this example, which is the simplest offered by exact science, logical technique faces a new problem. Mathematical concepts, which arise through genetic definition, through the intellectual establishment of a *constructive* connection, are different from empirical concepts, which aim merely to be copies of certain factual characteristics of the given reality of things. While in the latter case, the multiplicity of things is given in and for itself and is only drawn together for the sake of an abbreviated verbal or intellectual expression, in the former case we first have to create the multiplicity which is the object of consideration, by producing from a simple act of construction (*Setzung*), by progressive synthesis, a systematic connection of thought-constructions (*Denkgebilden*). There appears here in opposition to bare "abstraction," an act of thought itself, a free production of certain relational systems. It can easily be understood that the logical theory of abstraction, even in its modern forms, has frequently attempted to obliterate this opposition, for it is at this point that questions as to the value and inner unity of the theory of abstraction must be decided. But this very attempt leads at once to a transformation and disintegration of the theory, in whose favor it was undertaken. The doctrine of abstraction loses either its universal validity or the specific logical character that originally belonged to it.

Mill's analysis of mathematical concepts. Thus Mill, for instance, in order to maintain the unity of the supreme principle of experience, explains mathematical truths and concepts as also mere expressions of concrete physical matters of fact. The proposition that $1 + 1 = 2$ merely describes an experience which has been forced upon us by the process of joining things together. In another sort of world of objects, in a world, for example, in which by the combination of two things, a third always came into being of itself, it would lose significance and validity. The same is true of the axioms concerning spatial relations; a "round square" is only a contradictory concept for us because it has been shown in experience without exception, that a thing loses the property of having four corners the moment it assumes the property of roundness, so that the beginning of one impression is inseparably connected with the cessation of the other. According to this mode of explanation, geometry and arithmetic seem again resolved into mere statements concerning certain groups of presentations. But this interpretation fails when Mill further attempts to justify the value and peculiar significance, inherent in these special experiences of numbering and measuring, in the whole of our knowledge. Here, first of all, reference is made to the accuracy and trustworthiness of the images, which we retain of spatial and temporal relations. The reproduced presentation is, in this case, similar to the original in all details, as a varied experience has shown; the image that the geometrician frames corresponds perfectly to the original impression. In this way it can be conceived that, in order to reach new geometrical or arithmetical truths, we do not need each time renewed perceptions of physical objects; the memory-image, by virtue of its clarity and distinctness, is able to supplant the sensible object itself. However, this explanation is at once crossed by another. The peculiar "deductive" certainty, which we ascribe to mathematical propositions, is now traced back to the fact that in these propositions we are never concerned with statements. about concrete facts, but only with relations between *hypothetical* forms. There are no real things which precisely agree with the definitions of geometry; there are no points without magnitude, no perfectly straight lines, no circles whose radii are all equal. Moreover, from the standpoint of our experience, not only the actual reality, but the very possibility of such contents must be denied; it is at least excluded by the physical properties of our planet, if not by those of the universe. But psychical existence is denied no less than physical to the objects of geometrical definitions. For in our mind we never find the presentation of a mathematical point, but always only the smallest possible sensible extension; also we never "conceive" a line without breadth, for every psychical image we can

form shows us only lines of a certain breadth.[8]   It is evident at once
that this double explanation destroys itself.   On the one hand,
all emphasis is laid upon the similarity between mathematical ideas
and the original impressions; on the other, however, it is seen that
this sort of similarity does not and cannot exist, at least for those
forms which alone arc defined and characterized as "concepts"
in the mathematical sciences.   These forms cannot be attained by bare
selection from the facts of nature and presentation, for they possess no
concrete correlative in all of these facts.   "Abstraction," as it has
hitherto been understood, does not *change* the constitution of consciousness and of objective reality, but merely institutes certain
limits and divisions in it; it merely divides the parts of the sense-
impression but adds to it no new datum.   In the definitions of pure
mathematics, however, as Mill's own explanations show, the world of
sensible things and presentations is not so much reproduced as
transformed and supplanted by an order of another sort.   If we
trace the method of this transformation, certain forms of relation,
or rather an ordered system of strictly differentiated intellectual
functions, are revealed, such as cannot even be characterized, much
less justified, by the simple schema of "abstraction."   And this
result is also confirmed if we turn from the purely mathematical
concepts to those of theoretical physics.   For in their origin the
same process is shown, and can be followed in detail, of the transformation of the concrete sensuous reality,—a process which the
traditional doctrine cannot justify.   These concepts of physics also
are not intended merely to produce copies of perceptions, but to
put in place of the sensuous manifold another manifold, which agrees
with certain theoretical conditions.[9]

The defect of the psychological theory of abstraction.   Neglecting the
nature of abstract concepts, however, we find that the naïve view of
the world, to which the traditional logical conception especially
appeals and upon which it rests, conceals within itself what is ultimately the same problem.   The concepts of the manifold species
and genera are supposed to arise for us by the gradual predominance
of the similarities of things over their differences, *i.e.*, the similarities
alone, by virtue of their many appearances, imprint themselves upon

---

[8] *Cf.* Mill, *A System of Logic*, Ed. 7, London, 1868, Book II, Ch. 5, and Book
III, Ch. 24.
[9] *Cf.* more particularly Ch. IV.

the mind, while the individual differences, which change from case to case, fail to attain like fixity and permanence. The similarity of things, however, can manifestly only be effective and fruitful, if it is understood and judged *as such*. That the "unconscious" traces left in us by an earlier perception *are* like a new impression in point of fact, is irrelevant to the process implied here as long as the elements are not *recognized* as similar. By this, however, an act of identification is recognized as the foundation of all "abstraction." A characteristic function is ascribed to thought, namely, to relate a present content to a past content and to comprehend the two as in some respect identical. This synthesis, which connects and binds together the two temporally separated conditions, possesses no immediate sensible correlate in the contents compared. According to the manner and direction in which the synthesis takes place, the same sensuous material can be apprehended under very different conceptual forms. The psychology of abstraction first of all has to postulate that perceptions can be ordered for logical consideration into "series of similars." Without a process of arranging in series, without running through the different instances, the consciousness of their generic connection—and consequently of the abstract object—could never arise. This transition from member to member, however, manifestly presupposes a *principle* according to which it takes place, and by which the form of dependence between each member and the succeeding one, is determined. Thus from this point of view also it appears that all construction of concepts is connected with some definite form of construction of series. We say that a sensuous manifold is conceptually apprehended and ordered, when its members do not stand next to one another without relation but proceed from a definite beginning, according to a fundamental generating relation, in necessary sequence. It is the *identity* of this generating relation, maintained through changes in the particular contents, which constitutes the specific form of the concept. On the other hand, whether from the retention of this identity of relation there finally evolves an abstract *object*, a general *presentation* in which similar features are united, is merely a psychological side-issue and does not affect the logical characterization of the concept. The appearance of a general image of that sort may be excluded by the nature of the generating relation, without the definitive moment in the clear deduction of each element from the preceding being thereby removed.

In this connection, the real weakness of the theory of abstraction is apparent in the one-sidedness of its selection, from the wealth of possible principles of logical order, of merely the principle of similarity. In truth, it will be seen that a series of contents in its conceptual ordering may be arranged according to the most divergent points of view; but only provided that the guiding point of view itself is maintained unaltered in its qualitative peculiarity. Thus side by side with series of similars in whose individual members a common element uniformly recurs, we may place series in which between each member and the succeeding member there prevails a certain degree of difference. Thus we can conceive members of series ordered according to equality or inequality, number and magnitude, spatial and temporal relations, or causal dependence. The *relation of necessity* thus produced is in each case decisive; the concept is merely the expression and husk of it, and is not the generic presentation, which may arise incidentally under special circumstances, but which does not enter as an effective element into the definition of the concept.

The forms of series. Thus analysis of the theory of abstraction leads back to a deeper problem. The "comparison" of contents, here referred to, is primarily only a vague and ambiguous expression, which hides the difficulty of the problem. In truth, very different *categorical functions* are here united under what is merely a collective name. And the real task of logical theory with regard to any definite concept consists in setting forth these functions in their essential characteristics and in developing their formal aspects. The theory of abstraction obscures this task since it confuses the categorical forms, upon which rests all definiteness of the content of perception, with *parts* of this very content itself. And yet even the most simple *psychological* reflection shows that the "likeness" between any contents is not itself given as a further content; that similarity or dissimilarity does not appear as a special element of sensation side by side with colors and tones, with sensations of pressure and touch. The ordinary schema of the construction of concepts, therefore, calls for a thorough-going transformation, even in its outer form; for in it the qualities of things and the pure aspect of relation are placed on the same level and fused without distinction. Once this identification has taken place, it can indeed appear as if the work of thought were limited to selecting from a series of perceptions $a\alpha$, $a\beta$,

$a\gamma$ . . . . the common element a. In truth, however, the connection of the members of a series by the possession of a common "property" is only a special example of logically possible connections in general. The connection of the members is in every case produced by some general *law of arrangement* through which a thoroughgoing rule of succession is established. That which binds the elements of the series $a, b, c,$ . . . . together is not itself a new element, that was factually blended with them, but it is the rule of progression, which remains the same, no matter in which member it is represented. The function $F(a, b)$, $F(b, c)$, ... , which determines the sort of dependence between the successive members, is obviously not to be pointed out as itself a member of the series, which exists and develops according to it. The unity of the conceptual content can thus be "abstracted" out of the particular elements of its extension only in the sense that it is in connection with them that we become conscious of the specific rule, according to which they are related; but not in the sense that we construct this rule *out* of them through either bare summation or neglect of parts. What lends the theory of abstraction support is merely the circumstance that it does not presuppose the contents, out of which the concept is to develop, as *disconnected particularities*, but that it tacitly thinks them in the form of an ordered manifold from the first. The concept, however, is not deduced thereby, but presupposed; for when we ascribe to a manifold an order and connection of elements, we have already presupposed the concept, if not in its complete form, yet in its fundamental function.

The place of the thing-concept in the system of logical relations. There are two different lines of consideration in which this logical presupposition is plainly evident. On the one side, it is the category of the whole and its parts; on the other, the category of the thing and its attributes, of which application is made in the customary doctrine of the origin of the generic concept. That objects are given as organizations of particular attributes, and that the total groups of such attributes are divided into parts and sub-parts, which are common to several of them, is here taken as the self-evident, basic principle. In truth, however, the "given" is not thereby merely described, but is judged and shaped according to a certain conceptual contrast. But as soon as this is recognized it must become evident that we stand here before a mere beginning that points beyond itself. The categorical acts (*Akte*), which we characterize by the concepts of the whole and its parts, and of the thing and its attributes, are not isolated but belong to a *system* of logical categories, which moreover they by no means exhaust. After we have conceived the plan of this system in a general logical theory of relations, we can, from this standpoint, determine its details. On the other hand, it is not possible to gain a view of all possible forms of connection from the limited standpoint of certain relations emphasized in the naïve view of the world. The category of the thing shows itself unsuited for this purpose in the very fact that we have in pure mathematics a field of knowledge, in which things and their properties are disregarded in principle, and in whose fundamental concepts therefore, no general property of things can be contained.

III

The negative process of "abstraction." At this point, a new and more general difficulty arises to threaten the traditional logical doctrine. If we merely follow the traditional rule for passing from the particular to the universal, we reach the paradoxical result that thought, in so far as it mounts from lower to higher and more inclusive concepts, moves in mere *negations*. The essential act here presupposed is that we drop certain determinations, which we had hitherto held; that we abstract from them and exclude them from consideration as irrelevant. What enables the mind to form concepts is just its fortunate gift of forgetfulness, its inability to grasp the individual differences everywhere present in the particular cases. If all the memory images, which remained with us from previous experiences, were fully determinate, if they recalled the vanished content of consciousness in its full, concrete and living nature, they would never be taken as completely similar to the new impression and would thus not blend into a unity with the latter. Only the inexactness of reproduction, which never retains the whole of the earlier impression but merely its hazy outline, renders possible this unification of elements that are in themselves dissimilar. Thus all formation of concepts begins with the substitution of a generalized image for the individual sensuous intuition, and in place of the actual perception the substitution of its imperfect and faded remainder.[10]

[10] *Cf.*, Sigwart, *Logik*, Ed. 2, p. 50 f., also, H. Maier, *Psychologie des emotionalen Denkens*, Tübingen, 1908, pp. 168 ff.

If we adhere strictly to this conception, we reach the strange result that all the logical labor which we apply to a given sensuous intuition serves only to separate us more and more from it. Instead of reaching a deeper comprehension of its import and structure, we reach only a superficial schema from which all peculiar traits of the particular case have vanished.

The mathematical concept and its "concrete universality." But from any such conclusion we are once more safeguarded by consideration of that science in which conceptual definiteness and clarity have reached their highest level. It is at this point, indeed, that the mathematical concept appears most sharply distinguished from the ontological concept. In the methodological struggle over the limits of mathematics and ontology, which took place in the philosophy of the eighteenth century, this relation was incidentally given happy and significant expression. In his criticism of the logic of the Wolffian school, Lambert pointed out that it was the exclusive merit of mathematical "general concepts" not to cancel the determinations of the special cases, but in all strictness fully to retain them. When a mathematician makes his formula more general, this means not only that he is *to retain* all the more special cases, but also be able *to deduce* them from the universal formula. The possibility of deduction is not found in the case of the scholastic concepts, since these, according to the traditional formula, are formed by neglecting the particular, and hence the reproduction of the particular moments of the concept seems excluded. Thus abstraction is very easy for the "philosopher," but on the other hand, the determination of the particular from the universal so much the more difficult; for in the process of abstraction he leaves behind all the particularities in such a way that he cannot recover them, much less reckon the transformations of which they are capable.[11] This simple remark contains, in fact, the germ of a distinction of great consequence. The ideal of a *scientific* concept here appears in opposition to the schematic general presentation which is expressed by a mere *word*. The genuine concept does not disregard the peculiarities and particularities which it holds under it, but seeks to show the *necessity* of the occurrence and connection of just these particularities.

---

[11] Lambert, *Anlage zur Architektonik oder Theorie des Einfachen und des Ersten in der philosophischen und mathematischen Erkenntnis*, Riga, 1771, §193 ff. *Cf.* my *Erkenntnisproblem in der Philosophie und Wissenschaft der neuern Zeit*, Vol. II, p. 422 f.

What it gives is a universal *rule* for the connection of the particulars themselves. Thus we can proceed from a general mathematical formula,—for example, from the formula of a curve of the second order,—to the special geometrical forms of the circle, the ellipse, etc., by considering a certain parameter which occurs in them and permitting it to vary through a continuous series of magnitudes. Here the more universal concept shows itself also the more rich in content; whoever has it can deduce from it all the mathematical relations which concern the special problems, while, on the other hand, he takes these problems not as isolated but as in continuous connection with each other, thus in their deeper systematic connections. The individual case is not excluded from consideration, but is fixed and retained as a perfectly determinate step in a general process of change. It is evident anew that the characteristic feature of the concept is not the "universality" of a presentation, but the universal validity of a principle of serial order. We do not isolate any abstract part whatever from the manifold before us, but we create for its members a definite relation by thinking of them as bound together by an inclusive law. And the further we proceed in this and the more firmly this connection according to laws is established, so much the clearer does the unambiguous determination of the particular stand forth. Thus, for example, the intuition of our Euclidian three-dimensional space only gains in clear comprehension when, in modern geometry, we ascend to the "higher" forms of space; for in this way the total axiomatic structure of our space is first revealed in full distinctness.

The criticism of the theories of abstraction. Modern expositions of logic have attempted to take account of this circumstance by opposing,—in accordance with a well-known distinction of Hegel's,—the abstract universality of the concept to the concrete universality of the mathematical formula. Abstract universality belongs to the genus in so far as, considered in and for itself, it neglects all specific differences; concrete universality, on the contrary, belongs to the systematic whole (*Gesamtbegriff*) which takes up into itself the peculiarities of all the species and develops them according to a rule. "When, *e.g.*, algebra solves the problem of finding two whole numbers, whose sum is equal to 25, and of which one is divisible by 2 and the other by 3, by expressing the second by the formula $6z + 3$, in which z can only have the values 0, 1, 2, 3, and from which of itself $22-6z$ follows as a formula of the first, these formulae possess concrete universality. They are universal because they represent the law which determines all the numbers sought; they are also concrete because, when z is given the four above-mentioned values, the numbers sought for follow from these formulae as species of them. The same is true in general of every mathematical function of one or more variables. Every mathematical function represents a universal law, which, by virtue of the successive values which the variable can assume, contains within itself all the particular cases for which it holds."[12] If, however, this is once recognized, a completely new field of investigation is opened for logic. In opposition to the logic of the generic concept, which, as we saw, represents the point of view and influence of the concept of substance, there now appears the *logic of the mathematical concept of function.* However, the field of application of this form of logic is not confined to mathematics alone. On the contrary, it extends over into the field of the *knowledge of nature;* for the concept of function constitutes the general schema and model according to which the modern concept of nature has been molded in its progressive historical development.

Before we proceed to trace the construction of functional concepts in science itself and thus to verify our new conception of the concept by concrete examples, we may indicate the meaning of the problem by citing a characteristic turn recently taken by the theory of abstraction. Everywhere a new motive is apparent, which, if systematically developed and carried through, will raise logical questions that extend beyond the traditional point of view. An indication of this motive is to be found in the first place in Lotze's skeptical comments on the traditional doctrine of abstraction. As he explains, the real practice of thought in the formation of concepts does not follow the course prescribed by this doctrine; for it is never satisfied to advance to the universal concept by neglecting the particular properties *without retaining an equivalent for them.* When we form the concept of metal by connecting gold, silver, copper and lead, we cannot indeed ascribe to the abstract object that thus comes into being the particular color of gold, or the particular luster of silver, or the weight of copper, or the density of lead; however, it would be no less inadmissible if we simply attempted to deny all these particular determinations of it. For the idea obviously does not suffice as a characterization of metal, that it is neither red nor yellow, neither of this or that

[12] Drobisch, *Neue Darstellung der Logik,* p. 22.

specific weight, neither of this or that hardness or resisting power; but the positive thought must be added that it is colored in *some* way in every case, that it is of *some* degree of hardness, density and luster. And analogously, we would not retain the general concept of animal, if we abandoned in it all thought of the aspects of procreation, of movement and of respiration, because there is no form of procreation, of breathing, etc., which can be pointed out as common to all animals. It is not, therefore, the simple neglect of the "marks" p_1p_2, q_1q_2, that are different in the different species, which is the rule of abstraction; but always, in the place of the neglected particular determinations, the general "marks" P and Q must be set up, the particular species of which are p_1p_2 and q_1q_2. The merely negative procedure, on the contrary, would lead in the end to the denial of all determination, so that our thought would find no way of return from the logical "nothing" which the concept would then signify.[13] We see here how Lotze, on the basis of psychological considerations, approaches the problem which Lambert had clearly and definitely formulated, using the example of the mathematical concept. If we carry through the above rule to the end, it obliges us to retain, in place of the particular "marks" which are neglected in the formation of the concept, the systematic totality (*Inbegriff*) to which those marks belong as special determinations. We can abstract from the particular color only if we retain the total series of colors in general as a fundamental schema, with respect to which we consider the concept determined, which we are forming. We represent this systematic totality (*Inbegriff*) when we substitute for the *constant* particular "marks," *variable* terms, such as stand for the total group of possible values which the different "marks" can assume. Thus it becomes evident that the falling aside of the particular determinations is only in appearance a purely negative process. In truth, what seems to be cancelled in this way is maintained in another form and under a different logical category. As long as we believe that all determinateness consists in constant "marks" in things and their attributes, every process of logical generalization must indeed appear an impoverishment of the conceptual content. But precisely to the extent that the concept is freed of all thing-like being, its peculiar functional character is revealed. Fixed properties are replaced by universal rules that permit us to survey a total series of possible determinations

[13] Lotze, *Logik*, Ed. 2, Leipzig, 1880, p. 40 f.

at a single glance. This transformation, this change into a new form of logical being, constitutes the real positive achievement of abstraction. We do not proceed from a series $a\alpha_1\beta_1$, $a\alpha_2\beta_2$, $a\alpha_3\beta_3$.... directly to their common constitutive a, but replace the totality of individual members α by a variable expression x, the totality of individual members β by a variable expression y. In this way we unify the whole system in the expression $a\ x\ y\ \ldots$ which can be changed into the concrete totality (*Allheit*) of the members of the series by a continuous transformation, and which therefore perfectly represents the structure and logical divisions of the concept.

Objects of the "first" and "second" orders. This turn of thought can be traced even in those expositions of logic that, in fundamental tendency, retain the traditional theory of abstraction. It is significant of this tendency, for instance, that Erdmann, after completing his psychological theory of the concept, finds himself forced by his consideration of the mathematical manifold to the introduction of a new point of view and a new terminology. The *first* phase of every construction of concepts, he now teaches, does indeed involve the separating out of a certain universal on the basis of the uniformity with which its content recurs amid varying particulars; but this uniformity of the given, though perhaps the original, is not the *sole* condition which enables us to mark off the objects of our presentations. In the progress of thought, the consciousness of uniformity is rather supplemented by the consciousness of necessary connection; and this supplementation goes so far that ultimately we are not dependent upon number of repetitions to establish a concept. "Wherever in developed presentation a composite object is found in our perception, which takes its place as a well-defined member of a series of presentations, such as a new shade in the series of colors, a new chemical compound in a series of known compounds of similar constitution, there a single occurrence suffices to fix it in its definite character as a member of the series, even in case we never perceive it again."[14] In contrast to objects of sense-perception, which we can designate as "objects of the first order," there now appear "objects of the second order," whose logical character is determined solely by the form of connection from which they proceed. In general, wherever we unify the objects of our thought into a single object, we create a new "object of the second order," whose total content is expressed in the

[14] B. Erdmann, *Logik*, Ed. 2, p. 158. f.

relations established between the individual elements by the act of unification. This type of thought, to which Erdmann declares he was led by the problems of the modern theory of groups, breaks through the old schema of the formation of concepts; for instead of the community of "marks," the unification of elements in a concept is decided by their "connection by implication." And this criterion, here only introduced by way of supplement and as a secondary aspect, proves on closer analysis to be the real logical *prius;* for we have already seen that "abstraction" remains aimless and unmeaning if it does not consider the elements from which it takes the concept to be from the first arranged and connected by a certain relation.

The variety of objective "intentions." In general, as the purely logical aspect of the concepts of relation and of the manifold becomes clearer, a greater need is felt for a new psychological foundation. If the *objects* with which pure logic deals are not identical with the individual contents of perception, but possess their own structure and "essence," then the question must arise as to how this peculiar "essence" comes to our consciousness and by what acts it is grasped. It is clear that the mere sensuous experiences, however much we heap them up and however much we complicate them, can never suffice for this purpose. For sensuous experience is concerned exclusively with a particular object or with a plurality of such objects; no summation of individual cases can ever produce the specific unity which is *meant* in the concept. The theory of attention, therefore, as the truly creative faculty in the formation of concepts, loses all application in a deeper phenomenology of the pure thought processes. For attention only separates or connects elements already given in perception; it can give these elements no new meaning and invest them with no new logical function. It is such a change of function, however, which first transforms the contents of perception and presentation into concepts in the logical sense. From the standpoint of purely descriptive analysis of conscious process also, it is something different when I grasp this or that particular property of a thing, as when for example I select from the perceptual complex of a house its special red color, than when I contemplate "the" red as a species. There is a difference between making valid mathematical judgments concerning the number "four," thereby placing it in an objective connection of relations, and directing consciousness upon a concrete group of things or presentations of four elements. The logical determination of "four" (in the first case) is given by its place in an ideal and therefore timelessly valid whole of relations, by its place in a mathematically defined number-system; but sensuous presentation, which is necessarily limited to a particular 'here' and 'now,' is unable to reproduce this form of determination. Here the psychology of thought strives to make a new advance. By the side of what the content *is* in its material, sensuous structure, there appears what it *means* in the system of knowledge; and thus, its meaning develops out of the various logical "acts" which can be attached to the content. These "acts," which differentiate the sensuously unitary content by imprinting upon it different objectively directed "intentions," are psychologically completely underived; they are peculiar forms of consciousness, such as cannot be reduced to the consciousness of sensation or perception. If we are still to speak of abstraction as that to which the concept owes its being, nevertheless its meaning is now totally different from that of the customary sensationalistic doctrine; for abstraction is no longer a uniform and undifferentiated attention to a given content, but the intelligent accomplishment of the most diversified and mutually independent acts of thought, each of which involves a particular sort of *meaning* of the content, a special direction of objective reference.[15]

The serial form and the members of the series. Thus the circle of our subject is complete, since we are led, from the side of "subjective" analysis, from the pure phenomenology of consciousness, to the same fundamental distinction, the validity of which has been shown in the "objective" logical investigation. While the empiristic doctrine regards the "similarity" of certain contents of presentation as a self-evident psychological fact which it applies in explaining the formation of concepts, it is justly pointed out in opposition that the similarity of certain elements can only be spoken of significantly when a certain "point of view" has been established from which the elements can be designated as like or unlike. This identity of reference, of point of view, under which the comparison takes place, is, however, something distinctive and new as regards the compared contents themselves. The difference between these contents, on the one hand, and the conceptual "species," on the other, by which we unify them, is an irreducible fact; it is categorical and belongs to the

[15] On this whole subject *cf.* Husserl, *Logische Untersuchungen*, Vol. II, No. II, Halle, 1901: *Die ideale Einheit der Species und die neuern Abstractionstheorien.*

"form of consciousness." In fact, it is a new expression of the characteristic contrast between the *member of the series* and the *form of the series*. The content of the concept cannot be dissolved into the elements of its extension, because the two do not lie on the same plane but belong in principle to different dimensions. The meaning of the *law* that connects the individual members is not to be exhausted by the enumeration of any number of instances of the law; for such enumeration lacks the generating *principle* that enables us to connect the individual members into a functional whole. If I know the relation according to which $a\ b\ c\ .\ .\ .$ are ordered, I can deduce them by reflection and isolate them as objects of thought; it is impossible, on the other hand, to discover the special character of the connecting relation from the mere juxtaposition of a, b, c in presentation. (*Cf.* above pp. 16 ff.) In this conception there is no danger of hypostasizing the pure concept, of giving it an independent reality along with the particular things. The serial form $F(a, b, c\ .\ .\ .)$. which connects the members of a manifold obviously cannot be thought after the fashion of an individual a or b or c, without thereby losing its peculiar character. Its "being" consists exclusively in the logical determination by which it is clearly differentiated from other possible serial forms ϕ, ψ . . . .; and this determination can only be expressed by a synthetic act of definition, and not by a simple sensuous intuition.

These considerations indicate the direction of the following investigation. The totality and order of pure "serial forms" lies before us in the system of the sciences, especially in the structure of exact science. Here, therefore, the theory finds a rich and fruitful field, which can be investigated with respect to its logical import independently of any metaphysical or psychological presuppositions as to the "nature" of the concept. This independence of pure logic, however, does not mean its isolation within the system of philosophy. Even a hasty glance at the evolution of "formal" logic would show how the dogmatic inflexibility of the traditional forms begins to yield. And the new form that is beginning to take shape, is also a form for a new content. Psychology and criticism of knowledge, the problem of consciousness and the problem of reality, both take part in this process. For in fundamental problems there are no absolute divisions and limits; every transformation of the genuinely "formal" concept produces a new interpretation of the whole field that is characterized and ordered by it.

CHAPTER II

The Concept of Number

I

Among the fundamental concepts of pure science the concept of number stands in the first place, both historically and systematically. It is in connection with it that consciousness of the meaning and value of the formation of concepts first develops. In the thought of number all the power of knowledge seems contained, all possibility of the logical determination of the sensuous. If there were no number, nothing could be understood in things, either in themselves or in their relations to each other. This Pythagorean doctrine remains unchanged in its real import through all the changes in philosophical thought. The claim to grasp the substance of *things* in number has indeed been gradually withdrawn; but at the same time the insight has been deepened and clarified that in number is rooted the substance of rational knowledge. Even when the metaphysical kernel of the object is no longer seen in it, the concept of number remains the first and truest expression of rational method in general. In it are directly mirrored the differences in principle between the fundamental interpretations of knowledge. Through number the general ideal of knowledge gains a more definite form, in which for the first time it is defined with full clarity.

The sensationalistic deduction of number. Thus it is quite intelligible that we should meet upon the threshold of algebra the same typical opposition that was traceable in the field of logic. If we accept the traditional logical view, we should expect to find certain fundamental properties of objects revealed in the numerical concepts. The theory of abstraction provides, strictly speaking, for no other point of view. Just as objects are differentiated according to size and form, according to smell and taste, so also, on the theory of abstraction, they must have a certain property which gives them their numerical character. The concept of "two" or "three" would be abstracted from a plurality of objective groups, just as the concept of a certain color arises from the comparison of colored perceptual things. It is consistent, from this standpoint, to regard all assertions concerning numbers and numerical relationships as expressive of certain physical properties of objects. In the modern development of empiricism, this latent consequence has for the first time fully come to light. Thus, according to J. S. Mill, the proposition that $2 + 1 = 3$ represents no mere definition, no mere fixation of the meaning which we are to connect with the concepts of two and three, but it reports an empirical matter of fact which our spatial perception has hitherto always presented in the same way. We have always been able when we have seen three things before us in a certain arrangement,—for example, in the form 0^00,—to analyse them into partial groups of the sort 00, 0. Three pebbles do not make the same impression on our senses when they lie before us in two separate piles as when they are collected into a single pile. Hence the assertion, that the perception that arises in the first case can always be transformed through mere spatial re-arrangement of its parts into the second perception, is no mere identical proposition that says nothing, but an inductive truth learned through early experience and which has since been continually confirmed. Such truths constitute the foundation of the science of number. The appearance of ideality that attaches to this science must, therefore, disappear. The propositions of arithmetic lose their former exceptional position; they come to be on the same plane as other physical observations that we have made concerning separations and combinations in the world of bodies. For how can there be significant and valid judgments that have no reference to sensible facts? The concept of ten either means nothing or it means a certain uniform total impression that is always found in groups of ten bodies, ten tones or ten pulse-beats. And that the various impressions thus gained from objects form a system among themselves, in which certain constant relations prevail, is likewise a proposition possessing merely empirical certainty. A reality of another sort, a new physical environment into which we were thrust, could make the proposition that $2 \times 2 = 5$ just as familiar and self-evident to us as it now seems unintelligible and absurd.[1]

Frege's foundations of arithmetic. With this first step into the field of exact scientific problems, we can already see very clearly what real meaning and importance may be contained in what appear

---

[1] *Cf.* Mill, *System of Logic*, Bk. II, Ch. 6; *An Examination of Sir William Hamilton's Philosophy*, pp. 67 ff.

to be purely formal logical differences. For, however we may judge Mill's theory of the fundamental arithmetical principles, we must recognize that it follows with convincing necessity from his general interpretation of the concept. So much the more significant is it that the theory, when carried through, leads to a direct conflict with the *fact* of scientific arithmetic itself. Wherever an attempt has been made in modern mathematics to analyse and explain the fact of scientific arithmetic, the logical structure of pure number has been sharply distinguished from Mill's arithmetic of "pebbles and gingerbread nuts," thereby guarding against a possible illusion. Indeed, if Mill's deduction were correct, the arithmetical concepts would be deprived of the very determinateness which constitutes their peculiar value and import. The logical differences of numbers would be limited and restricted to whatever psychological power of differentiation we had attained in the apprehension of given groups of objects. The absurdity of this consequence can, however, be easily seen. The *number* 753,684 is just as definitely and clearly differentiated from the number that immediately precedes or follows it as three is from two or four; but who could point out the "impression" which differentiates the sensuous intuitions of the two concrete groups from each other? And in the same way that the characteristic content of the numerical concepts is lost here, so, on the other hand, we lose the scope and freedom of application that is essential to it. The synthesis of numbering can only take place, according to Mill, where the combining and separating instituted by it can actually be carried out with physical objects, where things themselves can be collected and separated into perceptible spatial groups. The varying images, which arise in us from the different groups, constitute the real and indispensable basis of all assertions concerning numerical relations. Hence beyond the field of spatial sensuous intuition, wherein these actual combinations and separations alone are possible, the real foundation of the numerical concepts would be lacking. But in truth, we speak not merely of the number of seeds in a pile, but of the number of categories, of the number of Kepler's laws, or of the number of factors of energy: all objects which cannot be arranged side by side and separated from each other like pebbles. "It would indeed be strange," remarks Frege, in his drastic and pertinent criticism of Mill's doctrine," if a property abstracted from outer things could be carried over without change of meaning to experiences, presentations and concepts. It would be precisely as if one were to speak of a fusible experience, of a blue presentation, of a salty concept or a sticky judgment. It is absurd that what is by nature sensible should present itself in connection with the insensible. When we see a blue surface, we have a peculiar impression which corresponds to the word blue; and this we recognize again when we look at another blue surface. If we assume that in the same way in looking at a triangle there is something sensuous which corresponds to the word "three" then this sensuous element must also be found in three concepts; something non-sensuous would have something sensuous as a property. It can be granted that there is a kind of sensuous impression corresponding to the word "triangular" in case we take the word as a whole. We do not see the three in it immediately, but we see something to which an intellectual activity can attach, which leads to a judgment in which the number 3 appears."[2]

The system of arithmetic. If the absurdities inevitably implied in the sensationalistic interpretation of number do not come to light directly in the first deduction, the reason lies in the fact that intellectual activities, processes of *judgment*, are not entirely excluded but are tacitly assumed. According to this theory, only the first truths of arithmetic, only the most elementary formulae, are to be the result of the immediate observation of physical facts, while the scientific system of algebra is not to rest upon the continually renewed influx of facts of perception, but upon a "generalization" of the original sensuous facts. But this conception involves all the riddles which the theory promised to solve. When we attempt to give such a conception clear and definite meaning, we find that it directly implies a plurality of different intellectual functions which partake in the construction of number. If it is to be possible to carry over observations that we have made with smaller complexes of objects progressively to larger and ever larger complexes, and to determine the properties of the latter after the analogy of the former, we must assume that some form of connection and dependence exists between the compared cases, by virtue of which the one can be deduced from the other. We would not have the right to extend any determination, which appears to us in any individual group, to groups of a

[2] Frege, *Die Grundlagen der Arithmetik.* Breslau, 1884, p. 31 f. For the whole matter *cf.* especially pp. 9 ff., 27 ff.

larger or smaller number of elements, if we did not comprehend them all as similar in "nature." This similarity, however, means nothing more than that they are connected by a definite rule, such as permits us to proceed from one manifold to another by continued identical application of the same fundamental relation. In fact, without the assumption of such a connection we should have to be prepared for the possibility that any unity, added to or substracted from a given group, would alter the total character of the group so that no inference would be possible from the relations of one group to those of any other. The new unities would then act like as many physical conditions or forces, which could completely transform the whole and cancel its fundamental characteristics. No generally applicable laws, no thorough-going relation would bind together the members of the realm of number; on the contrary, every arithmetical proposition would have to be specially proved by observation or perception for every individual number. The sensualistic theory is only able to avoid this consequence by an unnoticed deviation into another line of consideration. The demand for a generalization of primitive experiences of number contains, although concealed, that very function of the universality of number concepts, which was to have been set aside by the explanation. The road to a purely deductive construction of the realm of number is thereby reopened; for this the insight suffices, that the same intellectual procedure, which is indispensable to every theory in proceeding to the higher arithmetical forms, constitutes the necessary and adequate foundation for the determination of the elements themselves. In this consequence, which the sensualistic theory finally admits against its will, is offered the first view of a unified methodical deduction, deriving both its foundation and superstructure from a common principle.

Number and "presentation." At first, however, there seems to be another way to establish the desired relation between numerical propositions and the empirical existence of things. If we renounce the view that all arithmetical judgments are directed upon physical things and dependent upon them for their validity, there remains, nevertheless, another class of realities, in which we now seem to grasp the true original of the numerical concepts. Not external things, but "consciousness" itself in its peculiar and irreducible mode of being, is the source of these concepts; what they seek to encompass and represent is not a material but a mental being. The scope and universality of the numerical concepts seems here to be explained. Number, as presentation, as psychical reality, remains free from all the limitations that had to be placed upon it as long as it was taken as an expression of particular material existences and their relations. We can recognize how here, in connection with a special problem, the same intellectual transformation takes place that we met earlier in the field of general logical theory. The attempt to have the concept copy outer reality directly is abandoned; but in place of this outer reality there appears its phenomenal form in our mind. The act of enumeration does not give the relations of things in themselves, but only the way in which they are reflected in the comprehension of our ego.

However much this transformation may advance the problem, there still remains an element which it shares with the sensationalistic deduction. The doctrine of number here again fails to attain an independent logical foundation; it now forms an appendix to psychology, just as formerly it was a special case of physics. (*Cf.* above p. 9 ff.). For psychology, however, "presentation" means in the last analysis nothing else than a definite psychical content, which arises in the individual subject according to special circumstances, and can be destroyed again in the same way. Such a content is different in different individuals and, furthermore, with respect to the same subject, once it has disappeared, never returns in precisely the same form. Thus what is here given is always only a temporally limited and determined reality, not a state which can be retained in unchanging logical identity. It is the fulfillment of the demand for this latter, however, which constitutes all the meaning and value of the pure numerical concepts. The proposition that $7 + 5 = 12$ reports no connection of presentational experiences, either as it has occurred in the past or will occur in the future in thinking individuals; but it establishes a connection which, according to the Platonic expression, binds the seven and the five in themselves with the twelve in itself. The object upon which this judgment is directed has, in spite of its ideality, a fixity and definiteness which sharply differentiates it from the changing contents of presentation. The psychological image of two may, in the case of one person, be connected with an accompanying spatial presentation and, in the case of another, be without it; it may now be vividly grasped, now dimly; nevertheless, the arithmetical meaning of two is not affected by all these differences.[3] What the concept "is" and means can only be ascertained through understanding it as the bearer and starting-point of certain judgments, as a totality of possible relations. Concepts are identical when they can be substituted for each other in all the assertions into which they enter; when every relation which holds of the one can also be carried over to the other. If we apply this criterion, the whole difference between the *logical* meaning of the concept of number and the *psychological* conception of presentation is at once revealed. The characteristic relations which prevail in the series of numbers are not thinkable as properties of the given contents of presentation. Of a presentation it is meaningless to say that one is larger or smaller than another, the double or triple of it, that one is divisible by another, etc. And no less does the postulate of an infinity of numbers exclude any such conception, for all the "being" of a presentation is exhausted in its immediate givenness, in its actual occurrence. If the numbers are realities in individual consciousness, they can only be "given" in finite groups, *i.e.*, realized in this consciousness as particular elements.

The content of presentation and the act of presentation. Yet in establishing this opposition between the pure numerical concepts and the psychological contents of presentation, this criticism does not seem to have grasped the field of psychical being in its full significance and scope. It may with justice be objected that what is characteristic of number cannot be pointed out in any particular and isolated content of consciousness only because there is here a universal presupposition, which controls and guides the origin and formulation of contents in general. The act, by which we define any unity and the synthesis by which we join together such unities into new forms, constitute the only condition under which we can speak of a manifold of elements and their connection. The activity of differentiation and connection alone, not any particular content subsequently resulting from it, can be the desired psychological correlate of the numerical concepts. It is not with objects, either those of inner or outer reality, but with acts of apperception, that the numerical determination is connected, and to which it goes back for its real meaning. The "universality" of the pure numerical concepts can thus be understood and grounded from a new standpoint. Even sensationalism recognizes this universality; but it under-

---

[3] *Cf.* again Frege, *op. cit.*, p. 37.

stands it, according to its fundamental theory, as a thing-like "mark," which is uniformly found in a group of particular objects. "All numbers," says Mill, "must be numbers of something: there are no such things as numbers in the abstract. But though numbers must be numbers of something, they may be numbers of anything. Propositions, therefore, concerning numbers, have the remarkable peculiarity that they are propositions concerning all things whatever; all objects, all existences of every kind, known to our experience."[4] The mathematical property of the enumerability of things is thus ascertained here in the same manner as any physical property. Just as we learn by thorough-going comparison of individual cases that all bodies are heavy, so, by an analogous method, we discover their numerical determinateness. We recognize, however, that the assertion of the universality of number, in so far as it rests upon such a procedure as this, has in truth been gained surreptitiously; for nothing assures us that the cases beyond our experience show the same properties as those actually observed, and thus fall under the arithmetical laws. A new standpoint regarding the foundation of number is first reached through the deeper and more mature psychological deduction of the numerical concepts from the fundamental act of apperceptive connection and separation in general. From this standpoint, number is to be called universal not because it is contained as a fixed property in every individual, but because it represents a constant condition of judgment concerning every individual as an individual. The consciousness of this universality is not gained by running through an indefinite plurality of cases, but is already presupposed in the apprehension of every one of them; for the arrangement of these individuals into an inclusive whole is only rendered possible by the fact that thought is in a position to recognize and maintain a rule, in conceptual identity, in spite of all differences and peculiarities of application.

In this attempted deduction also, which goes back from the finished contents of presentation to the acts by which they are formed, the real logical problem of number is not so much solved as rather pushed back a step. For whatever constructive value we ascribe to the pure acts of thought, they remain, nevertheless, in their purely psychological sense, always occurrences which come and go in time. They thus belong to a certain individual stream of consciousness as

---

[4] Mill, *A System of Logic*, Bk. II, Ch. 6, 2.

it runs off here and now under the particular conditions of the moment. Here, however, the earlier question recurs. In the arithmetical judgments it is not the relation of temporally limited realities which is expressed and established; here thought reaches out beyond the whole field of thought-processes to a realm of ideal objects, to which it ascribes a permanent and unchanging form. It is by virtue of this fundamental form that every element of the numerical series is connected with every other according to a fixed systematic rule. But a psychological analysis of the acts of forming presentations cannot disclose how one is connected with two, or two with three, and how the entire logical complex of propositions contained in pure arithmetic arises according to this connection. The construction and objective foundation of this systematic connection belongs to a totally different method. (*Cf.* below, especially Ch. VIII.) In the beginning, indeed, this method is a mere postulate, the fulfillment of which must appear entirely problematical. For what way remains of grounding a concept, if we are to regard it neither as a copy of inner nor of outer being, neither of the psychical or the physical? This question, however, which constantly presses to the fore, is only an expression of a certain dogmatic view of the nature and function of the concept. The system of arithmetical concepts and propositions is not to be estimated in terms of this view; but considerations of formal logic, on the contrary, find a limit and standard in this system, which has gradually evolved out of its independent and immanent presuppositions.

II

The logical foundations of the pure concept of number. (*Dedekind.*) The development of scientific arithmetic in the last decades is characterized by the increasing demand for the deduction of the concept of number, in its full import, from purely logical premises. The science of space seems to belong to intuition, or perhaps even to empirical perception. On the other hand, the thought gains acceptance that all determinations of number are to be grounded, without any appeal to sensible objects or any dependence upon concrete measurable magnitudes, "by a finite system of simple steps of thought." In this deduction of arithmetic out of logic, however, the latter is presupposed in a new form. "If we trace exactly," Dedekind says in beginning his deduction of the concept of number, "what we do in counting a group or collection of things, we are led to the consideration of the power of the mind to relate thing to thing, to let one thing correspond to another thing, one thing copy another, a capacity in general without which thought is impossible. Upon this one, but absolutely inevitable, foundation the whole science of number must be erected. . . . ."[5] The starting-point here seems to be the traditional logical doctrine of a plurality of *things* and the power of the mind to copy them; nevertheless it becomes evident with deeper understanding that the old terms have here gained a new import and meaning. The "things," which are spoken of in the further deduction, are not assumed as independent existences present anterior to any relation, but they gain their whole being, so far as it comes within the scope of the arithmetician, first in and with the relations which are predicated of them. Such "things" are terms of relations, and as such can never be "given" in isolation but only in ideal community with each other. The procedure of "copying" has also undergone a characteristic transformation. For we are no longer concerned to produce a conceptual copy of outer impressions, such as shall correspond to them in some particular feature; for copying means nothing else than the intellectual arranging by which we bind otherwise totally diverse elements into a systematic unity. Here the question concerns merely the unification of members of a series by a serial principle, not their agreement in some factual constitutive part. After a certain starting-point has been fixed by an original assumption, all further elements are given by the fact that a relation (R) is given, which in continued application generates all the members of the complex. Thus arise systems and groups of systems in strict conceptual division without its being necessary that one element be connected with another by any sort of factual similarity. The "copying" does not produce a new thing, but a new necessary *order* among operations of thought and objects of thought.

The logic of relations. In his work, *Was sind und was sollen die Zahlen*, Dedekind has shown how the complete construction of arithmetic and the exhaustive exposition of its scientific content is possible on the basis of these simple principles. We shall not trace the mathematical development of this thought in its details, but shall

---

[5] Dedekind, *Was sind und was sollen die Zahlen?* Ed. 2, Braunschweig, 1893, p. VIII.

content ourselves merely with emphasizing its essential tendency, in as much as we are not interested in the concept of number for its own sake but only as an example of the structure of a pure "functional concept." The presuppositions of the deduction of the concept of number are given in the general *logic of relations*. If we consider the totality of possible relations according to which a series of intellectual constructions (*Denksetzungen*) can be arranged, there emerge in the first place certain fundamental formal determinations, such as uniformly belong to certain classes of relations and differentiate them from other classes of different structure. Thus if any relation between two members a and b is given, which we can symbolically represent by the expression aRb, it can in the first place be so constituted that it also holds in the same way between b and a, so that from the validity of aRb follows that of bRa. In this case we call the relation "symmetrical" and distinguish it, on the one hand, from the non-symmetrical relations, in which the validity of aRb indeed permits that of bRa but does not necessarily imply it, and on the other hand, from the asymmetrical relations, in which this sort of reversal is not possible, in which therefore aRb and bRa cannot exist together. A relation, furthermore, is called transitive when from the fact that it holds of each pair of members a and b, b and c, its validity follows for a and c. It is said to be non-transitive when this extension is not necessary, and intransitive when it is excluded by the nature of the relation under consideration.[6] These determinations, which have far-reaching application in the calculus of relations, come in for consideration here chiefly in so far as upon them rests the more exact definition of what we are to understand as the *order* of a given whole. It is, in fact, a naïve prejudice to regard the order which holds between the members of a manifold as something self-evident, as if it were immediately given through the bare existence of the individual members. In truth, it is not attached to the elements as such but to the serial relation by which they are connected, and all its determinate character and specific peculiarity are

[6] Russell, to whom these distinctions are due, illustrates them by different family relationships; the relation involved in "brothers and sisters" (*Geschwister*) is symmetrical and transitive; the relation "brother" is non-symmetrical and transitive; the relation "father" asymmetrical and intransitive, *etc.* On this and what follows see Russell, *The Principles of Mathematics*, I, Cambridge, 1903; *cf.* also my essay "*Kant und die moderne Mathematik,*" *Kant Studien* XII, p. 1 ff.

derived from this serial relation. Closer investigation shows that.it is always some transitive and asymmetrical relation that is necessary to imprint upon the members of a whole a determinate order.[7]

The concept of progression. If, now, we consider a series which has a *first* member, and for which a certain law of progress has been established, of such a sort that to every member there belongs an immediate successor with which it is connected by an unambiguous transitive and asymmetrical relation, that remains throughout the whole series, then, in such a "progression," we have already grasped the real fundamental type with which arithmetic is concerned. All the propositions of arithmetic, all the operations that it defines, are related solely to the general properties of a progression; hence they are never directed primarily upon "things" but upon the ordinal relation, which prevails between the elements of certain systematic wholes. The definitions of addition and subtraction, multiplication and division, the explanation of positive and negative, whole and fractional numbers, can be developed purely on this basis, and without especially going back to the relations of concrete measurable objects. According to this deduction, the whole "certitude" (*Bestand*) of numbers rests upon the relations, which they show within themselves, and not upon any relation to an outer objective reality. They need no foreign "basis" (*Substrat*) but mutually sustain and support each other in so far as the position of each in the system is clearly determined by the others. "When," says Dedekind in definition, "in the consideration of a simple infinite system N, arranged by the "copying" (*Abbildung*) ϕ, we totally abstract from the particular properties of the elements, retain merely their distinctness, and attend only to the relations in which they are placed to each other by the ordering "copying" ϕ, then these elements are called the natural numbers or the ordinal numbers or also simply numbers, and the fundamental element 1 is called the fundamental number of the numerical series N. With reference to this liberation of the elements from every other content (abstraction), we can correctly call numbers the free creation of the human mind. The relations or laws, which . . . . in all ordered simple infinite systems are always the same, whatever names may accidentally be given to the individual elements, form the primary object of the

---

[7] For more exact treatment *cf.* Russell, *op. cit.*, Chs. 24 and 25.

science of number or arithmetic."[8] From a logical standpoint, it is of special interest that here the concept and term "abstraction" is obviously applied in a new sense. The act of abstraction is not directed upon the separating out of the quality of a thing, but its aim is to bring to consciousness the meaning of a certain relation independently of all particular cases of application, purely in itself. The function of "number" is, in its meaning, independent of the factual diversity of the objects which are enumerated; this diversity must therefore be disregarded when we are concerned merely to develop the determinate character of this function. Here abstraction has, in fact, the character of a liberation; it means logical concentration on the relational connection as such with rejection of all psychological circumstances, that may force themselves into the subjective stream of presentations, but which form no actual constitutive aspect of this connection.

Number as ordinal number. It has occasionally been objected to Dedekind's deduction that according to it there remains in principle no distinguishing content for number, to mark its peculiarity as opposed to other serially ordered objects. Since in the determination of its concept only the general moment of "progression" is retained, everything that is here said of number is valid with regard to every progression in general; it is thus only the serial form itself that is defined and not what enters into it as material. If the ordinal numbers in general are to exist then they must, so it seems, have some "inner" nature and property; they must be distinguished from other entities by some absolute "mark," in the same way that points are different from instants, or tones from colors.[9] But this objection mistakes the real aim and tendency of Dedekind's determinations. What is here expressed is just this: that there is a system of ideal objects whose whole content is exhausted in their mutual relations. The "essence" of the numbers is completely expressed in their positions.[10] And the concept of position must,

[8] Dedekind, *op. cit.*, §6.—On the concept of *"Abbildung"* cf. above. On the definition of a "simple infinite system" cf. Dedekind, *op. cit.*, §5 and §6.

[9] *Cf.* Russell, *op. cit.*, §242.

[10] On the deduction of number as pure "serial number," *cf.* especially the exposition of G. F. Lipps (*Philosoph. Studien*, ed. by Wundt, Vol. III), and also the latest discussions of Natorp, which carry through these thoughts with special clarity and penetration. (*Die logischen Grundlagen der exakten Wissenschaften*, Leipzig, 1910, Chs. III and IV.)

first of all, be grasped in its greatest logical universality and scope. The distinctness required of the elements rests upon purely conceptual and not upon perceptual conditions. The intuition of pure time, upon which Kant based the concept of number, is indeed unnecessary here. True, we think the members of the numerical series as an ordered sequence; but this sequence contains nothing of the concrete character of temporal succession. The "three" does not follow the "two" as lightning the thunder, for neither of them possesses any sort of temporal reality but a merely ideal logical constitution. The meaning of the sequence is limited to the fact that two enters as a premise into the determination of three, that the meaning of one concept can only be explained by that of the other. The lower number is "presupposed" by the higher number; but this does not imply a physical or a psychological earlier and later, but a pure relation of systematic conceptual dependence. What characterizes the "later" position is the circumstance that it issues from the fundamental unity through a more complex application of the generating relation, and consequently takes up the elements that precede it as logical constitutive parts and phases into itself. Thus, time (if we understand by it the "concrete form" of the "inner sense") presupposes number, but number does not, conversely, presuppose time. Arithmetic can be defined as the science of pure time only when we remove from the concept of time (as Hamilton does, for instance), all special determination of character, and merely retain the moment of "order in progression."[11] It is just this that now proves to be the merit in fundamental method of the science of number: that in it the "what" of the elements of a certain progressive connection is disregarded and merely the "how" of this connection is taken into account. Here we meet for the first time a general procedure, which is of decisive significance in the whole formation of mathematical concepts. Wherever a *system of conditions* is given that can be realized in different contents, there we can hold to the form of the system itself as an *invariant*, undisturbed by the difference of the contents, and develop its laws deductively. In this way we produce a new "objective" form, whose structure is independent of all arbitrariness; but it would be uncritical *naïveté* to confuse the

---

[11] On William Hamilton's definition of algebra as "science of pure time or order in progression" and its relation to the Kantian concept of time, *cf.* my essay *"Kant und die moderne Mathematik,"* *Kant Studien*, Vol. XII, p. 34 f.

object which thus arises with sensuously real and effective things. We cannot read off its "properties" empirically, nor do we need to, for its form is revealed in all determinateness as soon as we have grasped in its purity the relation from which it develops.

The theories of Helmholtz and Kronecker. Fundamental as is the conceptual moment of order, yet it does not exhaust the whole content of the concept of number. A new aspect appears as soon as number, which has hitherto been deduced as a purely logical *sequence* of intellectual constructs, is understood and applied as an expression of *plurality*. This transition from the pure ordinal number to the cardinal number is made with general agreement by the various ordinal theories of arithmetic, as they have been developed by Dedekind and in particular by Helmholtz and Kronecker. Given any finite system, we can relate it to the previously developed totality of numbers in a clear and definite way by letting every element of the system correspond to one and only one position in this totality. In this way we are finally able, by following the fixed prescribed order of the positions, to coördinate the last member of the system with a certain ordinal number, n. This act of coördination, however, which concludes the process, contains in itself all the previous phases; for since the progress from 1 to n can only take place in one way, the number which we reach reproduces the total operation in its specific character. The number n, which was primarily gained as a characteristic of the last element, can thus be regarded, from another point of view, as a character of the total system: we call it the cardinal number of the system considered, and now say of the latter that it consists of n elements.[12] Here it is presupposed above all that there can be one and only one cardinal number of a given group, that thus the "position" we finally reach is independent of the order in which we successively regard and emphasize the members of the group. This presupposition, however, as Helmholtz in particular has shown, can be proved with all strictness from the premises of the ordinal theory without the assumption of any new postulate, provided we maintain the condition that the manifold considered forms a *finite* system. The definitions of the fundamental arithmetical operations can also be transferred without difficulty to the new sort of number. Thus the formation of the sum (a + b) means, with regard to the pure ordinal numbers, that starting from a, we "count

---

[12] *Cf.* especially, Dedekind, *Was sind und was sollen die Zahlen,* §161, p. 54.

on" b steps, that is, that we determine the place in the series which we reach when we coördinate the numbers following a, member for member, with the elements of the series 1, 2, 3, ... b. This explanation remains valid without modification when we go over to the addition of the cardinal numbers; it is evident that from the combination of the elements of two groups, which fall under the cardinal numbers a and b, there results a new group C, the number of whose members is given by the number (a + b) in the previously determined meaning. The consideration of the "cardinal numbers" thus occasions the discovery of no new property and no new relation, which could not have been previously deduced from the bare element of order. The only advantage is that the formulae developed by the ordinal theory gain a wider application, since they can henceforth be read in two different languages.[13]

Even though no actual new mathematical content is produced through this transition, nevertheless in the formation of the cardinal number a new logical function is unmistakably at work. As in the theory of the ordinal number the individual steps as such are established and developed in definite sequence, so here the necessity is felt of comprehending the series, not only in its successive elements, but as an ideal *whole*. The preceding moment is not to be merely set aside by its successor, but is to be retained in its entire logical import in the latter, so that the final step of the procedure contains in itself at once all the preceding steps and the law of their mutual connection. It is first in this synthesis that the bare sequence of the ordinal numbers is developed into a unitary, self-inclosed system, in which no member exists merely for itself alone but, on the contrary, represents the structure and formal principle of the whole series.

Criticism of the nominalistic deduction. If once these two fundamental logical acts at the basis of all differentiation and connection of numbers are recognized, then no further special presupposition is needed to determine the field of the operations of arithmetic. The demand is therewith realized for a purely rational deduction, which renounces all dependence on the empirical relations of physical objects. Indeed precisely this distinctive character has been frequently misunderstood in estimating the "ordinal" theory of number. The explanation of the theory, as given by Helmholtz for example,

[13] *Cf.* Helmholtz, *Zählen und Messen. erkenntnistheoretisch betrachtet* (*Philosoph. Aufsätze, Ed. Zeller gewidmet*, Leipzig, 1887, p. 33).

leads necessarily to the view that first of all concrete groups of objects are presupposed as given, and that the whole work of thought is exhausted in introducing a diversity of symbols corresponding to this diversity of things. However "symbols" are nothing themselves but groups of perceptible objects, visibly distinguishable from each other by shape and position. Hence we seem to be able to abstract from the immediate properties of things in assertions concerning numerical relations, only because we have substituted for the reality of things their sensible "copies." Thus the true beginning of the formation of numbers would be not an abstraction from physical objects, but on the contrary a solidification and concentration of their sensuous import. Every such interpretation, which different mathematicians have seemed to approach at times in their expositions of the ordinal theory of number, contradicts its essential and deeper logical tendency. The symbols produced would cease to be symbols, would lose their characteristic function, if they were judged merely according to what they sensuously *are*, and not according to what they intellectually *mean*. What would remain, in fact, would be only certain "images," which we could investigate as to form and size, position and color. But no mathematical "nominalism," however extreme, has ever actually attempted to transform the import of valid judgments about numbers into assertions of this kind. It is only the ambiguity in the concept of symbol, only the circumstance that under it can be understood, now the bare existence of a sensuous content, and now the ideal object symbolized by the latter, which makes possible this reduction to the nominalistic schema. Leibniz, whose entire thought was concentrated upon the idea of a "universal characteristic," clearly pointed out in opposition to the formalistic theories of his time, the fact which is essential here. The "basis" of the truth lies, as he says, never in the symbols but in the objective relations between ideas. If it were otherwise, we would have to distinguish as many forms of truth as there are ways of symbolizing. Among modern mathematicians, Frege especially has shown in penetrating, detailed criticism that the arithmetic of symbols is only able to keep itself in existence by being untrue to itself. In place of the empty symbols the meaning of the arithmetical concepts appears unnoticed in the course of the argument.[14]

[14] Frege, *Grundgesetze der Arithmetik*, Vol. II, Jena, 1903, pp. 69 ff., 139, *etc.*

In the theory of the pure ordinal numbers also, the nominalistic interpretation forms only an outer husk, which must be stripped off to reach the real kernel of the logical and mathematical thought. Once this is done, what we retain is the purely rational moment; for "order" is not something which can be immediately pointed out in sense-impressions but is rather something which belongs to them only by virtue of intellectual relations.   Thus the theory in its pure form does not require the assumption of a group of physically given, particular things, as has been urged against it.[15]   The manifolds which it makes fundamental are, not empirically present, but ideally defined, totalities, which are progressively constructed from an assumed beginning according to a constant rule.   In this rule also are rooted all genuinely "formal" properties, such as distinguish the series of numbers and make it the fundamental type of a conceptually intelligible connection in general.

III

Number and the concept of class.   If, however, we survey the actual modern development of the principles of mathematics, it seems as if the previous treatment of the subject had overlooked the essential moment which alone completes the logical characterization of number.   Wherever the attempt has been made to resolve the concept of number into pure "logical constants," *the concept of class* has been regarded as a necessary and sufficient presupposition. The analysis of number seems completed only when all its special import has been derived from the general function of the concept. But according to the dominant logical theory, here again, the formation of concepts means nothing but the collection of objects into species and genera by virtue of subsumption under general attributes.

Thus, in order to understand the concept of number, everything must first be removed from it which does not fit into this schema. But a fundamental difficulty here arises for the theory.   If we are considering not the thought of number in general, but the concept of this or that determinate number, we are not dealing with a logically universal concept but with an individual concept.   Our concern is not with giving a species, which can be found in any number of individual examples, but with the determination of a certain definite

---

[15] *Cf.* Couturat, *De l'Infini mathematique*, Paris, 1896, p. 318 ff.

position within a total system. There is only one "two," only one "four," and both of them possess certain mathematical properties and characteristics which they share with no other object. If the reduction of the concept of number to the concept of class is to be possible in spite of this, then another direction must be taken. To determine what a number "is," according to its pure essence, we do not attempt to analyse it directly into simpler constitutive parts, but we ask primarily what the *equality of numbers means*. As soon as it is established under what conditions we shall regard two groups as of the same numerical value, the peculiarity of the "mark," which we assume to be identical in both, is thereby indirectly determined. The criterion of the numerical equality of two groups, however, consists in that a certain relation can be given, by virtue of which the members of the two groups can be mutually coördinated one to one. By virtue of this process of coördination, we establish among the infinitely numerous possible classes of objects, certain connections by uniting into a total complex those groups which can be coördinated in this way. In other words, we unite into a species all the manifolds for which there exists such a relation of "equivalence" or of one to one coördination; while we regard groups in which this condition is not fulfilled as belonging to different species. When this has been done, any individual group, by virtue of the character of equivalence, can be regarded as a perfect representative of the whole species: For since it can be shown that two groups which are equivalent to a third are also equivalent to each other, it is enough to prove of a given total *M* that it can be coördinated member for member with any group of the whole complex in order to establish the certainty that the same is true of all groups of the complex in question. Now if we abstract the common relation, which all the wholes of such a complex have to each other and consider it as a possible object of thought, we gain the moment which, in ordinary language, we call the *number* of these wholes. "The number which belongs to a concept *F*," says Frege, to whom we owe this deduction in its fundamental features, "is the extension of the concept: numerically equal to *F*." We grasp the number of a concept when we do not regard the objects which fall under it merely for themselves alone, but at the same time include all those classes whose elements stand in the relation of one to one coördination with those of the whole under consideration.

Russell's theory of cardinal numbers. Thus it is characteristic of this conception that it emphasizes what appears in the ordinary view simply as the *criterion* of numerical equality as the peculiar constitutive character, upon which rests the whole content of the concept of number. While the traditional view presupposes the individual numbers as given, as known, and decides on the basis of this knowledge as to their equality or inequality, here the reverse process holds. Only the relation, which is asserted in the equation, is known; while the elements entering into this relation are at first undetermined and are only determinable by the equation. Frege thus describes the general procedure, "Our purpose is to shape the content of a judgment, which can be interpreted as an equation, in such a manner that upon each side of this equation there is a number. We thus wish . . . . *to reach by means of the already known concept of equality that which is to be regarded as equal.*" Here a methodological tendency is clearly defined, which is fundamental in all construction of mathematical concepts; the "construct" is to gain its total constitution from the relations in which it stands. (*Cf.* above p. 40 f.). The only question that remains is whether, in the relation of the equivalence of classes, we really grasp a relation which is logically simpler than the totality of functions which, in the ordinal theory, leads to the systematic series of the ordinal numbers. An advance in analysis would necessarily result in our being able to abstract entirely from all these functions, and yet accomplish in a new way the complete construction of the realm of number and its laws. It is upon this point, therefore, that all critical investigation must be concentrated: Is the deduction of the series of numbers out of the concept of class really successful, or does this deduction move in a circle by tacitly presupposing concepts from the very field which it undertakes to deduce?[16]

The theory, which has here been developed, although in sharp conflict with the empirical interpretation of the nature of number,

[16] The problem, which is involved here, has been discussed in lively fashion in the modern logico-mathematical literature; for positive exposition of the theory *cf.* especially the writings of Frege, Russell and Peano; for criticism *cf.* B. Kerry, *Über Anschauung und ihre psychische Verarbeitung, Vierteljahrsschr. f. wissensch. Philos.* XI, 287 ff.   Husserl, Philosophie der Arithmetik, I, Halle, 1891, p. 129 ff.; Jonas Cohn, *Voraussetzungen und Ziele des Erkennens, Leipzig,* 1908, p. 158 ff.

agrees with it in one *formal* characteristic: it also understands number as a "common property" of certain contents and groups of contents. The basis of numerical assertions, however, as is especially emphasized, is not to be sought in sensuous physical things themselves but wholly in the concepts of things. Every judgment about numerical relations ascribes certain attributes, not to objects, but to their concepts, by which they are divided into classes with peculiar properties. "When I say: Venus has 0 moons, there is in fact no moon or aggregate of moons there of which anything could be said; but a property is thereby ascribed to the concept 'moon of Venus,' namely, that of including nothing under itself. When I say: the coach of the emperor is drawn by four horses, I thereby ascribe the number four to the concept 'horse that draws the coach of the emperor.'" This fact alone explains the universal applicability of numerical assertions, which equally covers the material and the immaterial, outer and inner phenomena, things as well as experiences and deeds. This apparent diversity of the field of what can be numbered proves upon closer consideration to be strict uniformity; for the enumeration never concerns the heterogeneous contents themselves but the concepts under which they are comprehended, and thus is always concerned with the same logical nature. The previous exposition has shown how this can be more precisely understood; a certain numerical determination is to be ascribed to concepts when they are collected into classes with other concepts with which they stand in the relation of possible one to one coördination of their elements in extension.

Criticism of "Class theories." Against this explanation, however, one objection necessarily arises. The theory, which is here defended, is by no means concerned to excogitate the concept of number arbitrarily, but to indicate the real function which number possesses in the actual whole of knowledge. Precisely this is emphasized in opposition to the interpretation which proceeds from the ordinal numbers, that the "logical" properties of number here deduced are the very ones that are definitive and essential in its "use in daily life." A natural deduction, such as shall do justice to the concrete applications of number, is to be opposed to the technical deduction, which merely keeps in view the purposes of scientific arithmetic. But a closer examination shows that this goal is not attained; for what is logically deduced here is in no way identical with the real meaning that we connect with numerical judgments in actual knowledge. If we limit ourselves merely to the previous postulates, we are indeed enabled by means of them to place different groups of elements together and to regard them as similar from a certain point of view; but no sufficient determination is thereby gained of their "number" in the ordinary sense of the word. Our thought could run through any number of "equivalent" groups and consider their mutual relations without any characteristic consciousness of the pure numerical concepts resulting from the process. The specific meaning of "four" or "seven" could never result from the bare placing together of any number of groups of "four" or "seven" elements; the individual groups must first be determined as ordered sequences of elements, thus as numbers in the sense of the ordinal theory. The "how many" of the elements, in the ordinary sense, can be changed by no logical transformation into a bare assertion concerning "just as many;" that remains an independent problem of knowledge. Consideration of this problem, however, leads back to a deeper opposition in method between the two interpretations of number. It is a fundamental characteristic of the ordinal theory that in it the individual number never means anything by itself alone, that a fixed value is only ascribed to it by its position in a total system. The definition of the individual number determines at once and directly the relation in which it stands to the other members of the field; and this relation cannot be eliminated without losing the entire content of the particular numerical concept. In the general deduction of cardinal numbers, which we are considering, this connection is eliminated. It is also necessary that this deduction erect and logically derive a fixed principle of arrangement of the individual numbers; however the meaning of the elements is to be established before this arrangement and independently of it. The members are determined as the common properties of certain classes before anything whatever has been established as to their relation of sequence. Yet, in truth, it is precisely in the element here at first excluded that the peculiar *numerical* character is rooted. The conceptual construction at the basis of number does not tend to abstract similarities, as must be the case according to the traditional doctrine of abstraction, but to separate out and maintain diversity. The consideration of groups, which can be mutually coördinated member for member, can only lead to the separating out of the identical "mark" in them; but this "mark" is in itself not yet "number" but is merely a logical property not further defined. Such a property only becomes number when it separates itself from other "marks" of the same logical character by appearing in a relation to them of "earlier" or "later," or "more" or "less." Those very thinkers, who have carried through the explanation of number by equivalent classes most strictly and consistently, emphasize therefore that this explanation is irrelevant to the methods of pure mathematics. What the mathematician considers in number are merely the properties on which rest the order of the positions. Number in itself may be what it will; for analysis and algebra it only comes into consideration in so far as it can be purely and completely developed in the form of a "progression."[17] Strictly speaking, once this is admitted, the dispute as to the methodological precedence of the ordinal number is over. For where can more certain information be gained as to the "nature" of number, in the sense of the criticism of knowledge, than in its most general scientific use?

The appeal to the meaning attached to the concept of number in pre-scientific thought also does not withstand criticism. Psychological analysis at least offers no support to the theory. All reflection on the actual state of thought, on the contrary, shows clearly the inner difference between the thought of equivalence and that of number. If number were what it alone should be according to this deduction, then it would be a peculiarly complicated and difficult task to point out the process by which such a concept would arise and be maintained in consciousness. For number means here a relation between two classes wholly heterogeneous in content, which are connected by no further moment than the mere possibility of mutual coördination. But what intellectual motive would there be in general to relate such dissimilar groups with each other; what meaning would there be, for example, in placing the class of moons of Jupiter beside that of seasons of the year, the group of pins in nine-pins beside that of the muses! Such a comparison is intelligible *after* the numerical value of these classes and an indirect agreement between them has been established in another way; but here, on the other hand, where this value is not presupposed but is to be gained from the comparison, the comparison itself lacks any fixed guide or standard. It has been urged against the theory of equivalence that it leads to an "extreme relativism" in so far as the determination of number is to be a

[17] *Cf.* Russell, *op. cit.*, §230. On the concept of progression, *cf.* above p. 49.

property, which does not belong to a group in itself, but merely in relation to other groups. This charge is at least ambiguous; for the concept of number can, in fact, be nothing in any form of deduction but a pure relational concept. Only here the field and thus the logical place of the relation is shifted; for while we are concerned in the ordinal theory with ideal constructions, which are mutually related to each other, here each individual construction is deduced from the relation of given "classes."

The logical definition of the zero and of unity. The presuppositions, that are assumed here, come to light very clearly as soon as we proceed to give a strictly logical definition of the individual numerical values from this point of view, and to determine the conditions under which we are to regard two of these numerical values as immediately successive. In fact, in the explanation of *null* there appear grave difficulties; for there is obviously no meaning in still speaking of the mutual one to one coördination of the members of different classes in the case in which these classes by definition contain no members. But, even if this difficulty could be removed by complicated logical transformations of the concept of equivalence,[18] the circle in the explanation again becomes clear as soon as we advance to the definition of "one." What it means to apprehend an object as "one" is here assumed to be known from the very beginning; for the numerical equality of two classes is known solely by the fact that we coördinate with each element of the first class one and only one of the second. This remark, simple and even trivial as it seems, has been frequently controverted. It has been objected that it is something different whether I take the number one in its strictly mathematical meaning or merely in the vague sense represented by the indefinite article: it is merely this last sense, which is presupposed, when I am requested to take any member of a class *u* and relate it to a member of a class *v*. "That every individual or every member of a class is, in a certain sense, one," says Russell, *e.g.* "is naturally incontestable, but it does not follow from this that the concept of "one" is presupposed, when we speak of an individual. We can rather conversely regard the concept of the individual as the fundamental concept, from

---

[18] *Cf.* on this point: Frege, *Grundlagen der Arithmetik*, p. 82 ff.; Russell, p. 113, along with the criticism of Kerry, *Viertel jahrsschr. f. wiss. Philos.* XI, p. 287 ff., and also of Poincaré, *Science et Méthode*, Paris, 1908, Bk. II.—For criticism of Frege, *cf.* Natorp, *op. cit.*, p. 112 ff.

which the concept of one is derived." From this point of view, the meaning of the assertion that a class u possesses "one" member (in the arithmetical sense) is determined by the fact that this class is not null and that if x and y are u's, then x is identical with y. A similar determination is then to fix the meaning of the concept of mutual one to one correlation between terms: R is such a relation that in the case that x and x' have the relation R to y, and x possesses the relation R to y and y', x and x' and also y and y' are identical.[19] It is, nevertheless, easy to see that here the logical function of number is not so much deduced as rather described by a technical circumlocution. In order to understand the explanations given here, it is at least necessary that we comprehend a term x as identical with itself, while at the same time we relate it to another term y and judge the former as agreeing with or differing from the latter, according to special circumstances. Now if we take this process of positing (*Setzung*) and differentiation as a basis, we have done nothing but presuppose number in the sense of the ordinal theory. Thus, *e.g.*, the class of *two* objects is defined by Russell by the conditions, that it in general possesses terms and that, if x is one of its terms, there is another term of the class, y, which is different from x; while further, if x, y are different terms of the class u, and z is different from x and y, every class that belongs to z is also different from u. We see how here in order to complete the explanation, the elements x, y, z, are produced in progressive differentiation and must, therefore, be indirectly distinguished as first, second, third . . . . members.

In general, in order to bring the different numbers into the form of a definitely ordered "progression" (and it is upon this form their meaning and scientific use primarily rest), we must have a principle that enables us, when any number n is given, to define the succeeding number. This relation between two numbers "being neighbors" is now determined, according to the theory, by comparing the corresponding classes u and v, with each other, by coördinating their elements member for member. If it is found here that in class (v) a member remains, which possesses no corresponding member in the other (u), then we designate v in relation to u as the "next higher" class. Here also it is postulated that we first grasp as a whole the part of v, which can be coördinated member for member with u,

---

[19] *Cf.* Russell, §124–126, §496. Frege, *Grundlagen*, p. 40 ff.

in order to grasp the member that remains unconnected in this form of connection, as a "second." Thus here the progress from unity to what next follows it is founded on the same intellectual synthesis in principle as that upon which it rests in the theory of the ordinal number; and the only difference in method consists in that these syntheses appear, in the ordinal theory, as free constructions while here they depend upon given classes of elements.[20]

The presupposition of the class concept. It follows, however, from a final decisive consideration that the logical order of concepts is inverted in this interpretation. The determination of number by the equivalence of classes presupposes that these classes themselves are given as a plurality. The concept of the "similarity" of classes, on which the meaning of the cardinal numbers is to be based, requires at least the consideration of two wholes connected by a certain relation. It has been emphasized that, for the establishment of this one to one relation, it is not necessary that the members of the two manifolds be previously numbered but that, on the contrary, the general law is satisfied if any element of the first manifold is placed in connection with any element of the second. But even if we give up, in accordance with this point of view, the prior enumeration of the individual classes which we compare, nevertheless the circumstance remains that we must oppose the classes as wholes to each other, and thereby understand them as "two" different units. It may be objected that *this* difference is immediately given by the purely logical difference of the class concepts, and thus neither needs nor is capable of any further deduction. We would therewith be led from the classes themselves back to the generating relations, upon which they rest and to which they owe their definition and character. The

---

[20] In order to explain the relation, in which two neighboring members of the series of natural numbers stand to each other, Frege, for example, starts from the proposition: "there is a concept F and an object x falling under it of such a sort that the number which belongs to the concept F is n, and the number which belongs to the concept falling under F but not x is m;" this is explained as equivalent to the proposition that n is the immediate successor of m in the natural number series. *Op. cit.*, p. 89. Thus here a distinction is drawn within the totality F in which a single member x is selected and opposed to the others: all these others are then used in the definition of the neighboring "just lower" number. We thus have here also only a circumlocution of the "popular" view, according to which each member of the series of numbers is distinguished from its neighbors by the addition or absence of a "unity."

difference in the systematic wholes reduces to the difference in the conceptual law, from which they have proceeded. From this point, however, it is evident that the system of the numbers as pure ordinal numbers can be derived immediately and without the circuitous route through the concept of class; since for this we need assume nothing but the possibility of differentiating a sequence of pure thought constructions by different relations to a certain fundamental element, which serves as a starting-point. The theory of the ordinal number thus represents the essential minimum, which no logical deduction of number can avoid; at the same time, the consideration of equivalent classes is of the greatest significance for the application of this concept, yet does not belong to its original content.

The generic concept and the relational concept. Thus the conflict of mathematical theories here again combines with the general questions of logical principle that were our starting-point. In the different interpretations of the concept of number, there is repeated the general conflict between the logic of the generic concept and the logic of the relational concept. If the attempt to derive the concept of number from that of class were successful, the traditional form of logic would gain a new source of confirmation. The ordering of individuals into the heirarchy of species would be, now as before, the true goal of all knowledge, empirical as well as exact. In the attempts to ground the logical theory of the cardinal numbers, this connection has occasionally come clearly into view. According to Russell, if I grasp the thought "two men" I have thereby formed the logical product of the concepts "man" and the concept "couple;" and the proposition that there are two men says only that a complex is given which simultaneously belongs to the class "man" and the class "couple."[21] It becomes evident at this point that the theory has not carried through the fundamental critical ideas from which it started. Frege and Russell regard as the decisive merit of their doctrine, that in it number does not appear as a property of physical things but as an assertion concerning certain properties of classes; that in it therefore objects as such do not form the basis of numerical judgments but rather the concepts of these objects. (*Cf.* above p. 29 ff.) It is incontestable that, compared with the sensationalistic interpretation, an extraordinary liberation of thought and increase of depth is gained by this transformation. Nevertheless it does not

[21] *Cf.* Russell, *op. cit.*, §111.

suffice to emphasize the purely conceptual character of numerical assertions as long as thing-concepts and functional concepts are placed on the same plane. Number appears, according to this view, not as the expression of the fundamental condition which first renders possible every plurality, but as a "mark," that belongs to the given plurality of classes and can be separated from the latter by comparison. Thus the fundamental deficiency of the whole doctrine of abstraction is repeated: an attempt is made to view what guides and controls the formation of concepts, *i. e.*, a purely "categorical" point of view, as, in some way, a constitutive part of the compared objects. (*Cf.* above p. 24 ff.) The theory finally shows itself to be a subtle and extended attempt to deal, by means of the general schema of the generic concept, with a problem which belongs in its meaning and scope to a new field and presupposes another concept of knowledge.[22]

IV

Extension of the concept of number. The previous attempts to establish the character of number and the principle of its formation, however, have not grasped the question in that universality and breadth, which it has gained in the development of modern mathematics. It is to number in its most primitive form and meaning that the

[22] Indeed not merely logical views but also more special mathematical reasons led to the explanation of number through the equivalence of classes. Only on this foundation did it seem possible to produce a theory, which would not be limited from the very beginning to the finite numbers, but would include and characterize the "finite" and "infinite" numbers in a single deduction. The aspect of mutual one to one coördination of groups seemed of fundamental significance, for it remained when one abstracted from the finitude and therewith the enumerability of the groups,—according to the ordinary interpretation of enumeration as the successive advance from unity to unity. Fruitful as the general point of view of "power" (*Mächtigkeit*), which arises in this connection, has shown itself to be, it has nevertheless in no way been proved to be identical with the concept of number. The purely mathematical significance of the concept of "power" is obviously unaffected by whether we regard it as the original principle of number or as only a derived result that presupposes another explanation of number. The properties, which are *common* to the finite and transfinite numbers, by no means contain as such the essential element for the construction of number in general: the "summum genus" in the sense of the logic of the generic concept is here also not identical with the conceptual origin of knowledge. (On the problem of the transfinites, *cf.* below p. 80 ff.)

attempted deduction of the class theory as well as that of the ordinal theory applies. In principle, the standpoint of the Pythagoreans is not yet left behind; number, in the narrower sense of the whole number, still forms the only real problem. The scientific system of arithmetic, however, is first completed in the extensions, which the concept of number undergoes through the introduction of the oppositions of positive and negative, whole and fractional, rational and irrational numbers. Are these extensions, as prominent mathematicians have asserted, merely technical transformations, which can only be explained and justified as applications, or are they expressions of the same logical function that characterizes the first institution of numbers?

Gauss' theory of the negative and imaginary numbers. The difficulties encountered in the introduction of every new type of number, —of the negatives and the irrationals as well as of the imaginaries,— are easily explained if we consider that, in all these transformations, the real basis of numerical assertions seems more and more to disappear. Enumeration, in its most fundamental sense, could be immediately shown to be "real" by means of sensible objects and therefore valid. The meaning of "two" or "four" forms, as it seems, no serious problem, for the empirical world of things everywhere offers us groups of two and four things. With the first generalization and extension of the concept of number, however, this reference to things, upon which the naïve interpretation especially rests, disappears. The concept and designation of the "imaginary" number is the expression of a thought, that is effective in principle in each of the new types of number and gives them their characteristic stamp. Judgments and assertions concerning the "unreal" here lay claim to a definite, indispensable cognitive value. This connection and the general principle, to which all the different methods of the extension of number go back, is stated by Gauss with complete clarity and distinctness in a passage in which he sets himself the task of grounding the true "metaphysics of the imaginary." "Positive and negative numbers," it is here stated, "can only find application where what is enumerated has an opposite, to be united with which is identical to annihilation. In strictness, this presupposition is only realized where not substances (things conceivable by themselves) but relations between two objects are enumerated. It is thereby postulated that these objects are ordered in some way into a series, *e.g.*, A, B,

C, D, ... and that the relation of A to B can be regarded as the same as that of B to C. Now in this case nothing more belongs to the concept of opposition than the reversal of the relation; thus, if the relation (and thus the transition) between A and B is represented by $+1$, the relation of B to A is represented by -1. In so far as such a series is unlimited in both directions, every real whole number represents the relation of any member arbitrarily chosen as a beginning to some definite member of the series." The deduction of imaginary numbers rests, further, upon the fact that the objects we are investigating are no longer regarded as ordered in a single series, but as requiring for their arrangement consideration of a *series of series*, and thus the introduction of a new unity $(+i, -i)$. Here, if we abstract from all details of the deduction, the dominating logical view stands out clearly. The meaning of the generalized concepts of number cannot be grasped as long as we try to indicate what they mean with regard to substances, with regard to objects conceivable *in themselves*. But the meaning at once becomes intelligible when we regard the concepts as expressing pure relations through which the connections in a constructively produced series are governed. A negative substance, which would be at once being and not-being, would be a *contradictio in adjecto;* a negative *relation* is only the necessary logical correlate of the concept of relation in general, for every relation of A to B can also be represented and expressed as a relation of B to A. If we consider, therefore, the generating relation (R), on which rests the transition from a member of the series of numbers to that immediately following, we thereby postulate also a relation of the following member to the preceding, and thus define a second direction of progress, which we can understand as the converse of the first, or as the inverse relation $(\breve{R})$. The positive and negative numbers $(+a, -a)$ now appear merely as expressions of progress (*Fortgang*) in these two directions of the relations $(R^a, \breve{R}^a)$. From this fundamental conception, all the operations of calculation within the thus extended field of numbers can be deduced, for all these operations are founded on the character of pure number as relational number and express this with increasing clarity.[23]

The irrational numbers. Once more we shall not trace the development in all its particular phases but shall merely study *typical*

---

[23] *Cf.* here especially the penetrating exposition and proof of this connection given by Natorp, *op. cit.*, Chs. III and IV.

examples in which the logical tendency of the thought is most clearly
expressed. The new principle is verified, above all, in the deduction
of the irrational numbers. There seem at first two ways in which a
deduction of the irrationals can be attempted. We could start either;
from the relations between given geometrical extensions or from the
postulate of the solubility of certain algebraic equations. The first
method, which was almost exclusively accepted up to the time of
Weierstrass and Dedekind, bases the new number upon space and
thus upon relations found between measurable objects. Here again
it seems to be experiences of physical-spatial objects which control
the process of the formation of mathematical concepts and prescribe
its direction. Nevertheless it becomes evident that at least the appeal
to the relations of concrete empirical things must fail at this point.
The relative magnitudes of things are known to us only through
observation, and thus only within the limits imposed by the personal
equation (*Beobachtungsfehler*). To demand an absolutely exact
determination in this field is to mistake the nature of the question
itself. Thus the ordinary system of fractional numbers is obviously
an intellectual instrument adequate in every respect to accomplish
all the tasks that can arise in this field. For within this system there
is no smallest difference, for between any two elements, however near,
there can always be given a new element belonging to the system;
thus a conceptual differentiation is offered here, which is never
reached in the observable relations of things, to say nothing of being
surpassed. The determinations of size, to which we are led by external experience, can thus never force us to the concept of the irrational
in its strict mathematical significance. On the contrary, this concept must arise and be grounded from within the circle of the postulates, upon which rests the systematic connection of mathematical
cognitions. In any case, not the bodies of physical reality but the
purely ideal extensions of geometry can afford the desired basis for
the derivation of the irrationals. The new problem does not develop
out of the apprehension of given, factually presented magnitudes,
but out of the laws of certain geometrical *constructions*. Once this
is recognized, the further demand must arise that the construction,
indispensable in any attempted deduction, shall proceed and justify
itself from the fundamental principle of number itself. The shifting
of the question from number to space would destroy the unity and
completeness of the system of algebra itself.

The ordinary algebraical method, which introduces the irrational values as the solutions of certain equations is indeed inadequate, for it confuses the *erection* of a postulate with its *fulfillment*. For even if we abstract from the fact that there are infinitely many irrational values such as cannot be represented as the roots of algebraical equations, in any case, such an explanation does not decide whether the object produced by them is definitely determined or whether there are several different values which satisfy the stated conditions. An adequate definition must not characterize the ideal object upon which it is directed merely by some particular "mark" that belongs to it, but must comprehend and determine it in its full characteristic individuality, by which it is distinguished from all others. This individuality, however, is completely determined for any numerical value when its position in the total system is given with its deduction, and its relation thus determined to all the other known members of the realm of number. From the first, this relative position includes within it all other properties that can ever be ascribed to the individual numbers, for all these properties follow from it and are based on it.

Dedekind's explanation of the irrational numbers. This guiding thought appears in its purest form in Dedekind's well-known explanation of the irrational numbers as "cuts." If we first take as given the totality of rational fractions, a fraction being defined as a proportion (*Verhältniszahl*) and derived without appeal to measurable and divisible magnitudes from the consideration of the pure ordinal relations,[24] then every individual element *a*, which we can select from this totality, divides the totality itself into two classes 𝔄 and 𝔅. The first of these classes includes all numbers which are smaller than *a* (*i.e.*, which precede *a* in the systematic order of the whole); the second, all numbers which are "larger" than *a* (*i.e.* which follow *a*). If, however, the designation of any individual fraction implicitly involves such a division of the total system, the converse of this proposition does not hold; for not every strictly defined, unambiguous division, that can be made intellectually, corresponds to a definite rational number. For example, if we consider any positive whole number D, which however is not the square of a whole number, then it will always lie between two squares so that a positive whole number Λ can be pointed out such that

[24] More particularly *cf.*, *e.g.*, Russell, *op. cit.*, §144 ff., §230.

$\Lambda^2 < D < (\Lambda + 1)^2$. If now we unite all the numbers whose squares are smaller than D into a class $\mathfrak{A}$, while we think of all the numbers whose squares are greater than D as united in a class $\mathfrak{B}$, then any possible rational value belongs to one of these classes, so that the division here made completely exhausts the system of rational numbers. Nevertheless, there is evidently no element in this system, which produces this separation, and which would thus be greater than all the numbers of the class $\mathfrak{A}$, and smaller than all numbers of the class $\mathfrak{B}$. We have thus by means of a conceptual rule (by the side of which any number of others could be placed) reached an entirely sharp and clear relation between classes of numbers, which is nevertheless represented by no individual numerical value in the manifold as hitherto defined. It is this circumstance which now occasions the introduction of a new "irrational" element; an element which has no other function and meaning than to represent conceptually this *determinateness of division* itself. The new number, in this form of derivation, is thus not arbitrarily conceived, nor is it introduced as a mere "symbol;" but it appears as the expression of a complex whole of relations, which were first deduced with strict logic. It represents, from the beginning, a definite logical system of relations and into such can be again resolved.

The objection has frequently been raised against Dedekind's deduction, both from the side of philosophy and from that of mathematics, that it involves an indemonstrable assumption. The existence of one and only one numerical element, in every case of complete division of the system of rational numbers, is not proved but merely asserted on the ground of a general postulate. In fact, Dedekind's exposition suggests this consideration in so far as it takes its start from geometrical analogies, with the purpose of clarifying the fundamental thought. The continuity of the straight line, it is explained, is expressed by the fact that when all the points of a straight line are divided into two classes, in such a manner that every point of the first class lies to the left of every point of the second class, there exists one and only one point of the straight line which produces this division of all the points, this cutting of the line into two pieces.[25] The assumption of this property of a line is characterized by Dedekind himself as an *axiom*, by which we first recognize the continuity of the

---

[25] Dedekind, *Stetigkeit und irrationale Zahlen*, Ed. 2, Braunschweig, 1892, p. 9 ff.

line, by which we read continuity into it. "If, in general, space has a real existence, then it is not necessarily continuous; innumerably many of its properties would remain the same even if it were discontinuous. And even if we were sure that space were discontinuous, there would be nothing to hinder us from making it continuous by intellectually filling in its holes, if we chose to do so; this filling in, however, would consist in the creation of new individual points and would be done according to the above principle."[26] Such an opposition of "ideal" and "real" can indeed occasion the thought that no *conceptual determination*, which forces itself upon us in understanding the realm of number, need thereby involve a *determination of being*. The advance from an ideal systematic connection to the *existence* of a new element seems to involve a μετάβασις εἰς ἄλλο γένος. In truth, however, our concern here is not with an unjustified transition, for in the realm of number, at least, the whole dualistic separation of ideal and real being, of "essence" and "existence," is irrelevant. Even if, in the case of space, such a separation between the product of free geometrical construction and what space "is" in the nature of things can possibly be maintained, nevertheless in the field of pure number it loses all meaning. No number (the whole numbers as little as the fractional and irrational numbers) "is" anything other than it is made in certain conceptual definitions. The assumption that wherever there is a complete division or "cut" (*Schnitte*) of the rational number system there "exists" one and only one number, which corresponds to it, thus implies no questionable meaning. What is here determined with absolute definiteness is primarily the division itself. When the rational system is divided into two classes 𝔄 and 𝔅 by any sort of conceptual rule, we can decide with absolute certainty regarding any of its elements whether it belongs to one or the other class; and further show that this alternative leaves no member out of account, *i.e.*, that the resulting division is complete and exhaustive. The "cut" as such has thus indubitable logical "reality," which does not have to be conferred upon it by a postulate. Furthermore, the order in which the different "cuts" follow each other is not arbitrary but is definitely prescribed for them by their original concept. We call the first of the two "cuts" (𝔄, 𝔅) larger than the second (𝔄̄, 𝔅̄) when an element α can be indicated that belongs to class 𝔄 of the first division and class

$\mathfrak{B}$ of the second. Thus a fixed and universal criterion exists for determining the serial order of the individual "cuts." Thereby the forms thus produced receive the character of pure number. For number, in its original meaning, possesses no specific character but is merely the most general expression of the form of order and series in general; thus wherever such a form exists the concept of number finds application. The "cuts" may be said to "be" numbers, since they form among themselves a strictly ordered manifold in which the relative position of the elements is determined according to a conceptual rule.

Thus in the creation of the new irrational elements, we are not concerned with the notion that somehow "between" the known members of the system of rational numbers the *being* of other elements is supposed or assumed. This way of stating the question is, in fact, in itself meaningless and unintelligible. But we are concerned with the fact that there arises, on the basis of the originally given totality, another more complex system of serially arranged determinations. This system includes the previous totality and takes it up into itself; for the characteristic mark of succession which belongs to the "cuts" holds directly of the rational numbers themselves, which can all be understood and represented as "cuts." Thus here an inclusive point of view is found from which the relative position of all members of the old, as well as of the new system, is determined. We see how the fundamental thought of the ordinal theory of number is here verified. The notion must be given up that number arises from the successive addition of unities and that its true conceptual nature is based on this operation. Such a process contains indeed *one* principle from which ordered wholes result, but in no way contains the only principle for the production of such wholes. The introduction of the irrationals is ultimately nothing but the general expression of this thought: it gives to number the whole freedom and scope of a method for the production of order in general, by virtue of which members can be posited and developed in ordered sequence, without limiting it to any special relation. The conceptual "being" of the individual number disappears gradually and plainly in its peculiar conceptual "function." On the ordinary interpretation, with which Dedekind's deduction is at first connected, although a certain number, given and at hand, *produces* a definite "cut" in a system, none the less the process is finally reversed, for this *production* comes to be the necessary and sufficient condition of our speaking of the *existence* of a number at all.   The element cannot be separated from the relational complex, for it means nothing in itself aside from this complex, which it brings to expression, as it were, in concentrated form.

The problem of the transfinite numbers.   The general thought, on which the formation of number rests, takes a new turn when we pass from the field of finite number to that of transfinite number.   Here specifically philosophical problems accumulate; for the concept of the infinite, which is here the center of discussion, has always belonged no less to the domain of metaphysics than to that of mathematics. Thus, when Cantor, in the course of his decisive investigations, created the system of transfinite numbers, he conjured up again all the scholastic oppositions of the potential and actual infinite, of the infinite and the indefinite.[27]   We seem here finally to be forced from the question of the pure meaning of the concept for knowledge to the problem of absolute *being* and its properties.   The concept of the infinite seems to mark out the limits of logic and the point at which it comes in contact with another field that lies outside of its sphere.

The concept of "power."   Nevertheless the problems, which lead to the creation of the realm of transfinite numbers, issue with absolute necessity from purely mathematical presuppositions.   They arise as soon as we generalize the fundamental concept of "equivalence," which from the first forms the criterion for the numerical equality of finite groups, in such a way as to make it applicable to infinite wholes. Two wholes,—whether the number of their elements is limited or unlimited,—are equivalent or of the same "power" (*Mächtigkeit*), when their members can be mutually coördinated one to one.   The application of this criterion can manifestly not proceed, in the case of infinite groups, by coördinating the elements individually with each other but rather presupposes that a general rule can be given, by which a complete correlation is established that can be surveyed at a glance.   Thus we are sure that for every even number $2n$ there corresponds an odd number, $2n + 1$, and that, if we let n assume all the possible values of whole numbers, the two groups, of even and odd numbers, are exhaustively correlated in a one to one manner. The concept of "power," which is thus introduced, first gains a

[27] *Cf.* especially Cantor, *Zur Lehre vom Transfiniten.   Gesammelte Abhandl. aus der Zeitschr. f. Philosophie u. philos. Kritik.*   Halle a. S.   1890.

specifically mathematical interest when it is seen that it is itself capable of differences and degrees. If we say that all wholes whose elements can be coördinated in one to one fashion with the series of natural numbers belong to the first "power," then the question arises whether the totality of possible manifolds is exhausted in them or whether there are groups of another character, as regards the specified property. The latter is the case, in fact, as can be proved. For, while the advance from the positive whole numbers to the totality of rational numbers produces no change in "power," and the same is true when we go from the system of rational numbers to the system of algebraic numbers, nevertheless, the system assumes a new character when we add the totality of transcendental numbers to it and thus extend it to the group of real numbers in general. This manifold thus represents a new level, which rises above the former level; for, on the one hand, it includes systems of the first "power" within itself, while, on the other hand, it goes beyond them, since when we attempt to coördinate its elements with those of the series of natural numbers there always remains an infinity of unconnected elements.[28] The introduction of the transfinite numbers α_1 and α_0 is merely to maintain this characteristic difference. The new number here means nothing else than a new point of view, according to which infinite systems can be arranged. A more complex group of distinguishing characteristics results when we place along side of the transfinite cardinal numbers, which are limited to giving the "powers" of infinite groups, the coördinate system of ordinal numbers, which arises when we no longer compare the groups in question merely with regard to the *number* of their elements, but also with regard to the *position* of the elements in the system. We ascribe to the well-ordered groups[29] M and N the same ordinal number or the same "type of order" when the elements of both can be mutually co-ordinated one to one with each other, the sequence which holds for both being retained. Thus if E and F are elements of M, and E_1 and F_1 are the corresponding elements of N, the relative position of

---

[28] For more detailed exposition, *cf.* my essay, *"Kant und die moderne Mathematik"* (*Kant-Studien* XII, 21 ff.); for all particulars, the literature there cited should be consulted as well as Cantor's presentation in the *Mathemat. Annalen.*

[29] For the definition of "well-ordered groups," *cf.* Cantor, *Grundlagen einer allgemeinen Mannigfaltigkeitslehre*, §2.

E and F in the succession of the first group is in agreement with the relative position of E_1 and F_1 in the succession of the second. In other words, if E precedes F in the first group, then E_1 must precede F_1 in the second group.[30] Thus, while in the comparison of the "powers" of two manifolds use can be made of any possible arrangement of their members, in establishing their "type of order" we are bound to a certain prescribed kind of succession. If now we say that all series, that can be correlated in one to one fashion according to this condition with the sequence of natural numbers, belong to the "type of order" ω, then we can, by adding to such series in their totality 1, 2 or 3 members, form series of the types $\omega + 1$, $\omega + 2$, $\omega + 3$; and furthermore by uniting two or more systems of the type ω we can create the type of order 2ω, 3ω, ... $n\omega$, so that by further application of this procedure, we can go on to the production of types ω^2, ω^3 ... ω^η, or indeed to ω^ω, ω^{ω^ω}, etc. And these are by no means introduced here as mere arbitrary symbols but are signs of conceptual determinations and differences, that are actually given and can be definitely pointed out in the field of infinite groups. The form of *enumeration* also is only an expression of a necessary logical differentiation, which first gains clear and adequate interpretation in this form.

The production of transfinite numbers. In this type of deduction, the metaphysical problems of the actual infinite fall completely into the background. It has been rightly emphasized[31] that our concern with the new forms of number is not so much with "infinite numbers" as with "numbers of something infinite;" that is, with mathematical expressions, which we create in order to grasp certain distinctive characteristics of infinite totalities. The conflicts that result from the connection of the concepts "infinity" and "reality" are not involved here, where we move entirely in the realm of ideal constructions. These conflicts may be represented in two-fold form, according as they are regarded from the side of the object or from that of the subject, from that of the world or of the activity of the knowing ego. From the first point of view, the impossibility of the actual infinite is shown by the fact that the objects, upon which the act of enumeration is directed and which, as it seems, it must pre-

[30] Cantor, *op. cit.*, §2, p. 5.

[31] S. Kerry, *System einer Theorie der Grenzbegriffe*, Leipzig und Wien, 1890, p. 68 f.

suppose, can only be given in finite numbers. No matter what breadth and scope we may ascribe to abstract number, whatever is counted must always be thought as enclosed within certain limits, for it is not accessible to us otherwise than by experience, which advances from individual to individual. Viewed from the other point of view, it is the psychological synthesis of the act of enumeration, which would exclude the actual infinite; no "finite understanding" can actually survey an unlimited number of unities and add them successively to each other. But both objections are unjustified with reference to the "transfinite," when we limit the latter to its strictly mathematical meaning. The "material" of enumeration at our disposal is unlimited, for it is not of an empirical but of a logico-conceptual nature. It is not assertions concerning things that are to be collected, but judgments concerning numbers and numerical concepts; thus the "material," which is presupposed, is not to be thought of as outwardly given but as arising by free construction. Just as little is there demand for the psychological processes of particular, isolated acts of presentation and their subsequent summation. The concept of the transfinite implies rather the opposite thought: it represents the independence of the purely logical import of number from "enumeration" in the ordinary sense of the word. Even in the grounding of the irrational numbers, we could not avoid considering infinite classes of numbers, such as could be represented and surveyed in the totality of their elements only by a general conceptual rule and could not be counted off member for member. The new category of number gives this fundamental distinction its most general significance. Cantor expressly distinguishes the "logical function," on which the transfinite is based, from the process of successive construction and synthesis of unities. The number ω is not the result of such a perpetually renewed addition of particular elements, but is meant to be merely an expression for the fact that the whole unlimited system of the natural numbers, in which there is no "last member," "is given in its natural succession according to its law." "It is indeed permissible to think of the newly created number ω as the *limit* toward which the numbers 1, 2, 3, 4, . . . ν . . . tend, if we thereby understand nothing else than that ω is to be the *first* whole number that follows upon all the numbers ν; that is, is to be called greater than any of the numbers ν . . . The logical function, which gives us ω is obviously different from the first principle of generation; I call it the second principle of generation of real whole numbers and define the same more closely thus: when any definite succession of defined real whole numbers is given, among which there exists no one that is largest, a new number is produced, on the basis of this second generating principle, which is to be regarded as the limit of those numbers, *i.e.*, as the first number greater than all of them."[32]

The second "principle of generation" of number (Cantor). This "second principle of generation" is ultimately only permissible and fruitful because it does not represent an absolutely new procedure, but merely carries further the tendency of thought that is unavoidable in any logical basing of number. Consideration of the properties of external things, such as those of particular psychical contents and acts of presentation, proved them to be incapable not only of constructing the series of the "natural" numbers in their lawful order but of rendering them intelligible. Even here it was not the bare addition of unity to unity, which controlled the formation of concepts; it rather appeared that the individual members of the series of numbers could only be deduced in their total extent by considering one and the same generating relation, held to be identical in content through all variations in particular application. It is this same thought, which now attains sharper formulation. Just as the endless multiplicity of natural numbers is ultimately posited by *one* concept, *i.e.*, according to a universal principle, so its content can be drawn together again in a single concept. For mathematical thought, the fundamental relation, that includes within itself all the members that proceed from it, becomes itself a new element, a kind of fundamental unity, from which a new form of number-construction takes its start. The whole endless totality of natural numbers, in so far as it is "given by a law," *i.e.*, in so far as it is to be treated as a unity, becomes the starting-point for a new construction. From the first order there arise other and more complex orders, which use the former as material basis. Once more we see the liberation of the concept of number from the concept of *collective unity*. To seek to understand and represent the "number" ω as an aggregate of individual unities would be non-sensical, and would negate its essential concept. On the other hand, the ordinal view is verified here: for in the concept of a new construction following upon all elements of the series of natural numbers,

---

[32] Cantor, *Grundlagen*, §11, p. 33.

there is no contradiction, as long as it is remembered that this totality is to be logically surveyed and exhausted in a single concept.

The problem of the infinity of time can also be left out of consideration at first. For the meaning of "succession" in a series is independent of the concrete temporal succession. Thus three does not follow upon two in the sense of a succession of events, but the relation merely points out the logical circumstance that the definition of three "presupposes" that of two; the same holds in a still stricter sense for the relation between the transfinite and the finite numbers. That the number ω is "after" all the finite numbers of the series of natural numbers, means ultimately merely this same sort of conceptual dependence in the order of grounding. The judgments, into which the transfinite enters, prove to be complex assertions, which are reduced by analysis to relative determinations of infinite systems of "natural" numbers. In this sense, a thorough-going conceptual continuity prevails between the two fields. The new constructions are "numbers" in so far as they possess in themselves a prescribed serial form, and hence obey certain laws of connection for purposes of calculation, which are analogous to those of the finite numbers, though they do not agree with them in all points.[33]

Thus the new forms of negative, irrational and transfinite numbers are not added to the number system from without but grow out of the continuous unfolding of the fundamental logical function that was effective in the first beginnings of the system. A new direction in principle comes in, however, as soon as we advance from the complete and closed system of the real numbers to the more complex number systems. According to the "metaphysics of the imaginary," which Gauss founded and developed, we are no longer concerned here with the establishment of the most general laws of the order of a single series, but rather with the unification of a plurality of series, of which each one is given by a definite generating relation. With this transition to a multidimensional manifold, there appear logical problems, which find their complete formulation beyond the limits of the pure doctrine of numbers in the field of general geometry.

[33] For the arithmetic of the transfinite, *cf.* more particularly Russell, *op. cit.*, §286, §294.

CHAPTER III

THE CONCEPT OF SPACE AND GEOMETRY

I

The logical transformations of the concept of number are dominated by a general motive, which constantly gains more definite expression. The meaning of number is first grasped completely when thought has freed itself from seeking in concrete experience a correlative for each of its constructions. In its most general meaning, number reveals itself as a complex intellectual determination, which possesses no immediate sensible copy in the properties of the physical object. Necessary as it is in modern analysis and algebra to carry through this development, still it might appear that it represents only a technical detour of thought and not the original and natural principle of the scientific construction of concepts. This principle seems only to come to light in its purity where thought does not, as in the realm of number, act exclusively according to self-created laws but finds its value and support in intuition. Precisely here then lies the critical decision for every logical theory of the concept. A conceptual construction may be spun out ever so subtly and consistently from presuppositions; nevertheless it seems empty and meaningless so long as it does not deepen and enrich our intuition. But if we adhere to the criterion of intuition, the opposition of the logical theories to each other appears in a new light. The model, which theory must follow, lies henceforth not exclusively in algebra but, in purer form, in geometry. Not number, but the concepts of space, because of their immediate relation to concrete reality, must serve as the type.

Concept and form. In the historical beginnings of logic this fact is most evident. *Concept* and *form* are synonyms; they unite without distinction in the meaning of *eidos.* The sensuous manifold is ordered and divided by certain spatial forms, which appear in it and run through all diversity as permanent features. In these forms we possess the fixed schema by which we grasp in the flux of sensible things a system of unchanging determinations, a realm of "eternal being." Thus the geometrical *form* becomes at once the expression and the confirmation of the logical *type*. The principle of the logic of the generic concept is confirmed from a new angle; and this time it is neither the popular view of the world nor the grammatical structure of language, but the structure of a fundamental mathematical science upon which it rests. Just as we recognize as identical the outline of the visible form in whatever sensuous material or size it may appear, so in general we are able to grasp the highest genera, to which existence owes its uniform structure and the constant recurrence of its definite features.

The method of ancient geometry. These relations have not been exclusively significant with regard to logical problems; they have also been decisive in the scientific evolution of geometry. The synthetic geometry of antiquity is dominated by the same fundamental view that is given general expression in formal logic. The genera of existence can only be clearly grasped when they are strictly distinguished from each other and limited to a fixed circle of content. Thus also each geometrical form possesses an isolated and unchanging character. The proof at first is not directed so much upon the unity of the forms as upon their strict differentiation. The view, indeed, that the problem of change was in general alien to the mathematical spirit of the Greeks has been more and more completely refuted by investigation of the historical sources. Not only did they grasp the concept of number, including that of the irrational, in all sharpness, but the *Ephodion* of Archimedes also shows with great clarity how completely Greek thought, where it advanced freely in methodic discovery, was penetrated by the concept of continuity and thus anticipated the procedure of the analysis of the infinite itself.[1] But precisely when we realize this, the difference that remained between the method of *discovery* and that of scientific *exposition* becomes the more manifest. The scientific exposition is influenced by certain logical theories, from which it cannot entirely free itself. As circle and ellipse, ellipse and parabola do not belong to the same visible type, it seems that they cannot, in a strict sense, fall under the unity of a concept. However much the geometrical judgments, which we can make about the two fields, meet and correspond in content, nevertheless there is here only a secondary similar-

---

[1] *Cf.* the exposition of Max Simon, *Geschichte der Mathematik im Altertum in Verbindung mit antiker Kulturgeschichte*, Berlin, 1909, especially pp. 256, 274 ff., 373.

ity and not an original logical identity. The proofs of the two types of assertion are in each case strictly separate; they gain their validity and necessity only because they are individually derived from the concept in question and its specific structure. Every difference in the arrangement of the given and sought lines of a problem presents a new problem in regard to the proof; to every difference in the total sensuous appearance of a figure corresponds a difference in interpretation and deduction. A problem, which modern synthetic geometry solves by a single construction, was analysed by Apollonius into more than eighty cases, differing only in position.[2] The unity of the constructive principles of geometry is hidden by the specialization of its particular forms of which each is conceived as irreducible.

The concept of space and the concept of number. The transformation of geometry in modern times begins with an insight into the *philosophical* defect of this procedure. It was not by chance that the new form of geometry, although anticipated, especially by Fermat, first gained definitive formulation through Descartes. The reform of geometry could only be carried out after a new ideal of method had been clearly conceived. The method of Descartes is everywhere directed toward establishing a definite order and connection among all particular expressions of thought. It is not the content of a given thought that determines its pure cognitive value, but the necessity by which it is deduced from ultimate first principles in unbroken sequence. The first rule of all rational knowledge is then that cognitions be so arranged that they form a single self-contained series within which there are no unmediated transitions. No member can be introduced as an entirely new element, but each must issue step by step from the earlier members according to a certain rule. Whatever can be an object of human knowledge is necessarily subject to this condition of continuous connection, so that there can be no question, however remote, which we are not able to master fully by this methodical progress from step to step. This simple thought, on which the *Discours de la Méthode* is founded, at once demands and conditions a new conception of geometry. Geometrical knowledge in the strict sense is not found where the particulars are studied

[2] See Reye, *Die synthetische Geometrie im Altertum und in der Neuzeit* (*Jahresberichte der Deutschen Mathematik.*—Vereinigung. XI. (1902) p. 343 ff. *Cf.* also my work, *Leibniz' System in seinen wissensch. Grundlagen*, Marburg, 1902, p. 220 ff.

as isolated objects, but only where the totality of these objects can be constructively generated according to a given process. Ordinary synthetic geometry violates this postulate; for its object is the isolated spatial form whose properties it grasps in immediate sensuous intuition but whose systematic connection with other forms it can never completely represent. At this point, we are led with inner philosophical necessity to the thought of the completion of the concept of space through the concept of number. The notebook of Descartes, which reveals the development of his fundamental thought, contains a characteristic expression of this: "The sciences in their present condition are masked and will only appear in full beauty when we remove their masks; whoever surveys the chain of the sciences will find them no more difficult to hold in mind than the series of numbers."[3] This is the goal of the philosophical method: to conceive all its objects with the same strictness of systematic connection as the system of numbers. From the standpoint of the exact sciences in the time of Descartes, this is the only manifold which is built up from a self-created beginning according to immanent laws, and thus can conceal within itself no question in principle insoluble for thought. The demand that spatial forms be represented as forms of number and be wholly expressed in the latter, may appear strange when regarded from the standpoint of the Cartesian ontology; for in this, "extension" signifies the true substance of the external objects and is thus an original and irreducible condition of being. But here the analysis of being must be subordinated to the analysis of knowledge. We can only bring space to exact intelligibility by giving it the same logical character as hitherto belonged only to number. Number is not understood here as a mere technical instrument of measurement. Its deeper value consists in that in it alone is completely fulfilled the supreme methodological postulate, which first makes knowledge knowledge. The conversion of spatial concepts into numerical concepts thus raises all geometrical enquiry to a new intellectual level. The substantial form-concepts of ancient geometry, which remain opposed to each other in bare isolation, are changed by this into pure "serial concepts" which can be generated out of each other by a certain fundamental principle. The scientific discovery of analytic geometry rests, in fact, upon a true philosophical "revolution." Traditional logic seems impregnable as long as

[3] Descartes, *Oeuvres inédites*, pub. by Foucher de Careil, Paris, 1859, p. 4.

ancient synthetic geometry is at its side as an immediate confirmation and embodiment of its principles; the extension of geometry first makes room for a new logic of manifolds that extends beyond the limits of the syllogism.

The fundamental principle of analytic geometry. This connection appears even more clearly when we consider the special development which analytic geometry received from Descartes. Here it appears that the apparently individual form of exposition contains features of universal significance which, in another guise, run through the whole philosophical history of geometry. The fundamental concept upon which Descartes founds his considerations is the concept of movement. From the standpoint of the older interpretation, there is a problem even in this. For only the individual figure, which stands before us in fixed, closed limits, seems accessible to exact intellectual treatment, while the transition from one form into another seems to force us back into the chaos of mere perception, into the sensuous realm of "becoming." It might at first appear that recognition of the concept of motion introduced a not entirely rational element into the Cartesian geometry, contrary to its real tendency. Motion leads at once to the question of the moving "subject;" but does not this subject presuppose a material body, thus a purely empirical element? This doubt vanishes, however, when we analyse in detail the function here ascribed to the concept of motion. The various forms of plane curves arise by our prescribing to a given point, which we make a fundamental element, various kinds of movement relative to a vertical and a horizontal axis. From the unification of these types of movement must be completely deduced the characteristics of lines generated in this way as "paths" of points. Here, as we see, movement does not signify a concrete, but merely an ideal process; it is an expression of the synthesis by which the successive manifold of positions, that are connected by any law, are brought into the unity of a spatial form. Here the concept of motion, as previously the concept of number, serves simply as an example of the general concept of series. The individual point of the plane is first determined by its distance from two fixed lines, and gains hereby its fixed systematic place within the totality of possible positions. These point-individualities, which are characterized by definite numerical values, do not merely stand next to each other but are differently related, according to various complex

rules of arrangement, and are thus brought to unified forms. The
representation of the "movement" of the points is only the sensuous
symbol of this logical act of arrangement. The intuitive geometrical line is thus resolved into a pure succession of numerical values,
connected with each other by a certain arithmetical rule. All the
sensible properties by which we distinguish one line from another,
as for example, constancy or change in direction and curvature, must,
in so far as they can be given exact conceptual expression, be expressible as peculiarities of these series of values. The concept of motion
thus does not serve the purpose of pictorial representation but rather
that of progressive rationalization; the given form is destroyed that
it may rise anew from an arithmetical serial law. How strictly this
general postulate is maintained is seen in characteristic fashion in
Descartes' exposition, for it is this postulate that defines and limits
the field of geometry itself. "Transcendental" curves are excluded
because in their case,—with the technical means at Descartes'
disposal,—the postulated deduction from the relations of purely
numerical rules, seems impossible. These curves, which in their
intuitive construction have no exceptional position, are ruled out of
geometry because they can not be brought under the new definition
of the geometrical concept, by which the concept is ultimately
reduced to a system of elementary operations of calculation.

The infinitesimal geometry. Here, however, appear the limits of
Cartesian geometry, which had to be transcended in later historical
development. A new ideal of knowledge was affirmed; but this
ideal was not able to include all the scientific questions that had
hitherto been united under the name of geometry. Rigor of conceptual construction had to be purchased through the exclusion of
important and far-reaching fields. The path of logical progress was
thus clearly prescribed. The resolution of spatial concepts into
serial concepts remained the guiding standpoint; but the system of
serial concepts had to be deepened and refined so that not merely a
narrow selection but the whole field of possible spatial forms could be
surveyed and mastered. Because of this demand, Cartesian geometry develops with inner necessity into infinitesimal geometry. Here
in infinitesimal geometry first appears in more perfect form the
new conceptual construction which can be recognized in its general
outlines in analytic geometry. The procedure here also starts from
a fundamental series x_1 x_2 ... x_n, which is coördinated, ac-

cording to a definite rule, with another series of values y_1 y_2 ... y_n. But this coördination is no longer limited to the ordinary algebraic procedures, to the addition and subtraction, multiplication and division of numbers or groups of numbers, but includes all possible forms of the dependency of magnitudes according to law. The concept of number is fulfilled in the general concept of function, which permeates it; and their coöperation first enables us to develop the whole content of geometry in logical perfection. In the advance to differential geometry there appears, however, a new decisive moment. It is an *infinite* manifold of coördinations from whose unification the curve first results as a conceptual whole. The method of infinitestimal analysis first clearly shows that this infinity of determinations does not destroy all determinateness, but that it is possible rather to unify the determinations again in a geometrical conception. As in analytic geometry the individual point of the plane is essentially determined by the numerical values of its coördinates x and y, so, by the differential equation $f(x, y, y') = 0$, to every point thus given is coördinated a certain direction of movement, and the problem consists in reconstructing from the system of these directions the whole of a given curve with all peculiarities of its geometrical course. The integration of the equation signifies only the synthesis of these infinitely numerous determinations of direction into a unitary and connected structure. In the same way, a differential equation of the second order $f(x, y, y,' y'') = 0$ coordinates to each point and its direction of movement a certain radius of curvature, whereby the problem arises of deducing the form of the curve as a whole from the totality of values of curvature thus gained.[4] The elements, which are here indicated geometrically by the concept of direction and curvature, are in their most general expression obviously nothing else than simple serial principles, which we comprehend in their totality and in their transformation according to law.

If we consider, in the sense of infinitesimal analysis, the space measured by a moving body as represented by the integral of its velocities, the process that we apply consists in reading into every moment of the actually progressing movement a certain law of progress, by which the transition to the following points of space

---

[4] Compare, F. Klein, *Einleitung in die höhere Geometrie, Autographierte Vorlesung,* Göttingen, 1893, I, 143 ff.

is to be exactly determined. The "velocity," which a body possesses at a given point of its path at a given moment of time, can only be conceived and represented by the comparison and reciprocal relation of a series of space values and time values. Logically considered, velocity is no absolute property of the moving thing but merely an expression of this reciprocal relation of dependence. We assume that the body, if at a given point all outer influence upon it ceased, would move forward uniformly in a definite way, that is, that after a certain time t_1 had elapsed the distance s_1 would have been traversed, after the time $t_2 = 2t_1$ had elapsed the distance $2s_1$ would have been traversed, etc. In all this we are not concerned with indicating the real movement of the body by giving the particular places through which it passes, but with construing its path purely ideally according to the various laws of the possible coördination of points of space and points of time. The individual values within the various series are never factual, for the uniformity of movement is never actually realized; nevertheless, thought necessarily uses these hypothetical values and series of values in order to render the complex whole, i.e., the real path, completely intelligible. The same holds true of the procedure used in the analysis of the infinite in geometry. The curve is here also conceived as a certain order of points; but this order, which, as immediately given, is a highly involved serial form, is conceptually analysed as a manifold of simpler laws of serial order, that mutually determine each other. The concrete form is analysed into a system of *virtual* grounds of determination, which are assumed to be different from point to point. The geometrical form, which, from the standpoint of direct intuition and that of elementary synthetic geometry, seems to be something absolutely known and immediately comprehensible, appears here as a mediated result. The form is as if resolved into manifold strata of relations, which are superimposed upon each other and which, by the definite type of dependence among them, finally determine a single whole.

Magnitudes and functions. Herewith, however, is revealed a problem of comprehensive significance. The construction of curves out of the totality of their tangents, as shown in infinitesimal geometry, is only an example of a procedure of more general applicability. All mathematical conceptual construction sets itself a double task, in fact, the task of the analysis of a certain relational complex into elementary types of relation and the synthesis of these simpler types and laws of construction into relations of higher orders. The analysis of the infinite is logically a first and complete expression of this intellectual tendency. For even here mathematical investigation advances beyond the mere consideration of magnitudes and turns to a general theory of functions. The "elements" here joined into new unities are themselves not extensive magnitudes which are combined as "parts" of a whole, but are forms of function which reciprocally determine each other and unite into a system of dependencies. However, before we can follow this development, which gives mathematics its peculiar character, we must turn back to the special problems of geometry; for in the philosophical struggles concerning the geometrical methods, there clearly appear the beginnings of a new and universal formulation of the logical question.

II

Intuition and thought in the principles of the geometry of position. Modern geometry first attains a strictly logical construction of its field and true freedom and universality of method in advancing from the geometry of measure to the geometry of position. In relation to the analytic geometry of Descartes, this step seems to signify a reaction. Intuition again asserts its claim as in ancient synthetic geometry. It is not when we limit intuition as much as possible and seek to replace it by mere operations of calculation, that a truly logical and strictly deductive construction of the science of space results, but when we restore intuition to its full scope and independence. Thus the development leads back from the abstract concept of number to the pure concept of form. That in this there lies a new motive, in the philosophical sense, Descartes himself discovered and asserted. He saw in the methods of Desargues, which contained the first approach to a projective treatment and conception of spatial forms, an indication of a general "metaphysics of geometry."[5] If we follow this "metaphysics" further, it seems in immediate conflict with his own tendencies and deductions. In fact, the new interpretation could be carried through only by a stubborn struggle against the supremacy of the analytic methods. Criticism of these methods

[5] See Descartes' letter to Mersenne of Jan. 9, 1639, *Correspondance*, ed. Adam-Tannery, II, 490.

begins with Leibniz and is brought to a first conclusion with his founding the analysis of position. It is charged that analysis is not able to establish the universal principle of order, upon which it prides itself, within the field to be ordered, but that it is obliged to have recourse to a point of view external to the object considered. The reference of a spatial figure to arbitrarily chosen coördinates introduces an element of subjective caprice into the determination; the conceptual character of the form is not established on the basis of properties purely within itself, but is expressed by an accidental relation, which may be different according to the choice of the assumed system of reference. Whether from among all the various equations which can be applied, according to this process, in the expression of a spatial figure, the relatively simplest is chosen depends upon the individual skill of the calculator, and thus upon an element which the strict progress of method seeks to exclude. If this defect is to be avoided, a procedure must be found which is equal to the analytic methods in conceptual rigor but which accomplishes the rationalization wholly within the field of geometry and of pure space. The fundamental spatial forms are to be grasped as they are "in themselves" and understood in their own laws without translation into abstract numerical relations.[6]

Steiner and Poncelet. Nevertheless from this standpoint,—and this is philosophically characteristic and significant,—there is no possibility of return to the point of view of ancient elementary geometry. The reversion to the intuitive aspect of the figure produces only an apparent connecting link; for the content itself, that is now understood under geometrical "intuition," has been deepened and transformed. If, in order to gain a fixed criterion in the philosophical conflict of opinion, we enquire of the scientific founders of modern geometry concerning the meaning they attach to the concept and term "intuition," there appears at first a peculiarly conflicting result. On the one hand, Jakob Steiner, following his teacher and model Pestalozzi, is never tired of praising the logical justification and fruitfulness of pure intuition. Steiner and his disciples see the defect of ordinary synthetic geometry in that it teaches us to use intuition only in a limited sense, and not in the whole freedom and

---

[6] For further particulars regarding Leibniz' sketch of *"Analysis situs"* cf. *Leibniz' System in s. wiss. Grundlagen*, Ch. III.

scope of its meaning.[7]   On the other hand, in the chief work of Poncelet we find the opposite logical tendency expressed.   The value of the new method is found in that through it geometrical deduction can proceed entirely unhindered, in that, without being narrowed to the limits of possible sensuous representation, it especially takes account of imaginary and infinitely distant elements that possess no individual geometrical "existence," and thus first attains completeness of rational deduction.   The opposition here in the formulations of the thought is removed as soon as we follow the exposition on both sides more closely.   In the case where the geometry of position is founded purely on intuition, the meaning is not that of adherence to the sensuously given figure but is the free constructive generation of figures according to a definite unitary principle.   The various sensuously possible cases of a figure are not, as in Greek geometry, individually conceived and investigated, but all interest is concentrated on the manner in which they mutually proceed from each other.   In so far as an individual form is considered, it never stands for itself alone but as a symbol of the system to which it belongs and as an expression for the totality of forms into which it can be transformed under certain rules of transformation.   Thus "intuition" is never concerned with the particular figure with its accidental content, but is, according to Jakob Steiner, directed to the mediation of the dependency of geometrical forms upon each other.[8]   The particular terms are here also subor-

[7] *Cf.*, for example, B. Reye, *Die synthetische Geometrie im Altertum und in der Neuzeit, Op. cit.*, p. 347.

[8] *Cf.* the preface to J. Steiner's work: *Systemat. Entwicklung der Abhängigkeit geometrischer Gestalten voneinander*, Berlin, 1832: "The present work seeks to reveal the organization by which different kinds of phenomena are bound together in the spatial world.   There are a small number of simple fundamental relations, wherein the schema is expressed from which the remaining propositions develop logically and without any difficulty.   Through proper appropriation of a few fundamental relations, we master the whole subject; order comes out of chaos, and we see how all the parts naturally fit together, forming into series in the most perfect order, and the allied parts united into well-defined groups.   In this way we succeed, as it were, in gaining possession of the elements from which nature starts, by which, with the greatest possible economy and in the simplest way, innumerable properties can be ascribed to figures.   In this, neither the synthetic nor the analytic method constitutes the essence of the matter, which consists in the discovery of the dependency of forms on each other and the manner in which their properties are continued from the simpler figures to the more complex."

dinated to the systematic relation that unites them. The deduction itself of the fundamental forms expresses this in so far as, for example, the particular straight line is not defined in itself, but as an element in a pencil of rays, or the particular plane as an element in a sheaf of rays. In general, we see that the fundamental methodological standpoint, which led to the discovery of analytic geometry, is not set aside here but retained and brought to a new fruitful application in the field of the spatial itself. The motive of number is excluded; but the general motive of series stands forth the more clearly. We saw how, with Descartes, number was not characterized as the fundamental principle because of its own content, but was only retained because it represented the purest and most perfect type of a logically ordered manifold in general. Strictness of deductive connection seemed only able to be carried over to space by the mediation of number. (*Cf.* above p. 71 f.) We can understand that henceforth there must arise a new, important task continuous with the achievements of analytic geometry. The construction of the spatial forms from original fundamental relations remains as an inviolable postulate, but this postulate must now be satisfied by purely geometrical means and without the introduction of the concepts of measure and of number.

The concept of "correlation" and the principle of continuity. The evolution, which begins here, is characterized and guided in detail by logical views. This is especially manifest in the case of Poncelet who, in the conflict in which he engaged regarding the principles of his discipline, refers with increasing definiteness to the philosophical foundations. In opposition to the criticism which the Parisian Academy, especially Cauchy, made on the philosophical presuppositions of his work, Poncelet urges with emphasis that in these presuppositions we are not dealing with a secondary matter, but with the real root of the new view. He appropriates the words of Newton that in geometry the *method of discovery* means everything, so that when this is once found and established, results occur of themselves as the fruit of the method.[9] The theory of the projective properties is to be no mere material extension of the field of geometry, but is to

---

[9] *Cf.* Poncelet, *Traité des propriétés projectives des figures*, 2nd edition, Paris, 1865, I, p. 365; II, p. 357.

introduce a new principle of investigation and discovery.[10] The first and necessary step is to free geometrical thought from the narrowness of the sensuous view with its close adherence to the particularities of the directly given, individual figure. Descartes charged ancient mathematics with not being able to sharpen the intellect without tiring the imagination by close dependence on the sensuous form, and Poncelet maintains this challenge throughout. The true synthetic method cannot revert to this procedure. It can only show itself equal in value to the analytic method, if it equals it in scope and universality, but at the same time gains this universality of view from purely geometrical assumptions. This double task is fulfilled as soon as we regard the particular form we are studying not as itself the concrete object of investigation but merely as a starting-point, from which to deduce by a certain rule of variation a whole system of possible forms. The fundamental relations, which characterize this system, and which must be equally satisfied in each particular form, constitute in their totality the true geometrical object. What the geometrician considers is not so much the properties of a given figure as the net-work of correlations in which it stands with other allied structures. We say that a definite spatial form is correlative to another when it is deducible from the latter by a continuous transformation of one or more of its elements of position: yet in which the assumption holds that certain fundamental spatial relations, which are to be regarded as the general conditions of the system, remain unchanged. The force and conclusiveness of geometrical proof always rests then in the invariants of the system, not in what is peculiar to the individual members as such. It is this interpretation, which Poncelet characterizes philosophically by the expression *principle of continuity*, and which he formulates more precisely as the *principle of the permanence of mathematical relations*. The only postulate that is involved can be formulated by saying that it is

[10] "La doctrine des propriétés projectives, celle de la perspective-relief, le principe ou la loi de continuité, enfin la théorie des polaires réciproques et la théorie des transversales étendue aux lignes et surfaces courbes, ne forment pas simplement des classes plus ou moins étendues de problèmes et de theorèmes, mais constituent proprement, pour la Géométrie pure des principes, des méthodes d'investigation et d'invention, des moyens d'extension et d'exposition, dans le genre de ceux qu'on a nommé principes d'exhaustion, méthode des infiniment petits, *etc.*" *Op. cit.*, p. 5.

possible to maintain the validity of certain relations, defined once for all, in spite of a change in the content of the particular terms, *i.e.*, of the particular *relata*. We thus begin by considering the figure in a general connection (*Lage*), and do not analyse it in the beginning into all its individual parts, but permit changes of them within a certain sphere defined by the conditions of the system. If these changes proceed continuously from a definite starting-point, the systematic properties we have discovered in a figure will be transferable to each successive "phase," so that finally determinations, which are found in an individual case, can be progressively extended to all the successive members.

The transference of relations distinguished from induction and analogy. There is clearly manifest in these elucidations of Poncelet a tendency toward an exact and universal expression of the new thought. Above all, he is concerned to guard the transference of relations, which he assumes as basic, from any confusion with merely analogical or inductive inference. Induction proceeds from the particular to the universal; it attempts to unite hypothetically into a whole a plurality of individual facts observed as particulars without necessary connection. Here, however, the law of connection is not subsequently disclosed, but forms the original basis by virtue of which the individual case can be determined in its meaning. The conditions of the whole system are predetermined, and all specialization can only be reached by adding a new factor as a limiting determination while maintaining these conditions. From the beginning, we do not consider the metrical and projective relations in the manner in which they are embodied in any particular figure, but take them with a certain breadth and indefiniteness, which gives them room for development.[11] It may seem at first surprising and paradoxical that this indefiniteness of the starting-point is held to be the ground of the fruitfulness of the new procedure and of its superiority over the ancient methods. However, it soon appears that the expression of the thought here suffers from the ambiguity of the traditional logical terminology, in which concept and image are not strictly distinguished, and in which the identical and clearly defined meaning of a conceptual rule always threatens to dissolve into an abstract and schematic generic image. What appears as indefiniteness from the standpoint of image, because it neglects the individual features

[11] *Traité des propriétés projectives*, p. XIII f., XXI f.

of the picture (*Bild*), appears from the standpoint of the concept as the basis of all exact determination since in it is contained the universal rule for the construction of the individual. Between the "universal" and "particular" there subsists the relation which characterizes all true mathematical construction of concepts; the general case does not absolutely neglect the particular determinations, but it reveals the capacity to evolve the particulars in their concrete totality entirely from a principle. (See above p. 18 f.) As Poncelet emphasizes, it is never the mere properties of the particular kind but the properties of the genus, from which the projective treatment of a figure takes its start; the "genus," however, here signifies merely a connection of conditions by which everything individual is ordered, not a separated whole of attributes which uniformly recur in the individuals. The inference proceeds from the properties of the connection to those of the objects connected, from the serial principles to the members of the series.

Projection and the imaginary in geometry. The peculiarity of the method appears most clearly in its fundamental procedure. The most important form of correlation, by which different figures are connected, is found in the procedure of projection. The essential problem consists in separating out those "metrical" and "descriptive" elements of a figure which persist unchanged in its projection. All the forms, that can issue in this way from each other, are considered as an indivisible unity; they are, in the sense of the pure geometry of position, only different expressions of one and the same concept. Here it is immediately evident that to belong to a concept does not depend on any generic similarities of the particulars, but merely presupposes a certain principle of transformation, which is maintained as identical. The forms, which we unite in this way into a "group," can belong to totally different "types" in their sensuous intuitive structure; indeed, they can be deprived of any reference to such a type in so far as there is no geometrical existence, in the sense of direct intuition, corresponding to them. The new criterion of the geometrical construction of concepts is shown here in its general significance; for it is this criterion upon which the admission of the imaginary into geometry ultimately rests. In general, according to Poncelet, three different forms of the procedure of "correlation" can be distinguished. We can transform a certain figure, which we choose as a starting-point, into another by retaining all its parts as well as their mutual arrangements, so that the difference consists exclusively in the absolute magnitude of the parts. In this case, we can speak of a *direct* correlation, while in the case in which the order of the individual parts is exchanged or reversed in the deduced figure, we can speak only of an "indirect" correlation. Finally,—and this is methodologically the most interesting and important case—the transformation can proceed in such a manner that certain elements, which could be indicated in the original form as real parts, entirely disappear in the course of the process. If we consider, for example, a circle and a straight line that intersects it, then we can transform this geometrical system by continuous displacements in such a manner that the straight line finally falls wholly outside of the circle, so that the intersections and the directions of the radii corresponding to them are to be expressed by imaginary values. The coördination of the deduced figure with the original figure no longer connects elements that are actually present and observable, but merely intellectual elements; it has resolved itself into a purely *ideal correlation*.

But it is precisely these ideal correlations that cannot be spared if geometry is to be given a unitary and self-contained form. The defect of the ancient methods consists in that they neglect this fundamental logical instrument, and thus consider only magnitudes of an absolute and quasi-physical existence. The new view is obliged to break with this procedure, since from the beginning it defines as the real object of geometrical investigation not the individual form in its sensuous existence, but the various species of dependency that can subsist between forms. (*Cf.* p. 77.) From this point of view, real and imaginary elements are essentially similar; for the latter also are the expression of perfectly valid and true geometrical relations. That under definite conditions certain elements of a figure disappear and cease to exist, this is in itself no merely negative knowledge but contains a fruitful and thoroughly positive geometrical insight. Further, the imaginary intermediate members always serve to make possible insight into the connection of real geometrical forms, which without this mediation would stand opposed as heterogeneous and unrelated. It is this ideal force of logical connection, that secures them full right to "being" in a logico-geometrical sense. The imaginary subsists, in so far as it fulfills a logically indispensable function in the system of geometrical propositions. The only "reality," which we can intelligibly expect and demand of it consists in the truth it contains, in its relation to the valid propositions and judgments which it brings to expression. Here in the realm of geometry is repeated the same process that we are able to follow in the realm of number; from the retention of definite relations arise new "elements," essentially similar and equal to the earlier, since these also have no deeper or firmer basis than consists in the truth of relations. (*Cf.* above p. 60 ff.)

If we consider a simple example from ordinary geometry, *viz.*, of two circles in a plane, then in the case in which they intersect, there is given in the straight line, that connects their two points of intersection, a new structure of definite properties. The points of this straight line, which we call the "common chord" of the two circles, are distinguished by the fact that the tangents drawn from them to the circle are equal to each other. The geometrical relation thus established can also be traced and expressed in the case in which the circles no longer intersect, but fall completely outside of each other. In this case, also, there always exists a straight line,—the so-called "radical axis" of the two circles,—which satisfies the essential condition previously mentioned, and in this sense can be called the ideal common chord of the two circles, which contains their two "imaginary" points of intersection. Thus a certain intuitive element is expressed here and completely replaced by certain conceptual properties belonging to it; and this logical determination is also retained after the substratum, in connection with which it was first discovered, has disappeared. We proceed from the persistence of the relation and create by definition in the imaginary "points" the "subjects," of which the relation is predicated. The fruitfulness of this procedure is seen in that a systematic connection between forms is thereby established, such as permits us to carry over propositions, which are discovered and proved in connection with one form, to another in connection with which they were not immediately obvious.[12] Along with the particular relations of content, there are

[12] Thus, *e.g.*, if any three circles in a plane are given, and we construct for each two of them the "radical axes," it can easily be shown that the three lines so arising intersect in one point; from this it results further that the same also holds, in the special case of three common chords of really intersecting circles, *etc.* The real properties of real common chords are thus discovered and grounded by reference to the "ideal" chords. *Cf.* Chasles, *Aperçu historique sur l'origine et le développement des méthodes en Géométrie*, 2nd. ed., Paris, 1875, p. 205 ff.; also Hankel, *Die Elemente der projektivischen Geometrie*, Leipzig, 1875, p. 7 ff.

above all certain universal formal determinations by which the "unreal" elements, which geometry produces, are connected with the "real" points. The principle of the "permanence of formal laws," even before it was used in algebra as the justification of the generalized concepts of number, was introduced and founded by Poncelet from a purely geometrical standpoint. The infinitely distant point in which, according to the projective theory of space, two parallels intersect, and the infinitely distant straight line in which two parallel planes intersect, are logically justified conceptual constructions, not only because they represent concentrated assertions about definite relations of position, but because these new constructions are completely subjected to the geometrical axioms, as can be shown, in so far as they do not refer to relations of measure. Here a higher point of view is found, which is equally just to the "real" and "unreal" points. The new elements, Poncelet says with distinctness, are paradoxical in their object but they are nevertheless thoroughly logical in their structure, in so far as they lead to strict and incontestable truths.[13]

Metrical and projective geometry, and the quadrilateral construction of Staudt. The development of projective geometry, which cannot be followed here in detail, has thus brought the philosophical principles upon which it is founded to more and more explicit expression. To the extent that the geometry of position is built up from independent assumptions, the general logical character and meaning of the new method becomes evident. The constructive process, by which we generate in strict deduction the whole of projective space from the simple concepts of the point and the straight line, begins with the consideration of harmonic pairs of points. Thus in the first phase of projective geometry, the harmonic position of four points on a straight line is at first introduced exclusively by means of the concept of the double proportion: the points a, b, c, d form a harmonic sequence when the relation of the distances $a\,b$ to $b\,c$ is the same as that of the

---

[13] For the whole matter, compare Poncelet, *Considérations philosophiques et techniques sur le principe de continuité dans les lois géométriques*, section III. (*Applications d'Analyse et de Géométrie*, Paris 1864, p. 336 ff.), as well as *Traité des propriétés projectives* I, p. XI ff., 66 ff. For the designation of the principle of continuity as the "permanence of geometrical relations, "*cf. Applic.* p. 319; *Traité* II, 357; the same thought is expressed by Chasles in his "*Principe des relations contingentes*" with another turn of expression. (*Aperçu historique* p. 204 ff., 357 ff., 368 ff.)

distances *a d* to *c d*. This explanation obviously presupposes the measurement and comparison of certain distances, and is thus in essence of purely metrical nature; that it is, nevertheless, made the basis of the geometry of position rests on the fact that it represents a metrical relation, which remains unchanged in every projective transformation of a given figure. The concept of measure is here not excluded, but is taken up into the foundations as an underived element. Projective geometry only gains an independent and strictly unified exposition when this last limit is also set aside, when the determination, which is characterized metrically as the double proportion, is derived in a purely descriptive way. The decisive method for this is given in the known quadrilateral construction of Staudt. We determine the fourth harmonic point *d* to three given co-linear points *a b c* by constructing a quadrilateral of such a sort that two opposite sides pass through *a*, the diagonal through *b*, and the two other opposite sides through *c;* the point of intersection of the second diagonal of the quadrilateral with the straight line *a b c* is the desired point *d*, which is definitely determined by this method, since it can be proved that the construction indicated always gives the same result no matter what quadrilateral is taken as a basis, so long as it satisfies the conditions.[14] Thus without any application of metrical concepts, a fundamental relation of position is established by a procedure which uses merely the drawing of straight lines. The logical ideal of a purely projective construction of geometry is thus reduced to a simpler requirement; it would be fulfilled by showing the possibility of deducing all the points of space in determinate order as members of a systematic totality, by means merely of this fundamental relation and its repeated application.

Projective metric (*Cayley and Klein*). The demonstration of this is furnished in the formulation which projective geometry has gained through Cayley and Klein. We here gain a general procedure that enables us to coördinate all the points of space that can be generated from a given starting-point by progressive harmonic constructions, with certain numerical values, and thus to give them a fixed position within a general serial order. If we start with three points *a*, *b*, *c*, in a straight line to which we coördinate the values *0, 1,* ∞ then by means of the Staudtian quadrilateral construction, we can find their

---

[14] *Cf.* Staudt, *Geometrie der Lage*, Nürnberg 1847, §8, p. 43 ff.; Reye, *Die Geometrie der Lage*, 4th. ed., Leipzig 1899, I, p. 5.

fourth harmonic point to which we let the number 2 correspond, and
we can further determine a new point, which forms with the points
1, 2, ∞ a harmonic fourth, and give this the value 3, until finally by
virtue of this method we gain an infinite manifold of simple determinations of position, to each of which a whole number is coördinated.
This manifold can be further completed by becoming a universally
"dense" group in which every element corresponds to a definite
rational positive or negative number. The transition to the point-
continuum takes place on the basis of a further intellectual postulate,
which is analogous to the postulate by which Dedekind in his theory
introduces the irrational numbers as "cuts" (*Schnitte*). We thus
gain a complete scale on the basis of which a unified projective
metric can be evolved in which the elementary operations, such as
addition and subtraction, multiplication and division of distances,
are defined purely geometrically. Also the advance to structures of
higher dimensions offers no difficulty in principle; it results when we
take into consideration, instead of the points of one straight line,
those of two or more straight lines.[15]

The concept of space and the concept of order. The working out of
this thought is chiefly of technical mathematical interest; but a general
philosophical result, anticipated from the beginning of modern
geometry, also is evident here. The inclusion of the spatial
concepts in the schema of the pure serial concepts is here finally
accomplished. The designation of the individual points of space by
corresponding numbers might indeed at first cause the illusion that
concepts of magnitude, of length and distance, are applied in this
deduction. In truth, however, number is only used here in its most
general logical meaning: not as an expression of the measurement
and comparison of magnitudes but as the expression of an ordered
sequence. We are not concerned with the addition or division of
distances and angles, but only with the differentiation and gradation
of the members of a certain series, the elements of which are defined
as pure determinations of position. Here we find verification of the
fact that, in our general logical deductions, number was evolved as

---

[15] For all particulars regarding this method, of which only the principle can
be suggested here, *cf.* F. Klein, *Vorlesungen über Nicht-Euklidische Geometrie*,
2nd. impression, Göttingen 1893, p. 315 ff., 338 ff.; as well as Math. Annalen, IV,
p. 573 ff. On projective *metric*, see also Weber-Wellstein, *Encyklopädie de*
Elementar-Mathematik, Vol. II, §18.

pure ordinal number and kept free from all connection with measurable magnitudes. The demand, which Descartes made, is thus satisfied in a new way. The order of points of space is conceived in the same manner as that of numbers. True, the two fields remain strictly separated in essence; the "essence" of the figure cannot be immediately reduced to that of number. But precisely in this relative independence of the elements, as in the independence of their fundamental relation, is manifest the connection in general deductive method. As in the case of number we start from an original unit (*ursprüngliche Einheitssetzung*) from which, by a certain generating relation, the totality of the members is evolved in fixed order, so here we first postulate a plurality of points and a certain relation of position between them, and in this beginning a principle is discovered from the various applications of which issue the totality of possible spatial constructions. In this connection, projective geometry has with justice been said to be the universal "*a priori*" science of space, which is to be placed beside arithmetic in deductive rigor and purity.[16] Space is here deduced merely in its most general form as the "possibility of coexistence" in general, while no decision is made concerning its special axiomatic structure, in particular concerning the validity of the axiom of parallels. Rather it can be shown that by the addition of special completing conditions, the general projective determination, that is here evolved, can be successively related to the different theories of parallels and thus carried into the special "parabolic," "elliptical" or "hyperbolic" determinations.[17]

Geometry and the group theory. Thus in contrast to the multiplicity of geometrical methods, the single form of the geometrical concept stands out with increasing clarity. Its logical character persists through all change in particular application. This character can be brought to mind by considering the most general interpretation which the modern concept of geometry has attained. The addition of geometry to the theory of groups forms the final and decisive step for the whole interpretation. The very definition of "group" contains a new and important logical aspect, in so far as through it is brought to intellectual unity, not so much a whole of individual elements or structures, as a system of operations. A totality of operations forms a group, when with any two operations their

[16] *Cf.* Russell, *The Foundations of Geometry*, Cambridge 1897, p. 118.
[17] *Cf.* F. Klein, *Mathem. Annalen IV*, 575 ff.

combination is also found in the group, so that the successive application of different transformations belonging to the totality leads
only to the operations originally contained in it.   In this sense, a
group is formed by all the geometrical transformations which result
when we permit the elements any movements whatever in ordinary
three-dimensional space; for the result of two successive movements
can here always be represented by a single movement.[18]   In this
concept of the group, a general principle of classification is gained by
which the different possible kinds of geometry can be unified under a
single point of view and their systematic connection surveyed.   If we
raise the question as to what in general is to be understood as a
"geometrical" property, we find that we consider as geometrical only
such properties as remain unaffected by certain spatial transformations.   The propositions, which geometry evolves about a
certain structure, persist unchanged when we vary the absolute
position of this structure in space, when we increase or decrease the
absolute magnitude of its parts proportionately, or when finally we
reverse the arrangement of the individual parts, as when we substitute for the original figure another, which is related to it as its
image in a mirror.   The thought of independence from all these transformations must be added to the intuition of the individual form, that
serves as a starting-point, to give this form true universality and
therewith true geometrical character.   "Geometry is distinguished
from topography by the fact that only such properties of space are
called geometrical as remain unchanged in a certain group of operations."   If we adhere to this explanation, we gain a view of very
diverse possibilities for the construction of geometrical systems, all
equally justified logically.   For as we are not bound in the choice of
the group of transformations, which we take as the basis of our
investigation, but can rather broaden this group by the addition of
new conditions, a way is opened by which we can go from one
form of geometry to another by changing the fundamental system to
which all assertions are related.   For instance, if we take the ordinary
metrical geometry as characterized by the appropriate group of
spatial transformations (*i.e.*, by the specific operations of movement,
of similarity-transformation and of 'mirroring' (*Spiegelung*)), we
can broaden it to projective geometry by adding the system of all
projective transformations to this group, and considering the con-

[18] *Cf.* F. Klein, *Einleitung in die höhere Geometrie* II, p. 1 ff.

stant properties in this broadened sphere of transformations. As F. Klein has shown in detail, the most diverse kinds of geometry can be methodically grounded and similarly deduced, by proceeding from a given group to a more inclusive system by means of a definite rule. In general, in each of these geometries where a manifold containing a transformation-group is given, the problem is to develop the invariant-theory applicable to the group.[19]

The concepts of constancy and change in geometry. This universal procedure throws a bright light on the essential relation of the concepts of constancy and change in the foundations of geometry. We saw how from the beginnings of Greek mathematics, the philosophical question constantly reverts to this relation. If geometry were defined, in Platonic language, as what possesses "eternal being," if it were true that exact proof were only possible of that which always maintains itself in the same form, then change could be tolerated as an auxiliary concept, but could not be used as an independent logical principle. The field of becoming marked out a region within which pure mathematical thought possessed no force and which thus seemed given over to the indeterminateness of sensuous perception. This emphasis on permanence, which was intended to exclude all sensuous elements from the foundations of pure mathematical knowledge, finally proved to work within geometry in the opposite direction. The requisite rigid constancy of the intuitive spatial form narrowed the freedom of geometrical deduction; thought remained entangled in the particular figure instead of directing itself to the ultimate grounds of the connection of figures according to law. A new development begins only after the concept of change has been critically tested and confirmed by *analysis*. This development reaches its systematic conclusion in the theory of groups; for here change is recognized as a fundamental concept, while, on the other hand, fixed logical limits are given it. The Platonic explanation is now confirmed in a new sense. Geometry, as the theory of invariants, treats of certain unchangeable relations; but this unchangeableness cannot be defined unless we understand, as its ideal background, certain fundamental changes in opposition to which it gains its validity. The unchanging geometrical properties are not

---

[19] For all particulars, we must refer again to F. Klein's "*Erlanger Programm*" of 1872, *Vergleichende Betrachtungem über neuere geometrische Forschungen* (reprinted *Math. Annalen* 43, 1893, p. 63 ff.)

such in and for themselves, but only in relation to a system of possible transformations that we implicitly assume. Constancy and change thus appear as thoroughly correlative moments, definable only through each other. The geometrical "concept" gains its identical and determinate meaning only by indicating the definite group of changes with reference to which it is conceived. The permanence here in question denotes no absolute property of given objects, but is valid only relative to a certain intellectual operation, chosen as a system of reference. Here already a change appears in the meaning of the general category of substantiality, that must constantly grow clearer in the course of the enquiry. Permanence is not related to the duration of things and their properties, but signifies the relative independence of certain members of a functional connection, which prove in comparison with others to be independent moments.

III

The evolution of modern mathematics has approached the ideal, which Leibniz established for it, with growing consciousness and success. Within pure geometry, this is shown most clearly in the development of the general concept of space. The reduction of metrical relations to projective realizes the thought of Leibniz that, before space is defined as a *quantum*, it must be grasped in its original qualitative peculiarity as an "order of coexistence" (*ordre des coexistences possibles*). The chain of harmonic constructions, by which the points of projective space are generated, provides the structure of this order, which owes its value and intelligibility to the fact that it is not sensuously presented but is constructed by thought through a succession of relational structures.[20] We can still take the elemen-

[20] It is of historical interest that the logical problem of a metrical geometry based on pure projective relations was, as a matter of fact, grasped by Leibniz. Against Leibniz's definitions of space as an order of coexistence and time as an order of succession, Clarke, who advocated Newton's theory of absolute space and absolute time, raised the objection that they did not touch the essential import of the two concepts. Space and time are first of all *quantities*, which position and order are not. Leibniz replied that also within pure determinations of order, determinations of magnitude are possible, in so far as a preceding member is distinguished from a succeeding member and the "distance" between them can be conceptually defined. "Relative things have their magnitudes just as well as absolute things; thus, *e.g.*, in mathematics, relations or proportions have magnitudes, which are measured by their logarithms;

tary contents of geometry: the point, the straight line and the plane, from intuition; but all that refers to the connection of these contents must be deduced and understood conceptually. In this sense, modern geometry seeks to free a relation, such as the general relation of "between," which at first seems to possess an irreducible sensuous existence, from this restriction and to raise it to free logical application. The meaning of this relation must be determined by definite axioms of connection in abstraction from the changing sensuous material of its presentation; for from these axioms alone is gained the meaning in which it enters into mathematical deduction. By this extension, we can make the concept of "between" independent of its original perceptual content and apply it to series in which the relation of "between" possesses no immediate intuitive correlate.[21]

Characteristic (Kombinatorik) as pure "doctrine of forms" (Leibniz). This interpretation, however, advances still further when it attempts to subsume the specific order of spatial externality under a universal system of possible orders in general. Again we are led to the Leibnizian conception of mathematics. According to this conception, mathematics is not the general science of magnitude but of form, not the science of quantity but of quality. Characteristic (*Kombinatorik*) thereby becomes the fundamental science; we do not comprehend under it the doctrine of the number of combinations of given elements, but the universal exposition of possible forms of connection in general and their mutual dependency.[22] Wherever a definite form of connection is given, which we can express in certain rules and axioms, there an identical "object" is defined in the mathematical

nevertheless they are and remain relations." Leibniz, *Hauptschriften zur Grundlegung der Philosophie*, I, Philos. Bibl. 107, p. 189 f. We see here a reference to a question, which has been repeated in the modern grounding of projective metrical geometry: for in the latter, in fact, the "distance" between two points is defined and measured by the logarithm of a certain double-relation. *Cf.* Klein, *Vorlesungen über Nicht-Euklidische Geometrie*, p. 65 ff.

[21] More especially see Pasch, *Vorles. über neuere Geometrie*, §1 and 9.

[22] "Hinc etiam prodit ignorata hactenus vel neglecta subordinatio Algebrae ad artem Combinatoriam, seu Algebrae Speciosae ad Speciosam generalem, seu scientiae de formulis quantitatem significantibus ad doctrinam de formulis, seu ordinis similitudinis relationis, etc., expressionibus in universum, vel scientiae generalis de quantitate ad scientiam generalem de qualitate, ut adeo speciosa nostra Mathematica nihil aliud sit quam specimen illustre Artis Combinatoriae seu speciosae generalis. Leibniz, *Math. Schriften*, Gerhardt, VII, 61.

sense. The relational structure as such, not the absolute property of the elements, constitutes the real object of mathematical investigation. Two complexes of judgments, of which the one deals with straight lines and planes, the other with the circles and spheres of a certain group of spheres, are regarded as equivalent to each other on this view, in so far as they include in themselves the same content of conceptual dependencies along with a mere change of the intuitive "subjects," of which the dependencies are predicated. In this sense, the "points" with which ordinary Euclidian geometry deals can be changed into spheres and circles, into inverse point-pairs of a hyperbolic or elliptical group of spheres, or into mere number-trios without specific geometrical meaning, without any change being produced in the deductive connection of the individual propositions, which we have evolved for these points.[23] This deductive connection constitutes a distinct formal determination, which can be separated from its material foundation and established for itself in its systematic character. The particular elements in this mathematical construction are not viewed according to what they are in and for themselves, but simply as examples of a certain universal form of order and connection; mathematics at least recognizes in them no other "being" than that belonging to them by participation in this form. For it is only this being that enters into proof, into the process of inference, and is thus accessible to the full certainty, that mathematics gives its objects.

Geometry as pure "doctrine of relations" (Hilbert). This interpretation of the methods of pure mathematics receives its clearest expression in the procedure which Hilbert has applied in the exposition and deduction of the geometrical axioms. In contrast to the Euclidian definitions, which take the concepts of the point or the straight line as immediate data of intuition, from which fixed content they proceed, the nature of the original geometrical objects is here exclusively defined by the conditions to which they are subordinated. The beginning consists of a certain group of axioms, which we assume, and their compatibility has to be proved. From these rules of connection, that we have taken as a basis, follow all the properties of the elements. The point and the straight line signify nothing but structures which stand in certain relations with others of their kind, as these relations

[23] *Cf.* the very instructive examples and explanation given by Wellstein, *Encyklopädie der Elem. Mathematik.*, Vol. II, Bk. I, 2nd. sect.

are defined by certain groups of axioms. Only this systematic "complexion" of the elements, and not their particular characters, is taken here as the expression of their essence. In this sense, Hilbert's geometry has been correctly called a pure theory of relations.[24] In this, however, it forms the conclusion to a tendency of thought, which we can trace in its purely logical aspects from the first beginnings of mathematics. At first, it might seem a circle to define the content of the geometrical concepts exclusively by their axioms: for do not the axioms themselves presuppose certain concepts in their formulation? This difficulty is disposed of when we clearly distinguish the psychological beginning from the logical ground. It is true that, in the psychological sense, we can only present the meaning of a certain relation to ourselves in connection with some given terms, that serve as its "foundations." But these terms, which we owe to sensuous intuition, have no absolute, but rather a changeable existence. We take them only as hypothetical starting-points; but we look for all closer determination from their successive insertion into various relational complexes. It is by this intellectual process that the provisional content first becomes a fixed logical object. The law of connection, therefore, signifies the real $\pi\rho\delta\tau\rho\sigma\nu$ $\tau\tilde{\eta}$ $\varphi\upsilon\sigma\epsilon\iota$, while the elements in their apparent absoluteness signify only a $\pi\rho\delta\tau\epsilon\rho\sigma\nu$ $\pi\rho\delta s$ $\dot{\eta}\mu\tilde{a}s$. Intuition seems to grasp the content as an isolated self-contained existence; but as soon as we go on to characterize this existence in judgment, it resolves into a web of related structures which reciprocally support each other. Concept and judgment know the individual only as a member, as a point in a systematic manifold; here as in arithmetic, the manifold, as opposed to all particular structures, appears as the real logical *prius*. (*Cf.* above p. 68). The determination of the individuality of the elements is not the beginning but the end of the conceptual development; it is the logical goal, which we approach by the progressive connection of universal relations. The procedure of mathematics here points to the analogous procedure of theoretical natural science, for which it contains the key and the justification. (*Cf.* Ch. V.)

The syntheses of generating relations. From this point of view, we can understand how the center of gravity of the mathematical system has moved in a definite direction in its historical development. The circle of objects to which mathematics is applicable is extended;

[24] Wellstein, *op. cit.*, p. 116.

finally it becomes clear that the peculiarity of the method is bound and limited by no particular class of objects. The *"mathesis universalis,"* in the philosophical sense, which it had for Descartes, was to form the fundamental instrument for all problems relating to order and measure. Leibniz replaced this mere conjunction of two aspects by a relation of logical subordination; the doctrine of the different possible types of connection and arrangement was made the presupposition of the science of measurable and divisible magnitudes.[25] Modern mathematics clarifies this conception. The development of projective geometry realizes the ideal of mathematical exposition independently of the instrumentalities of measure and the comparison of magnitudes. Metrical geometry itself is deduced from purely qualitative relations, that merely concern the relative position of the points of space. The extension of mathematics beyond traditional bounds is still more striking in the case of the theory of groups; there the immediate object is not determinations of magnitude or position, but a system of operations, which are investigated in their mutual dependency. In the theory of groups, for the first time the supreme and universal principle is reached, from which the total field of mathematics can be surveyed as a unity. In its general meaning, the task of mathematics does not consist in comparing, dividing or compounding given magnitudes, but rather in isolating the generating relations themselves, upon which all possible determination of magnitude rests, and in determining the mutual connection of these relations. The elements and all their derivatives appear as the result of certain original rules of connection, which are to be examined in their specific structure as well as in the character that results from their composition and interpenetration. The various forms of calculus of modern mathematics, Grassmann's *Ausdehnungslehre,* Hamilton's theory of quaternions, the projective calculus of distances, are only different examples of this logically universal

---

[25] *Cf.* above p. 121, note 2; also Leibniz' *Hauptschriften* (Phil. Bibl. Vol. 107), Leipzig, 1904, p. 5, p. 50, p. 62. For the modern interpretation, see Russell, *Principles of Mathematics,* p. 158 and 419: "Quantity, in fact, though philosophers appear still to regard it as very essential to Mathematics, does not occur in pure Mathematics, and does occur in many cases not at present amenable to mathematical treatment. The notion which does occupy the place traditionally assigned to quantity is *order."* *Cf.* Gregor Itelson's definition of mathematics as the science of ordered objects. (*Revue de Métaphysique,* XII, 1904.)

procedure. The methodological merit of all these procedures consists precisely in that the "Calculus" here achieves completely free and independent activity (*Betätigung*), in that it no longer remains limited to the compounding of quantities, but is directly applied to the synthesis of relations.

We were able to trace this synthesis as the real goal of mathematical operations in the field of magnitude itself in the development of the analysis of the infinite. (*Cf.* above p. 73 f.) Now, however, the sphere of consideration is widened; for any arbitrarily chosen element can serve as a foundation in so far as a new structure can be made to issue from it by repeated application of a certain defined relation. It is merely this possibility of determination, which is retained in the calculus and constitutes its necessary and sufficient condition. The certainty of the deductive structure is bound to no particular element. We can deal with products of points or of vectors, as in Grassmann's geometrical characteristic and in the theory of quaternions; we can have points characterized not only by their different positions in space, but by different mass-values, as in Möbius' barycentric calculus; we can compound distances or triangular surfaces, forces or pairs of forces in any way with each other and establish the result by calculation.[26] In all these cases, we are not concerned in analyzing a given "whole" into parts similar to it, or in compounding it again out of these, but the general problem is to combine any conditions of progression in a series in general into a unified result. If an initial element is defined and a principle given by means of which we can reach a manifold of other elements by a regular progression, then the combination of several such principles will be an operation, which can be reduced to fixed systematic rules. Wherever such a transition from simple to complex series is possible, there a new field for deductive mathematical treatment is defined.

Grassmann's Ausdehnungslehre and its logical principles. It seems to have been this general thought, as it evolved in strict sequence from the philosophical ideal of Descartes and Leibniz of "*mathesis universalis*," which led also to one of the most weighty and fruitful conceptions of modern mathematics, *viz.*, to Hermann Grassmann's *Ausdehnungslehre*. The general considerations, which Grassmann

[26] *Cf.* more particularly regarding these methods of calculation, Whitehead, *Universal Algebra* I, Cambridge, 1898, as well as H. Hankel, *Theorie der komplexen Zahlensysteme*, Leipzig, 1867.

prefaced to his work, if regarded as mathematical definitions, might occasionally seem unsatisfactory and obscure; nevertheless, they signify a clear methodological project, whose significance is explained and confirmed by the further development of the problems.[27] The goal, which Grassmann set himself, was to raise the science of space to the rank of a universal science of form. The character of a pure science of form is defined by the fact that in it proof does not go beyond thought itself into another sphere, but remains entirely in the combination of different acts of thought. This postulate is fulfilled in the science of numbers; for all details in the field of number can be entirely deduced from the system of ordered postulations, to which the number series itself owes its being. But as "immediate" a "beginning" must now be gained for geometry as is already given and assured within arithmetic.[28] For this purpose, we must here also go back from the given extensive manifold to its simple "manner of generation," by virtue of which the manifold is first surveyed and grasped. In the ordinary account of the geometrical elements, we are accustomed to speak of the generation of the line out of the point, of the surface out of the line; but what is here meant as a mere picture, must receive a strictly conceptual interpretation in order to serve as the starting-point of the new science. The intuitive spatial relations may offer the first occasion for grasping pure conceptual relations; but they do not exhaust the real content of the latter. Instead of the point (*i.e.*, the particular place), we now assume the *element*, by which is meant only a pure particular grasped as different from other particulars. A specific content is thus not yet assumed: "there can be no thought here as to what sort of particular this really is—for it is the particular absolutely, without any real content; nor can there be question as to in what relation one particular differs from another,—for the particular is defined as different absolutely, without the assumption of any real content with reference to which it is different."[29] In the same way, we expressly abstract from all special characters of the changes, which we think of the fundamental element as undergoing, and merely retain the abstract thought of an

---

[27] *Cf.* especially V. Schlegel, *Die Grassmannsche Ausdehnungslehre. Ztschr. f. Mathem. u. Physik*, Vol. 41, 1896.

[28] Grassmann, *Die lineale Ausdehnungslehre: ein neuer Zweig der Mathematik* (1844). *Ges. mathemat. u. physikal. Werke*, Leipzig 1894, I, p. 10, p. 22.

[29] *Ausdehnungslehre, op. cit.*, p. 47.

original beginning, from which there issues by continuous repetition of one operation a multiplicity of members. Thus if the concrete working out of Grassmann's *Ausdehnungslehre* is primarily concerned with definite kinds of transformation, nevertheless, the total scheme from the beginning reaches further. Here we are occupied only with that aspect, which stands out as the most general function of the mathematical concept: with giving some qualitatively definite and unitary rule that determines the form of the transition between the members of a series. "The Different must evolve according to a law, if the result is to be definite. The simple form of extension is thus that which arises by a transformation of the generating element according to one law; the totality of the elements, which can be generated according to the same law, we call a system or a field."[30] Similarly, there arise systems on higher planes when we combine different transformations in such a way, that first a manifold evolves by a certain transformation from the initial element, and then the totality of its members is subjected to a new transformation. In as much as the fields, which we consider, are not *given* us from elsewhere, but are merely known and defined by the rule of their construction, it is clear that this rule must suffice to represent exhaustively all their properties.

These general considerations gain a more precise mathematical meaning, when Grassmann goes on to develop the various possible forms of connection in detail and to limit them from each other by the formal conditions to which they are subjected. There results a developed doctrine of the "addition" and "subtraction" of similar or dissimilar transformations, a theory of external and internal multiplication of distances and points, *etc*. All these operations agree with the algebraic procedures of the same names merely in certain formal peculiarities, such as subjection to the associative or the distributive law; but in and for themselves they represent entirely independent processes, by which a new structure can be definitely determined from any given elements. We advance from the relatively simple forms of "generation," which we have established by definition, to ever more complex ways of constructing a manifold out of certain fundamental relations. If an initial member α_0 is assumed and a plurality of operations $R_1 R_2 R_3$ . . . . indicated at the same time, which successively transform it into the different values

[30] *Ausdehnungslehre*, p. 28.

$\alpha_1 \ \alpha_2 \ \alpha_3, \ \alpha_1' \ \alpha_2' \ \alpha_3'$ etc., then the result of the compounding of these operations and the various possible types of this compounding is deductively determined. The considerations by which Grassmann introduces his work thus create a general logical schema under which the various forms of calculus, which have evolved independently of the *Ausdehnungslehre*, can also be subsumed; for they only show from a new angle that the real elements of mathematical calculus are not magnitudes but relations.

The forms of calculus, and the concept of the Source. If we survey the whole of these developments, we recognize at once how the fundamental thought of logical idealism has been progressively confirmed and deepened in them. More and more the tendency of modern mathematics is to subordinate the "given" elements as such and to allow them no influence on the general form of proof. Every concept and every proposition, which is used in a real proof and is not merely related to pictorial representation, must be fully grounded and understood in the laws of constructive connection. The logic of mathematics, as Grassmann understands it, is, in fact, in a strict sense "logic of the source" (*Logik des Ursprungs*). Cohen's *Logik der reinen Erkenntnis* developed its fundamental thought of the Source in connection with the principles of the infinitesimal calculus.[31] Here, in fact, is the first and most striking example of the general point of view, which leads from the concept of magnitude to the concept of function, from "quantity" to "quality" as the real foundation. In advancing to the other fields of modern mathematics, the logical principle here established gains new confirmation. However different these fields may be in content, in structure they all point back to the fundamental concept of the Source. The postulate of this concept is fulfilled wherever the members of a manifold are deduced from a definite serial principle and exhaustively represented by it. The most diverse forms of "calculus," in so far as they satisfy this condition, belong to one logical type, as also they agree in their fruitfulness for the problems of mathematical natural science. Thus Möbius applied his universal calculus to a strictly rational construction of statics, while Maxwell evolved the elements of mechanics from the fundamental concepts of vector analysis.[32] The systematic

[31] Cohen, *Logik der reinen Erkenntnis*, especially p. 102 ff.
[32] Möbius, *Lehrbuch der Statik* (T. I, 1837); *cf.* especially Hankel, *Theorie der komplexen Zahlensysteme*, VII; Maxwell, *Matter and Motion*.

connection of operations, once deduced, remains unchanged when we substitute forces for straight lines, pairs of forces for certain distance-products, and thus relate every geometrical proposition directly to a mechanical proposition. The subordination of the infinitesimal analysis to the more inclusive system of "analysis of relations" as such serves also to fix and limit its own problem. In spite of the protests of idealistic logic, the concept of the "infinitely small" has continually led to the misunderstanding that here magnitudes are not *understood* from their conceptual principle, but rather *compounded* from their disappearing parts. Thereby, however, the real question is mistaken and displaced; for we are not concerned with pointing out the ultimate substantial constitution of magnitudes, but merely with finding a new logical point of view for their determination. This point of view comes out sharply, however, when we place the other possible forms of mathematical "determination" by the side of the procedure of the infinitesimal calculus. For example, it would be nonsense to ascribe the ordinary "arithmetical" meaning to operations in the barycentric calculus, such as when simple points are added or the sum of two distances with direction is represented by the diagonal of the parallelogram constructed out of them,—or when we speak of the product of two or three points or of the product of a point and a distance. The relation of the "whole" to its component "parts" is here excluded and replaced by the general relation of the conditioned to the individual moments, which conceptually constitute it. The distinction, clearly emphasized by Leibniz, is unavoidable: in contrast to the "analysis into parts," there appears everywhere the "resolution into concepts," which as the universal instrument guarantees the certainty and the progress of pure deduction.

IV

The problem of metageometry. The extension, which the system of Euclidean geometry has undergone through metageometrical investigations and speculations, falls in point of content outside of the sphere of our enquiry. For we are not concerned with presenting the results of mathematics, significant and fruitful as they may be from the standpoint of the critique of cognition, but merely with determining the principle of the mathematical construction of concepts. But even from this limited point of view, we cannot avoid taking up the problem of metageometry; for it is the special distinction of this problem that it has not merely transformed the content of mathematical knowledge, but also the interpretation of its basis and source. The question necessarily arises as to whether the view, which has previously been gained of the mathematical concept, can be maintained in the face of the new problems arising in this connection. That there is here a justified extension of the original field of geometry is now beyond question with philosophers as with mathematicians; so much the more necessary is it to discover whether the new content breaks through the logical form of geometry or confirms it.

The attempt at an empirical grounding of geometry (*Pasch*). The answer of mathematics itself seemed for a time definitive; in general it was the *empirical* character of the geometrical concepts, that was deduced from the metageometrical researches. Veronese's *Fondamenti di geometria*, the first complete historical survey of all critical attempts to reform the theory of the principles of geometry, affirms as a common conviction of scientific investigators, that at least the ordinary geometry of tridimensional space is founded merely on experience.[33] If we examine the motives and reasons more closely, which have led individual investigators to this decision, we soon recognize that the agreement of interpretation is only apparent. It is as if geometry, on entering the field of philosophical speculation, had lost its characteristic privilege of applying its concepts in a strictly unambiguous sense. The whole indefiniteness belonging to the concept of experience in popular usage at once comes to light. An empirical grounding of the mathematical concepts would only be given, in the strict sense, where proof was adduced that their entire content was rooted in concrete perceptions, and deducible from them. Thus the one consistent empiristic system of mathematics has been constructed by Pasch, in so far as he attempts to introduce the elementary structures, such as the point and the straight line, not in exact conceptual form but merely in the meaning which they can possess for sensation. The fruitful application, which geometry continually receives in natural science and in practical life, Pasch explains, can only rest on the fact that its concepts originally correspond exactly to the actual objects of observation. Only secondarily is this original content overlaid with a net-

---

[33] Veronese, *Grundzüge der Geometrie von mehreren Dimensionen und mehreren Arten geradliniger Einheiten*, German ed., Leipzig, 1894, p. VIII, Note 1.

work of technical abstractions, by which indeed its theoretical construction is furthered, but nothing is added to the fundamental truth of its propositions. If we abandon these abstractions, if we turn resolutely back to the real psychological beginnings, geometry retains the character of a natural science and is only distinguished from the other natural sciences by the fact that it only needs to take a very limited number of concepts and laws directly from experience, and gains all the rest by the development of this once assumed material. The "point," according to this conception, is nothing but a material body, which proves to be not further divisible within the given limits of observation, while distance is compounded out of a finite number of such points. The validity of the geometrical principles is accordingly subject to certain limitations, which are demanded by the nature of the geometrical objects as mere objects of perception. Thus to the proposition, that between two points we can draw one and only one straight line, the reservation is to be added that the points considered must not be too close to each other. The theorem, that between two given points a third can always be inserted, remains in force only for these cases, while it loses its validity when we go beyond certain limits, which cannot indeed be clearly assigned.[34]

Ideal objects in empirical geometry. All these developments are consistent with the chosen starting-point; but it soon appears impossible to reach the structure of the total historical system of scientific geometry from it. In order to give the proofs true rigor and universality, we are forced from the assumption of "real" points, that represent actual objects of observation, to the assumption of "unreal" structures, which are ultimately nothing but a result of those ideal constructions, that we originally sought to exclude. The concepts of perfectly determinate points, straight lines and planes are used also, and serve as a basis for the definitions of those elements in which the geometrical idea is only imperfectly and approximately realized. Every geometry of approximation is obliged to operate with presuppositions taken from "pure" geometry; it cannot serve for the deduction of methods, of which it is rather only a special *application.*[35]

---

[34] Pasch, *Vorlesungen über neuere Geometrie*, p. 17 f.

[35] *Cf.* the criticism of Pasch's system by Veronese, p. 655 ff., and by Wellstein, *op. cit.*, p. 128 f.

Veronese's modification of empiricism. The search for an empirical foundation of geometry is thus led into a new path. Veronese, who at first approves the search, gives the thought a new turn, when he urges that geometrical "possibility" is not to be based merely on direct external observation, but also on "mental facts." The geometrical axioms are not copies of the real relations of sense perception, but they are postulates by which we read exact assertions into inexact intuition. The raw material of sense impressions must be worked over by our mind before it can be useful as a starting-point for mathematical considerations; and it is this "subjective" element which in pure mathematics, geometry and rational mechanics asserts its superiority over the "objective" element. Although geometry is defined here also as an exact experimental science, nevertheless, the logical rôle of experience has become entirely different. We start from "empirical considerations," from certain facts of sensuous intuition; but these facts serve, in Platonic language, only as the "spring-board" from which we ascend to the conception of universal systems of conditions with no sensuous correlate. The sensuous contents form indeed the first occasion, but express neither the limit nor the real meaning of the mathematical construction of concepts. They serve as the first incentive, but as such do not enter into the system of deductive proof, which is to be formed in strict independence. But in establishing this, the issue is already decided from the standpoint of the critique of knowledge; for such critique does not ask as to the origin of concepts, but only what they mean and are worth as elements of scientific proof.

Rationalism and empiricism. Thus we are obliged ultimately to appeal to a specific function of the intellect in the deduction of the geometries of more than three dimensions. In the system of Pasch, as Veronese remarks, multiple-dimensional geometry is not excluded *a posteriori* but *a priori*, *i.e.*, not factually but methodically. For the data of observation negate every attempt to enter a field, which lies beyond the possibilities of our spatial intuition. For this always demands a pure act of construction, a possible "intellectual activity" in which we go beyond the given, and in which the generated element is determined from the beginning by the fact that we subject it to certain general laws of relation. As the axioms, propositions and proofs of geometry cannot contain any undefined element of intuition, when we abandon intuition in general, there must at least remain a purely hypothetical connection of abstract truths, which is accessible to intellectual investigation. "If we are called," Veronese adds, "Rationalists or Idealists because of these ideas, we accept the title in distinction from those who would unjustifiably deny the greatest possible logical freedom to the mathematical and geometrical intellect, and who would enquire whether each new hypothesis possesses a possible perceptual representation, *e.g.*, in geometry a purely external perceptual representation. We accept the title, however, only under the condition that no really philosophical meaning be attached to it." The "really philosophical" meaning, that is here guarded against, is,—as the reference to P. du Bois-Reymond shows,[36]—only the hypostatization of mathematical ideals into a sort of absolute existences; their purely intellectual value as hypotheses is not being thereby affected.[37]

Mathematical space and sensuous space. The logical freedom here sought for geometrical concepts cannot, however, merely relate to those that apply to more than three-dimensional spaces; in so far as a true unity of principles is sought, it must be recognized in the methods of ordinary Euclidean geometry. If the "point" of this geometry were only the image of an object existing outside of thought, "because there are outer objects which directly (!) present or arouse in us the perception of a point without which there are no real so-called points,"[38] the continuity of the system of geometry would be broken; for what conceptual analogy and affinity subsists between elements, which are copies of presented things and elements that entirely result from "intellectual activities"? And conversely, if those intellectual procedures suffice to constitute the element of an *n*-dimensional manifold, what difficulty is there in gaining the element in the special case of three dimensions? In fact, it is precisely when we compare Euclidean space with other possible "forms of space" that its peculiar conceptual character stands forth sharply. If from the standpoint of metageometry, Euclidean geometry appears as a mere beginning, as given material for further developments, nevertheless, from the standpoint of the critique of knowledge, it represents the end of a complicated series of intellectual operations. The psychological investigation of the origin of the idea of space (including those which were

[36] *Cf.* more particularly Ch. IV, p. 162 ff.

[37] Veronese, *op. cit.*, p. VIII ff., XIII ff., p. 658, 687, *etc.*

[38] Veronese, *op. cit.*, p. VII, *cf.* p. 225 f.

undertaken with a purely sensationalistic tendency) have indirectly confirmed and clarified this. They show unmistakably that the space of our sense perception is not identical with the space of our geometry, but is distinguished from it in exactly the decisive constitutive properties. For sensuous apprehension, every difference of place is necessarily connected with an opposition in the content of sensation. "Above" and "below," "right" and "left" are here not equivalent directions, which can be exchanged with each other without change, but they remain qualitatively distinct and irreducible determinations, since totally different groups of organic sensations correspond to them. In geometrical space, on the contrary, all these oppositions are cancelled. For the element as such possesses no specific content, but all its meaning comes from the relative position it occupies in the total system. The principle of the absolute homogeneity of spatial points denies all differences, like the difference of above and below, which merely concern the relation of outer things to our bodies, and thus to a particular, empirically given object.[39]   Points are what they are only as starting-points of possible constructions, in which the postulate holds that the identity of these constructions can be recognized and retained through all diversity of the initial elements. The further moments of geometrical space, such as its continuity and infinity, rest upon a similar foundation; they are in no way given in spatial sensations, but rest upon ideal completions, which we assume in them. The appearance that the continuity of space is a sensuously phenomenal property has been definitely set aside by the deeper mathematical analysis of the continuum, which has been carried out through the modern theory of the manifold. The concept of the continuum used by the mathematician in his deductions is in no way to be gained from the indefinite image of space, that is offered us by sensuous intuition. This image can never represent precisely the ultimate deciding difference by which continuous manifolds are distinguished from other infinite totalities; no sensuous power of discrimination, however sharp, can discover any difference between a continuous and a discrete manifold in so far as

[39] Concerning the differentiation of "homogeneous" geometrical space from inhomogeneous and "anisotropic" physiological space, *cf.* more particularly Mach, *Erkenntnis u. Irrtum*, Leipzig, 1905, p. 331 ff. *Cf.* especially the exposition of Stumpf, *Zur Einteilung der Wissenschaften* (*Abhandl. der Berliner Akademie d. Wiss.*, 1906, p. 71 ff.).

the elements of the latter are "everywhere dense," *i.e.*, where between any two members, however close we may choose them, another member can be discovered belonging to the assemblage itself.[40] Just as the field of rational numbers is broadened by gradual steps of thought into the continuous totality of real numbers, so by a series of intellectual transformations, does the space of sense pass into the infinite, continuous, homogeneous and conceptual space of geometry.

Objections to the Kantian theory of geometry. It is thus a strange anomaly, when from the possibility of metageometry the empirical character of Euclidean space is inferred. Euclidean geometry does not cease to be a purely rational system of conditions and consequences, when it is shown that along with it other systems can be thought, which are capable of the same logical strictness of connection. It is to be noted that two opposite objections, based upon the *same* premises which are taken from metageometry, have been expressed against the Kantian theory of geometry. On the one hand, the pure apriority of space is contested on the basis of these premises, while on the other hand, it is objected that in Kant's own exposition, the *a priori* freedom of the mathematical concept and its possible separation from all sensuous representation is not satisfactorily expressed. Kant's view that the axioms were "given" in "pure intuition" can only be explained "by that residuum of sensualism which still attached to the Kantian idealism."[41] Of these two opposed objections, only the last possesses an entirely clear and consistent meaning. Not the empirical but the logical character of the fundamental concepts is confirmed and illumined in a new way by the modern extension of the field of mathematics. The rôle, which we can still ascribe to experience, does not lie in founding the particular systems, but in the selection that we have to make among them. It is reasoned that, as all the systems are equally valid in logical structure, we need a principle that guides us in their application. This principle can be sought only in reality, since we are not here concerned with mere possibilities, but with the concept and the problem of the real itself; in short, it can be sought only in observation and scientific experiment. Experience thus never serves as a

---

[40] For explanation and examples, *cf.* especially Huntington, "The Continuum as a type of order," *Annals of Mathematics*, 2 ser., VI and VII; compare my "Kant und die moderne Mathematik," *Kant Studien*, XII, 15 ff.

[41] Wellstein, *op. cit.*, p. 146.

proof or even as a support of the mathematical system of conditions, for such a system must rest purely in itself; but it points the way from the truth of concepts to their reality. Observation closes the gap left by purely logical determination; it leads from the many forms of geometrical space to the one space of the physical object.

Real space and experiment. This connection leads, however, beyond the bounds of pure mathematics and results in a problem, which can only be adequately solved by a critical analysis of the procedure of physics. The question of the method and value of the physical experiment itself now becomes of central importance. If one looks to experiment for the confirmation or refutation of a certain system of mathematical hypotheses, experiment is essentially understood in the Baconian sense of the *"experimentum crucis."* Experience and hypotheses belong accordingly to separate fields; each exists for itself and can function by itself. "Pure" experience, which is conceived as separated from any conceptual presupposition, is appealed to as a criterion of the value or lack of value of a certain theoretical assumption. The critical analysis of the concept of experience shows, on the contrary, that the separation here assumed involves an inner contradiction. Abstract theory never stands on one side, while on the other side stands the material of observation as it is in itself and without any conceptual interpretation. Rather this material, if we are to ascribe to it any definite character at all, must always bear the marks of some sort of conceptual shaping. We can never oppose to the concepts, which are to be tested, the empirical data as naked *"facta"*; but ultimately it is always a certain logical system of connection of the empirical, which is measured by a similar system and thus judged.[42] But if the measuring experiment is always bound in this way to a system of presuppositions, which include both purely geometrical assumptions concerning space and concrete physical assumptions concerning the relations of bodies, then it is clear that we can expect no clear decision from it with regard to the conflict of geometrical systems. Wherever a value gained by experiment contradicts the value demanded by deductive theory, the alternative is left open to us whether we shall restore the agreement of concept and observation by changing the *mathematical* part or the *physical* part of our abstract hypothesis. And thought would undoubtedly avail itself of this latter procedure. The possible

[42] *Cf.* here the detailed grounding in Ch. IV, especially Sect. IV.

variation of the conditions follows certain rules and is bound to a certain sequence. Before we would proceed on the basis of the results of astronomical measurements, to change from the geometry of Euclid to the geometry of Lobatschefski, we would first have to investigate as to whether we could take account of the new result by an altered conception of the system of physical laws, for example, by revising the assumption of the strictly rectilinear propagation of light. This state of affairs has been continually emphasized from the philosophical side in the controversies about the principles of geometry; but it seems that it was first through the expositions of Poincaré, which were decisive in this connection, that it became clear and gained wide recognition within mathematics. As Poincaré justly emphasizes, all our experiences are related only to the relations of bodies to each other and their physical interactions, but never to the relation of bodies to pure geometrical space, or the parts of this space to each other. It is thus vain to expect instructions about the "essence" of space from a procedure which, according to its whole tendency and disposition, is directed upon entirely different questions. Since the objects with which experience deals are of an entirely different sort from the objects of which the assertions of geometry hold, since the investigation of material things never directly touches the ideal circle or straight line, we never gain in this way a decision among the different systems of geometry.[43]

The conceptual principles of pure space. Thus, if the choice between the various systems is not to be surrendered entirely to subjective caprice, we must face the problem of discovering a rational criterion of difference. Logical consistency, such as belongs to all these systems, is merely a negative condition, which they all share among themselves. But within the group thus established the differences in fundamental structure and in relative simplicity of structure are not extinguished. While from the standpoint of the principles of identity and contradiction, the thought of the heterogeneity of space may be equivalent to that of homogeneity, nevertheless, there can be no doubt that within the rational system of knowledge the concept of uniformity, in the most diverse fields, always precedes that of non-uniformity. The non-uniform is always gained from the uniform in the process of constructive synthesis by the addition of a new condition, and thus represents a more complex intellectual structure.

[43] *Cf.* Poincaré, *La Science et l'Hypothèse*, Chs. 3-5.

The form of Euclidean space is thus in fact "simpler" than any other form of space, in the same sense that within algebra a polynomial of the first degree is simpler than a polynomial of the second degree.[44] In the order of knowledge, at least, there is here a necessary and definite sequence; but it is this order of knowledge by which, in the critique of cognition, we determine the order of objects. The differences between Euclidean space and the space represented in the hypothesis of Lobatschefski or of Riemann first become manifest when we compare parts of these spaces, which transcend a certain magnitude, with each other. If we limit ourselves, on the contrary, to the generating element of all these spaces, the difference disappears. The Euclidean standard holds without modification for measurements in infinitesimals, which thus proves it in principle really fundamental. It represents the first and fundamental schema with which all other constructions are connected and from which they are distinguished. The uniformity of Euclidean space is really only an expression of the fact that it is conceived merely as a pure relational and constructive space and that all further determination of content, which might lead to a difference in absolute magnitude and in absolute direction, is eliminated from it.[45] In so far as absolute determinations of magnitude as such are permissible in pure geometry, they always rest on a universal system of relations, which have been previously developed independently and which is only more closely determined in details by the addition of particular conditions.

Euclidean space and the other forms of mathematical space. Thus Euclidean space remains, indeed, a conceptual hypothesis in a system of possible hypotheses; but within this system, nevertheless, it possesses a peculiar advantage in value and significance. From a system of pure logico-mathematical forms, we select a manifold that corresponds to certain rational postulates, and attempt with the help of this manifold to render the character of the real intelligible. We do not thereby exclude that, along with the fundamental system, the more complex systems also possess a certain sphere of application in which they also gain concrete significance. In the first place,

---

[44] *Cf.* Poincaré, *op. cit.*, p. 61.

[45] *Cf., e.g.*, Grassmann, *Ausdehnungslehre* of 1844, §22: "The simplicity of space is expressed in the principle: space has the same properties in all places and in all directions, that is, in all places and in all directions, the same constructions can be produced."

the results of these systems are often themselves capable of an interpretation and translation which brings them, at least indirectly, to
intuitive representation.   As Beltrami has shown, the relations of
Lobatschefskian geometry find their exact correlate and copy in the
geometry of pseudo-spherical surfaces, itself a particular section
of ordinary Euclidean geometry; while the "elliptical geometry" of
planes, as developed by Riemann, corresponds to the geometry of
spherical surfaces within the Euclidean space of three dimensions.
And also, when we go over to systems of higher dimensions, this
possibility of referring back does not cease.   We can again select
within our intuitive space itself structures that are subject in all
their reciprocal relations to the abstract rules, which are deduced and
proved for any manifold of more than three dimensions.   Thus the
manifold of all spheres forms a linear manifold of four dimensions,
the form of which can be investigated and established in universal
geometry.[46]   But even where we lack this reduction to known spatial relations and problems, the possibility is not excluded of interpreting the propositions of non-Euclidean geometry so that a definite
concrete "meaning" corresponds to them.   For all these propositions
only express a system of relations, while they make no final determination of the character of the individual members, which enter into
these relations.   The points, with which they are concerned, are not
independent things, to which in and for themselves certain properties
are ascribed, but they are merely the assumed *termini* of the relation
itself and gain through it all their character.   (*Cf.* above p. 94 ff.)
Hence, where any system is found in accordance with the rules of
connection of any of these general theories of relation, a field of
application for the abstract propositions is indicated and defined, no
matter whether the qualitative character of the elements of the
system can be pointed out or whether it can be intuitively represented
in space.   In so far as physics offers us systems, which require a
plurality of means of determination for their complete exposition,
we can speak of a manifold of several "dimensions," to be judged and
treated according to the previously evolved deductive laws of these
manifolds, regardless of whether or not these means of determination permit a spatial interpretation.

Geometry and reality.   In any case, the result is that the purely
rational form of the geometrical construction of concepts, as the

---

[46] *Cf.* more particularly Wellstein, *op. cit.*, p. 102.

latter has been gradually established, is not threatened by the metageometrical considerations but is rather confirmed. Even if one heeds all the doubts that may be awakened through these considerations, still these doubts never concern the real ground of the concepts, but only the possibility of their empirical application. That experience in its present scientific form gives no occasion to go beyond Euclidean space is expressly admitted even by the most radical empiristic critics.[47] From the standpoint of our present knowledge, they also conclude, we are justified in the judgment that physical space "is to be regarded as positively Euclidean." Only we must not exclude the possibility that in a distant future perhaps changes will take place here also. If any firmly established observations appear, which disagree with our previous theoretical system of nature, and which cannot be brought into harmony with it even by far-reaching changes in the physical foundations of the system, then, all conceptual changes within the narrower circle having been tried in vain, the query may arise whether the lost unity is not to be reëstablished by a change in the "form of space" itself. But even if we take into account such possibilities, the proposition would only be thereby strengthened that, as soon as we enter the field of the determination of reality, no assertion, however indubitable it may appear, can lay claim to absolute certainty. It is only the pure system of conditions, which mathematics erects, that is absolutely valid, while the assertion, that there are existences corresponding to these conditions in all respects, possesses only relative and thus problematic meaning. The system of universal geometry shows that this sphere of problems does not affect the logical character of mathematical knowledge as such. It shows that the pure concept on its side is prepared and fitted for all conceivable changes in the empirical character of perceptions; the universal serial form is the means by which every order of the empirical is to be understood and logically mastered.

[47] Enriques, *Problemi della Scienza*, Bologna, 1906, p. 293 ff. (See Ch. VI of the Supplement on Einstein's theory of relativity. Tr.)

CHAPTER IV

The Concepts of Natural Science

I

The constructive concepts and the concepts of nature. The logical nature of the pure functional concept finds its clearest expression and most perfect example in the system of mathematics. Here a field of free and universal activity is disclosed, in which thought transcends all limits of the "given." The objects, which we consider and into whose objective nature we seek to penetrate, have only an ideal being; all the properties, which we can predicate of them, flow exclusively from the law of their original construction. But precisely here, where the productivity of thought unfolds most purely, its characteristic limit seems to come to light. The constructive concepts (*Konstruktionsbegriffe*) of mathematics may be fruitful and indispensable in their narrow field; but they seem to lack an essential element for serving as an example for the whole circle of logical problems, as typical of the properties of the concept in general. For however much logic limits itself to the "formal," its connection with the problems of being is never broken. It is the structure of being with which the concept and the logically valid judgment and inference are concerned. The Aristotelian conception and foundation of the syllogism assume this at all points: ontology gives the basic plan for the construction of logic. (*Cf.* above p. 4 ff.) If this, however, is the case, mathematics can no longer serve as the type and model, for since it remains strictly within the field of its self-created structures, it has in principle no concern with being. The difference between the "generic concept" in the sense of traditional logic and the constructive mathematical concept may be freely granted; but one might be tempted to explain this difference by the fact that within mathematics the final and conclusive function of the concept is not sought, and is accordingly not found. The voluntary limitation, which we assume in it, is justified; but it would be a failure in method, if we were to attempt to solve all logical problems from the narrow standpoint that we have here defined for ourselves. The decision as to the direction of logic cannot be gained by a type of consideration which remains one-sidedly in the ideal. Rather it is the genuine concepts of being, the assertions concerning things and their real properties, which must constitute the true standard. The question as to the meaning and function of the concept gains its final and definitive formulation only in the concepts of nature.

The concept of traditional logic and the scientific ideal of pure description. If we proceed, however, from this conception of the problem, the solution seems to turn at once in favor of the traditional logical view. The concepts of nature know and can know no other task than to copy the given facts of perception, and to reproduce their content in abbreviated form. Here truth and certainty of judgment rest only on observation; there remains no creative freedom and arbitrariness of thought, but the character of the concept is from the beginning prescribed by the character of the material. The more we free ourselves from our own constructions, from the "idols" of the mind, the more purely is the image of outer reality presented to us. It is passive surrender to the object, which here seems to secure to the concept its force and effectiveness. We thus stand again wholly within the general view, that has found its logical expression in the theory of abstraction. The concept is only the copy of the given; it only signifies certain features, which are present and can be indicated in the perception as such. (*Cf.* above p. 5.) The conception of the meaning and task of natural science also corresponds completely to this view. The whole meaning and certainty of the concept as found in natural science depends accordingly on the condition, that it contain no element which does not possess its precise correlate in the world of reality. In order to represent adequately a certain group of phenomena, theory may indeed assume and apply certain hypothetical elements; but in this case, also, the postulate holds that these elements must at least be validated in a *possible* perception. An hypothesis signifies only a gap in our knowledge; it means the assumption of certain data of sensation, that have hitherto been accessible to us in no direct experience, but which are nevertheless regarded as thoroughly homogeneous in their properties with the really perceived elements. Perfect knowledge could abandon this *asylum ignorantiae:* for it, reality would be clearly and completely given as a whole in actual perception.

The apparent logical ideal of physics.   The whole modern philosophy of physics appears at first glance merely as the increasingly rigorous and consistent working-out of this view.   In this view alone, does
the possibility seem given of sharply separating experience and
speculative philosophy of nature; and in it there seems to be indicated a necessary condition, by which the scientific concept of physics
is first defined and completed.   In opposition to the metaphysical
ideal of the explanation of nature, there now appears the more modest
task of *describing* the real completely and clearly.   We no longer
reach beyond the field of the sensible in order to discover the inexperienceable, absolute causes and forces, upon which rest the multiplicity and change of our world of perception.   The content of
physics is rather constituted merely by the phenomena in the form in
which they are immediately accessible to us.   Colors and tones,
smell and taste sensations, sensuous muscle-feelings and perceptions
of pressure and contact are the only material out of which the world
of the physicist is constructed.   What this world seems to contain in
addition, what is added in concepts, as atom or molecule, ether or
energy, is in truth no fundamentally new element, but only a peculiar
guise in which the data of sense appear.   Complete logical
analysis reduces these concepts to their significance, when it
recognizes them as symbols for certain impressions and complexes of
impressions.   The unity of the physical method seems thereby to
be secured for the first time; for now it is no longer compounded from
heterogeneous elements, but in the general concept of sensation the
common denominator is fixed, to which all assertions concerning reality must be ultimately reducible.   Whatever resists this reduction
thereby shows itself to be a factor arbitrarily introduced, which
must disappear in the final result.   The goal of this philosophy of
physics would be reached, if we resolved every concept, which enters
into physical theory, into a sum of perceptions, and replaced it by
this sum; that is, if we retraced the path from the intellectual abbre-
viation (which is what all concepts reveal themselves to be) to the
concrete fullness of the empirical facts.   The exclusion of all elements, which possess no direct sensuous correlate in the world of
perceptible things and processes, would be, accordingly, the true
logical ideal of physics.

Is this the true ideal of physics?   However we may judge concerning the justification of this ideal, its very conception contains an

ambiguity, which must first be set aside. The description of the actual status of physical theories is confused with a general *demand* that is made of these theories. Which of the two elements is the original and determining one? Is it merely the actual procedure of science itself that is here brought to its simplest and shortest expression, or, on the contrary, is this procedure measured by a general theory of knowledge and of reality, which decides concerning its value? In the latter case, whatever the final result might be, the method of consideration would not be changed in principle. Again it would be a certain metaphysics of knowledge which sought to point out the way to physics. The answer to this question can only be won by following the course of physical investigation itself and considering the function of the concept that is involved directly in its procedure. The same impartiality, that is demanded by the positivistic critic with regard to the facts of sense perception, must also be demanded with respect to the more complex facts of knowledge. Here also the first task is to grasp the "factual side" of scientific theory in its purity, before we decide as to the value or lack of value of the view of reality which it contains. Is this theory, as it is historically presented, really only a collection of observations strung together as if on a thread, or does it contain elements, which belong to another logical type and therefore demand another foundation?

II

Numbering and measuring as presuppositions. The first and most striking characteristic which forces itself upon us with regard to any scientific theory, involves a peculiar difficulty when we consider it from the standpoint of the general logical demand for *description of the given.* The theories of physics gain their definiteness from the mathematical form in which they are expressed. The function of numbering and measuring is indispensable even in order to produce the raw material of "facts," that are to be reproduced and unified in theory. To abstract from this function means to destroy the certainty and clarity of the facts themselves. However self-evident, indeed trivial, this connection may seem, it is highly paradoxical in principle when we look back over our general estimate of the principle of mathematical conceptual construction. It has become increasingly clear that all content belonging to the mathematical concept rests on a pure *construction*. The given of intuition forms merely the psychological starting-point; it is first known mathematically when it is subjected to a transformation, by which it is changed into another type of manifold, which we can produce and master according to rational laws. Every such transformation, however, must obviously be abandoned where we are merely concerned with grasping the given as *given* in its specific individual structure and properties. For the purposes of knowledge of nature, in the positivistic sense of the word, the mathematical concept is not so much a justified and necessary instrument to be applied along with experiment and observation as a constant danger. Does it not falsify the immediate existence, revealed to us in sensation, to subject this existence to the schema of our mathematical concepts, and thus to let the empirical determinateness of being disappear into the freedom and caprice of thought?

And yet this danger, however clearly it may be envisaged, is never to be avoided or set aside. No matter how penetratingly the physicist may portray it as empirical philosopher, he directly falls into it again as soon as he sets to work as scientific investigator. There is no exact establishment of a time-space fact, which does not involve the application of certain numbers and measures. One might overlook the difficulty in this, if it were merely a matter of the elementary concepts and structures of mathematics. Although the first of Kepler's laws of planetary motion makes use of the purely geometrical definition of the ellipse as a conic section, and the third of the arithmetical concepts of the square and the cube, at first no epistemological problem might be seen in this; to the naïve comprehension, number and form themselves appear as a sort of physical property, inhering in things precisely as do their color or their lustre and hardness. (*Cf.* above p. 28.) The more this appearance is destroyed in the advance of mathematical conceptual construction, the more strikingly the general question is thrown into relief. For it is precisely the complex mathematical concepts, such as possess no possibility of direct sensuous realization, that are continually used in the construction of mechanics and physics. Conceptions, which are completely alien to intuition in their origin and logical properties, and transcend it in principle, lead to fruitful applications within intuition itself. This relation finds its most pregnant expression in the analysis of the infinite, yet is not limited to the latter. Even so abstract an intellectual creation as the system of complex numbers offers a new example of this connection; Kummer, for instance, has developed the thought that the relations, which prevail within this system, possess their concrete substratum in the relations of chemical combination. "Chemical combination corresponds to the multiplication of the complex numbers; the elements, or more exactly the atomic weights of the same, correspond to the prime factors; and the chemical formulae for the analysis of bodies are exactly the same as the formulae for the analysis of numbers. Even the ideal numbers of our theory are found in chemistry, perhaps only too often, as hypothetical radicals, which have hitherto not been analysed, but which, like the ideal numbers, have their reality in compounds. . . . . The analogies here indicated are not to be regarded as a mere play of wit, but have their justification in the fact that chemistry, as well as the part of the number theory here considered, have both the same fundamental concept as their principle, namely, that of composition, although in different spheres of being."[1] The real problem is, however, precisely this transference of structures, whose whole content is rooted in a connection of purely ideal constructions, to the sphere of concrete factual being. Even here it appears that it is upon a peculiar interweaving of "real" and "not-real" elements, that every scientific theory rests. As soon as we take one step beyond the first naïve observation of isolated facts, as soon as we ask about the *connection* and *law* of the real, we have transcended the strict limits prescribed by the positivistic demand. In order even to indicate this connection and law clearly and adequately, we must go back to a system that develops only universal hypothetical connections of grounds and consequences, and which renounces in principle the "reality" of its elements. That form of knowledge, whose task is to describe the real and lay bare its finest threads, begins by turning aside from this very reality and substituting for it the symbols of number and magnitude.

Mechanism and the concept of motion. The first phase of the scientific theory of nature clearly expresses this. The exact concept of nature is rooted in the thought of mechanism, and can only be reached on the basis of this thought. The explanation of nature, in its later development, may attempt to free itself from

---

[1] Crelle's *Journal*, Vol. 35, p. 360. Cited by Hankel, *Theorie der komplexen Zahlensysteme*, p. 104.

this first schema and to replace it by one broader and more universal; nevertheless, motion and its laws remain the real problem in connection with which knowledge first becomes clear regarding itself and its task. Reality is perfectly understood as soon as it is reduced to a system of motions. This reduction, however, can never be accomplished as long as consideration remains in the sphere of mere data of perception. Motion, in the universal scientific sense, is nothing but a certain relation into which space and time enter. Space and time themselves, however, are assumed as members of this relation not in their immediate, psychological and "phenomenal" properties, but in their strict *mathematical* meaning. As long as we understand by space nothing else than a sum of various visual and tactual impressions, qualitatively different from each other according to the special physiological conditions under which they come into existence, no "motion" is possible in it in the sense of exact physics. This latter demands the continuous and homogeneous space of pure geometry as a foundation; continuity and uniformity, however, never belong to the coexistence of the sensuous impression itself, but only to those forms of manifold, into which we constructively transform it by certain intellectual postulates. (*Cf.* above p. 105.) Thus, from the beginning, motion itself is also drawn into this circle of purely conceptual determinations. It is only in appearance that it forms a direct fact of perception, indeed the fundamental fact, which all outer observation first presents us. Perhaps the change of sensations, *i.e.*, the qualitative difference of successive presentations may be conceived in this way; but this aspect alone is in no sense sufficient to ground the strict concept of motion, that is needed by mechanics. Here unity is demanded along with diversity, identity along with change; and this identity is never provided by mere observation, but involves a characteristic function of thought. The individual positions of Mars, which Kepler took as a basis, following the observations of Tycho de Brahe, do not in themselves alone contain the thought of the orbit of Mars; and all heaping up of particular positions could not lead to this thought, if there were not active from the beginning ideal presuppositions through which the gaps of actual perception are supplemented. What sensation offers is and remains a plurality of luminous points in the heavens; it is only the pure mathematical concept of the ellipse, which has to have been previously conceived, which transforms this discrete aggregate into a continuous system. Every assertion concerning the unitary path of a moving body involves the assumption of an infinity of possible places; however, the infinite obviously cannot be perceived as such, but first arises in intellectual synthesis and in the anticipation of a universal law. Motion is gained as a scientific fact only after we produce by this law a determination that includes the totality of the space and time points, which can be constructively generated, in so far as this determination coördinates to every moment of continuous time one and only one position of the body in space.

Thus, from a new angle it is revealed that even the first approach to mechanics depends upon presuppositions, which go beyond possible sensuous experience. The well-known definition of Kirchoff, which defined the task of mechanics as the complete and unambiguous description of the movements taking place in nature, may be entirely justified in the meaning which its author connected with it, yet without the philosophical consequences that are ordinarily drawn from it being thereby justified. Kirchoff himself leaves no doubt that the "description," at which he aimed, has the exact mathematical equations of motion as a presupposition, and in them is involved the concepts of the material point, of uniform and variable velocity as well as uniform acceleration. All these concepts may justifiably serve the mathematical physicist as fixed and immediate data; but they are in no way such for the epistemologist. For the latter, a "nature" exists in which movements are found as describable objects only as a result of a thorough intellectual transformation of the given. This mathematical transformation, which the physicist assumes to have taken place, constitutes the real and original problem. If the thought of the continuity and uniformity of space, as well as the exact concepts of velocity and acceleration, are grasped and grounded, then with the help of this logical material the totality of possible phenomena of motion can be completely surveyed and mastered in its form; but the question arises all the more urgently as to the intellectual means by which this result is reached.

The "subject" of motion. This ideal dependence stands out most sharply when we pass from the *process* of movement to the *subject* of movement. Again it seems as if this subject could be directly pointed out in perception; it is body, it is a complex of tangible and visible qualities, to which motion is ascribed as a property. Even at this point, however, sharper conceptual analysis meets peculiar difficulties. In order to serve as the subject of movement, the empirical body must be definitely determined, distinguished and limited from all other structures. As long as it is not enclosed in fixed and unchanging limits, by which it is separated from its surroundings and recognized as a whole of individual form,—so long it is unable to furnish a constant point of reference for change. Yet the bodies of our world of perception never satisfy this condition. They owe their determinateness merely to a first and superficial unification, wherewith we unite into a whole parts of space that seem to possess approximately the same sensuous properties. Where such a unity begins and ends can never be determined with absolute exactitude; a keener faculty of sensuous discrimination would show us, at the point where two different bodies seem to be in contact, a constant reciprocal exchange of parts and thus a continuous movement of the limiting surfaces. Only when we ascribe a strict geometrical form to the body and thus raise it from the sphere of the merely perceptible to the determinateness of the concept, does it attain that *identity*, which makes it useful as the "bearer" of motion. And as exact limitation of the body is required with respect to all elements of its outer surroundings, so on the other hand, it is to be demanded that it represent a strict unity in itself. As soon as we think of its individual parts as movable with relation to each other, the supreme condition of the definiteness of the point of reference is again abandoned; in place of the one movement have been substituted as many different movements as there are independently moving particles. Thus a system must be taken as a basis, which is closed off from the outside and also is in itself incapable of further differentiation and disintegration into a plurality of independently moving subjects. The "rigid" body of pure geometry has to be substituted for the perceptible body and its limitless changeability, if the grounding of the exact theory of motion is to be accomplished.

The "limiting concept" and its significance for natural science. (*Karl Pearson*). In fact, the necessity of such a transformation of the problem is recognized and emphasized by the adherents of the theory of "description" themselves. It is Karl Pearson above all, who has described this process with clearness and emphasis in his work, *The Grammar of Science.* As he explains, it is never the contents of perceptions as such that we can use as foundations for the judgments of pure mechanics, as points of application in the expression of the laws of motion. Rather, all these laws can only be asserted with meaning of the ideal limiting structures which we conceptually substitute for the empirical data of sense-perception. Motion is a predicate that is never immediately applicable to the "things" of the surrounding sense-world, but holds solely of that other class of objects, which the mathematician substitutes for them in his free construction. Motion is not a fact of sensation, but of thought; not of "perception" but of "conception." "Startling as it may, when first stated, appear, it is nevertheless true that the mind struggles in vain to clearly realize the motion of anything which is neither a geometrical point nor a body bounded by continuous surfaces; the mind absolutely rebels against the notion of anything moving but these conceptual creations, which are limits unrealizable, as we have seen, in the field of perception." Groups of sensuous impressions can change, can lose old parts and gain new, can form into new groups; but these changes in no way signify the real object of mechanics. "It is in the field of conception solely that we can properly talk of the motion of bodies; it is there, and there only, that geometrical forms change their position in absolute time—that is, move." The contradictions, in which mechanics often becomes involved and which have come to light especially in the attempts to apply the general mechanical laws to the movements of the ether, can be explained for the greater part by the fact, that the two spheres of knowledge here opposed have not been sharply and definitely separated from each other. These contradictions disappear as soon as we learn not to confuse immediately sensuous with conceptual elements, as soon as we give up trying to conceive an intellectual construction for the establishment of a scientific order of phenomena, as itself a particular phenomenal existence. What we can alone accomplish in physics is the construction of a world of geometrical forms; yet these, in the multiplicity of movements we ascribe to them, reproduce and represent with wonderful exactness the complex phases of our sensuous experience. As soon as we read this whole thought-world directly into the sense-world again, as soon as we transform its logical assumptions directly into parts of reality, which would thus be apprehended by sensation, we fall once more into all the antinomies that necessarily inhere in every type of dogmatism, physical as well as metaphysical.[2] All this exposition of Pearson's

[2] *Cf.* Pearson, *The Grammar of Science*, Second edition, London 1900, p. 198 ff., p. 239 ff., 282, 325 *etc.*

is admirable; but we ask in vain how, on these assumptions, mechanics can still be conceived as purely descriptive science. Can it be called a description of perceptual contents to substitute in place of them a system of geometrical ideals, such as are necessarily foreign to the world of our perceptions? If the task of a true "objective" description is to conceive the given as faithfully as possible, neither adding nor subtracting anything: then, on the contrary, it is precisely that sort of transformation of the initial experience, which constitutes the character and value of the intellectual procedure of physics. Instead of a mere passive reproduction, we see before us an active process, which transports what is at first given into a new logical sphere. It would be a strange way of describing what is presented, if for this purpose we concerned ourselves with bare concepts, which can themselves in no way be "presented."

P. du Bois-Reymond's theory of the limiting concept. The question as to the character of the fundamental concepts of natural science merges here into a more general problem. We saw how the first step in the formation of these scientific concepts was to introduce, in place of the members of a certain sensuous manifold, the ideal limit of this manifold. The justification of such construction of a limit cannot be perfectly demonstrated by natural science, as long as it remains purely within its sphere; yet the construction rests on general logical principles. The advantage to be gained from this reduction of the question, however, is slight as long as logic and epistemology themselves have not reached clarity on this point. Here more than anywhere else they both seem entangled in insoluble difficulties; and the only way out for clear thought seems to be not so much in resolving the antinomies, which appear at this point, as rather in understanding and recognizing them in their insolubility. In fact, this decision is expressly represented by a noted mathematician in more recent times. According to him, the consideration of the concept of the mathematical limit leads back to a fundamental metaphysical problem, which, like all problems of this species, is not to be solved according to strict objective criteria, but according to the subjective inclination of the individual investigator. The "general theory of functions," as Paul du Bois-Reymond develops it, illumines this dualism on all sides; but disclaims from the beginning any attempt to remove it. When we raise the question whether there exists an exact limit to a definite given sequence of presentations, as for example to the figures of a decimal fraction, such that the limit possesses the same existence as the members of the sequence themselves, the answer we give cannot be clearly determined by logical and mathematical considerations alone. The simple mathematical problem leads us into the conflict of two general views of the world, which stand irreconcilably opposed. We must choose between these two views of the world: either with empiricism we must assume as existent only what can be pointed out as an individual in the real presentation, or with idealism, affirm the existence of structures, which constitute the intellectual conclusion of certain series of presentations, but which can never themselves be directly presented. The mathematician is not in a position to grant the victory to either one of these fundamental views; all that he can and must do, in order to bring clarity into the foundations of analysis, is to follow them to their ultimate intellectual roots. The solution of the riddle is that it remains and will remain a riddle. "The most persistent observation of our thought-process," says du Bois-Reymond, "and of its relations to perception does not go beyond showing that there are two completely different interpretations, that have equal claim to serve as the foundations of exact science, because no absurd consequence of them is found, at least as long as we are concerned with pure mathematics. . . . . However, it remains a very strange phenomenon that, after the removal of everything which might conceal the truth and when at last one might expect to behold its image clearly and definitely, it appears before us in double form. He, who first noticed through a transparent crystal the double image of a single object, could not have showed it to his friends more moved, than I today, at the end of most careful and eager reflection, am forced to develop before the reader the double interpretation of the foundations of our science."[3]

The problem of existence. It is, in fact, worth while to seek out the origin of this peculiar result; for here we stand at a point, which represents the decisive turning-point of all critique of knowledge. The old question as to the relation of concept and existence, of idea and reality, here meets us once more in a characteristic and original form. Indeed the suspicion must at once arise as to whether the opposition found here between "empiricism" and "idealism" repre-

[3] Paul du Bois-Reymond, *Die allgemeine Funktionstheorie*, Tübingen 1882, p. 2 f.

sents a complete disjunction, whether it contains within itself all possible manners of thinking. In this case only, would the antinomy be insoluble; while if it could be shown that there are problems wholly removed from the opposition, which here serves as a starting-point, and which are thus wholly independent of its solution in their logical structure and validity, the antinomy would lose its sharpness. In truth, it appears even in the first arguments of du Bois-Reymond, that it is not the mathematician but the philosopher and psychologist, who is speaking here. What in the world could "the persistent observation of our thought-process and its relations to perception" contribute to the solution of any particular, specifically mathematical problem? Pure mathematics is precisely characterized by the fact that it abstracts completely from all such investigation of the thought-process and its subjective conditions, and merely directs itself upon the objects of thought as such and their objective logical connection. The manner, in which the concept of existence alone appears within mathematics, confirms this exclusive direction of interest. The student of algebra, who speaks of the "existence" of the numbers e and π, undoubtedly intends to signify no fact of outer physical reality; but just as little is it the presence of certain contents of presentation in any perceiving and thinking subjects, that is to be thereby affirmed. If this were the meaning of the assertion, the mathematical standpoint would lack any means of testing and verifying it; for only experiment and generalizing induction warrant us in making a decision concerning real events in the psychic life of individuals. The existence of the number e means nothing else than that, within the ideal number system, one and only one position is determined definitely and with objective necessity by the series, which we apply in its definition. If we assume the general rule $1 + \frac{1}{1} + \frac{1}{1.2} + \frac{1}{1.2.3} + \ldots$ (in inf.), then by it the system of rational numbers is analysed into two strictly divided classes, of which one contains all elements, that are ever exceeded by the series when it is carried far enough, while the other contains all those elements, with which this is not the case. By virtue of this complete division, which it effects in the field of rational numbers, the series gains a definite relation to the members of this field, since it stands to them in the relation of "before" and "after," and thus of "smaller" and "greater." The validity of all these relations alone justifies us in speaking of a "number" e, and constitutes the entire "being," the complete and

self-contained existence of this number. (*Cf.* above p. 61 f.) The,
determination, which arises in this way, is, although purely ideal,
nevertheless in principle of no other sort than the whole numbers and
fractions: since the value of *e* is just as strictly and sharply distinguished from that of any other number, however near *e* it may
lie, as the value of 1 is distinguished from the value of 1,000. Here
we do not appeal in any way to the faculty of separating presentations
and similar particular contents of perception in consciousness; we
are concerned on both sides with pure concepts, which are sufficiently
divided from each other by the logical conditions, which their definition imposes upon them.

The existence of the limiting point. It appears, indeed, to be otherwise when we turn from the algebraic meaning of limit to its geometrical meaning. The existence of a point seems to be verifiable, in
fact, only by a procedure, which allows us to point it out in intuition
and to distinguish it from other positions. Here, however, certain
limits to further advance are felt, on the basis of the psychological
principle of the threshold of discrimination. If we remain at the
standpoint of the "empiricist," if we hold to the belief that we are
only justified in assuming a particular "thing," where there is a
particular presentation for its representation at our command, then
we see that, according to this assumption, the existence of a limiting
point for any definite converging sequence of points can never be
proved from the consideration of the sequence itself. For example,
if we think of the individual numbers of a convergent series as represented by points on the abscissa, then all these points, as we advance
further in the series, will move nearer and nearer to each other, until
finally our intuition is unable further to separate them. After a
certain member, the terms become indistinguishable and flow into
each other; we are, accordingly, not in a position to decide finally
whether that point, which corresponds to the algebraic limit of the
series, exists as a particular geometrical individual, or whether only
those positions possess reality, which are algebraically expressed
by the members of the series itself. "We demand, in fact, what is
impossible," remarks du Bois-Reymond, "if we demand that the
sequence of points abstracted from the given points shall determine a
point not belonging to the given points. I hold this to be so incon-
ceivable that I affirm, that no intellectual labor will extort from a
brain such a proof for the existence of the limiting point, even if it

united Newton's gift of divination, Euler's clarity and the crushing power of the intellect of Gauss."[4]

Logical idealism on the problem of existence. It is entirely correct that all these powers would not suffice to produce the desired proof; for with the mere question under investigation we have already set ourselves outside the field of pure mathematics. To "prove" the existence of points, in the sense in which existence is here taken, will never be attempted by anyone who has ever made fully clear to himself even the critical refutations of the ontological argument. The deeper ground of all the misunderstandings and contradictions, however, lies here also in the indefiniteness and ambiguity, in which the concept of being is understood. The "being" of the geometrical point is not different in principle from that of the pure numbers, and belongs to no other logical sphere. The construction of the geometrical manifold takes place, it was seen, according to laws thoroughly analogous to those of the systematic development of the system of numbers. Here as there we start from an ideal postulation of unity, and here as there intellectual progress consists in our taking up into the system all elements, that are connected with the original by an unambiguous conceptual relation or a chain of such relations. We saw how the paradoxes of imaginary and infinitely distant points were solved from this standpoint: little as these points could claim for themselves any sort of mysterious "reality" in space, they proved themselves, on the other hand, an expression of valid spatial relations.[5] Their being is exhausted in their geometrical meaning and necessity. (*Cf.* above p. 83 ff.). It is this necessity which the true "idealism" can alone demand and claim for the structures of pure mathematics. On the other hand, the idealist in the sense of du Bois-Reymond, goes far beyond such a demand. "The fundamental view of the idealistic system," we read here, "is thus the real existence not only of that which is presented, but of the intuitions necessarily following from the presentations. . . . . The idealist believes in some sort of existence (*Vorhandensein*) of unpresentable, verbal conclusions of sequences of presentations, generated by our thought process."[6] Here speaks an "idealist,"

[4] *Allgemeine Funktionentheorie*, p. 66 f.

[5] *Cf.* Kerry's pertinent criticism of the doctrine of P. du Bois-Reymond, *System einer Theorie der Grenzbegriffe*, Lpz. and Wien, 1900, p. 175 ff.

[6] *Allgemeine Funktionentheorie*, p. 87. *Cf.* du Bois-Reymond's work, *Über die Grundlagen der Erkenntnis in den exakten Wissenschaften*, Tübingen, 1890, p. 91.

who has permitted his conception to be perverted by his opponent the "empiricist," as we can easily see, when as here he only recognizes the existent (*Vorhanden*) as true. The whole antinomy unfolded in the *Allgemeine Funktionentheorie* disappears as soon as we destroy this confusion of truth and reality, which is common to the advocates of both these.

Consequences of the confusion of truth and reality. The consequences of this confusion appear even more sharply in the interpretation of the fundamental concepts of natural science than in the purely mathematical discussion. These concepts are drawn into the same conflict; they also continually go beyond the given, but this unavoidable process cannot be critically justified and grounded. We cannot abandon the concepts of the absolutely rigid body, of the atom or of force at a distance, although, on the other hand, we must give up all hope of finding a direct verification of them in any part of the outer world of perception. The consciousness of the limits set to all our knowing by its nature and essence is felt increasingly here. Ever anew we find ourselves led to unpresentable elements, which lie behind the known and accessible world of sensuous appearance, and ever again it appears, when we attempt to grasp and analyse them, that no intelligible meaning can be gained. "Our thought is as if paralyzed, and makes no progress." The organ for reality is and remains denied to us. "We are enclosed in the box of our perceptions and for what is beyond them born blind. We cannot have a glimmer of light, for a glimmer is already light: but what corresponds in the real to light?"[7] This radical scepticism, in which the exposition of the foundations of exact knowledge here results, is a consistent and significant consequence. On the basis of this view, in fact, we possess no "organ" for the real; for the necessary *concepts*, which form the real organs for the logical interpretation and mastery of the manifold of sensations, are transformed into mysterious realities behind the phenomena.

The "idealization" of presentations. If this transformation is once understood, however, the mist is again dispelled, which threatened to settle more thickly around the image of scientific reality. Indeed, this image arises first through a process of idealization, in which the indefinite data of sensation are supplanted by their strict conceptual

---

[7] P. du Bois-Reymond. *Über die Grundlagen der Erkenntnis in den exakten Wissenschaften*, Abschn. VIII.

limits.   But the assertion of the objective validity of this process is not the same as the assertion of a new class of objects.   "Our field of thought," affirms du Bois-Reymond's "idealist," "contains not only the mosaic of the perceptible and the images and concepts deduced from it by the process of thought, thus by transformation and combination, but there dwells within us the indestructible conviction . . . . of the presence of certain things outside the system of presentations."[8] This proposition is undoubtedly correct, in so far as by "system of presentations" nothing else is understood than the mass of given perceptions, than the system of colors and tones, of tastes and odors, of pressure and contact sensations.   But the completion of this "mosaic of the perceptible" cannot take place by our simply inserting new "insensible" things into this first empirical reality; for the parts of the mosaic would thereby, indeed, be moved together more closely and densely, but in spite of this, no new form of connection, no deeper relation would be gained. The aggregate of sensuous things must be related to a system of necessary concepts and laws, and brought to unity in this relation. This process of thought, however, demands really more than the mere combination and transformation of parts of presentations; it presupposes an independent and constructive activity, as is most clearly manifest in the creation of limiting structures.   The "empiricist" also must accept this form of idealization; for, without it, the world of perception would not be merely a mosaic but a true chaos.   It is a mere misunderstanding when he affirms that he does not recognize that the absolutely straight line and the absolutely exact plane exist, but only more or less straight lines, more or less exact planes.   For this very discrimination of different stages of exactitude presupposes comparison with the *exact idea*, whose fundamental function is thus here throughout confirmed.   The "being" of the idea, however, consists in this function and needs no other support and no other proof.   Also the ideal concepts of natural science affirm nothing regarding a new realm of separate absolute objects, but they would only establish the inevitable, *logical lines of direction*, by which alone complete orientation is gained within the manifold of phenomena.   They only go beyond the given, in order to grasp the more sharply the systematic structural relations of the given.

---

[8] *Allgemeine Funktionentheorie*, p. 110 f.

As soon as the empiricist, along with du Bois-Reymond, character-izes idealization as throughout justifiable, and explains that he only refuses to accept the *ideal itself*, all conflict is removed in principle.[9] For the existence of the ideal, which can alone be critically affirmed and advocated, means nothing more than the objective logical neces-sity of idealization. That we are not here concerned with such a necessity, not with an arbitrary play of phantasy, becomes clearer the more deeply the concept of the object is analysed into its condi-tions. It is vain to interpret the ideal limits, ascribed to certain sequences on the basis of conceptual criteria, as mere verbal conclu-sions with no real or logical meaning corresponding to them. "The perfect," it is affirmed, "can in no way be grasped as a pictorial presentation. Nevertheless, as it enters our thought and finds application there, ... and as our thought consists in the succession of presentations, so it must be somehow a presentation, and it is, namely,—as a word. The sequence of objective presenta-tions of what is exact have, therefore, as their conclusion, a word for something that cannot be presented."[10] This nominalism, however, fails in the explanation of the concept of limit, as it has already failed in the explanation of the pure numbers. (*Cf.* above p. 43 ff.) For here precisely the characteristic meaning and the real function of the concept of limit is obviously excluded. Between the limit and the members of the series, certain relations hold, which are mathe-matically fixed and cannot be arbitrarily changed. The "number" e stands in certain numerical relations to the other numbers, that are gained from the partial sums of the defining series; it takes its place with them in a series, in which the position of each element, its earlier and later, is unalterably ascribed. Is there any meaning in asserting that there are such relations of order in the sequence, in the greater and smaller of elements, where *one* is taken as an actual, psychologically significant image, while its correlate is made to consist of a mere sound? There can only be valid mathematical relations between ideas and ideas, not between ideas and words.

The relation of the ideal and reality. From this connection with the logic of mathematics, we can explain and understand better why any attempt to interpret the concepts of natural science as mere aggre-

[9] *Allgemeine Funktionentheorie*, p. 118.

[10] *Grundlagen der Erkenntnis*, p. 80; cf. *Allgemeine Funktionentheorie*, p. 95.

gates of facts of perception must necessarily fail. No scientific theory is directly related to these facts, but is related to the ideal limits, which we substitute for them intellectually. We investigate the impact of bodies by regarding the másses, which affect each other, as perfectly elastic or inelastic; we establish the law of the propagation of pressure in fluids by grasping the concept of a condition of perfect fluidity; we investigate the relations between the pressure, tempera- ture and volume of gas by proceeding from an "ideal" gas and com- paring a hypothetically evolved model to the direct data of sensation. "Such extrapolations," says so convinced a "positivist" as Wilhelm Ostwald, "are a procedure very generally applied in science; and a very large part of the laws of nature, especially all *quantitative* laws, *i.e.*, such as express a relation between measurable values, only hold exactly for the ideal case. We thus stand before the fact that many and among them the most important laws of nature are asserted and hold of conditions, *which in reality in general are never found.*"[11] The problem here raised reaches further than appears in this initial formulation. If the procedure of natural science only consisted in substituting the ideal limiting cases for the directly observable phenomena, then we could attempt to do justice to this method by a simple extension of the positivistic schema. For the objects with which the theoretical consideration of nature is concerned, although they fall beyond the real field of empirical perception, seem to lie on the same line with the members of this field; and the laws, that we assert, do not seem to represent a transformation so much as a mere extension of certain perceptible relations. Yet, in sooth, the relation between the *theoretical* and *factual* elements at the basis of physics cannot be described in this simple way. It is a much more complex relation, it is a peculiar interweaving and mutual interpene- tration of these two elements, that prevails in the actual structure of science and calls for clearer expression logically of the relation between principle and fact.

III

The problem of the physical method and its history. In epistemologi- cal discussion of the foundations of natural science, we often meet the view that the ideal of pure description of the facts is a specifically modern achievement. It is thought that here for the first time

[11] Ostwald, *Grundriss der Naturphilosophie* (Reclam), p. 55.

physics has reached true clarity regarding its proper goal and intellectual instruments; while before, in spite of all the wealth of results, the way to these results remained in darkness. The separation of "physics" and "metaphysics," the exclusion in principle of all factors which cannot be empirically confirmed, is thought to be the decisive product of the critically philosophical and most modern research. This view, however, signifies a misapprehension of the continuous course by which physics has reached its present form. From the first scientific beginnings of physics, the problem of method has been continually and vitally important, and it was only in struggling with this problem that physics gained full mastery over its field of facts. Reflection and productive scientific work have never been strictly separated here, but have mutually assisted and illumined each other. And the further back one follows this reflection, the more clearly is a fundamental opposition of viewpoints discoverable in it. This opposition persists in the modern expositions unweakened; but it gains full sharpness and definiteness, when we trace it back to its general systematic and historical sources.

The problem of knowledge. (*Plato.*) As modern investigation has more and more destroyed the prejudice that the scientific use of experiment was unknown to the Greeks, so we can also indubitably recognize in ancient philosophy the theoretical controversy over the principles of empirical knowledge. The conflict, that begins here, affects the whole speculative view. It is expressed in an incomparable and unforgettable picture in the Platonic metaphor of the cave. For the human mind, there are two types of consideration and judgment regarding the phenomena of the world of sense, phenomena which pass across the mind like shadows. The one is satisfied merely with grasping the sequence of the shadows, with fixing their before and after, their earlier and later. Custom and practice gradually enable us to distinguish certain uniformities in the sequence of phenomena, and to recognize certain connections between them as uniformly recurrent, without the grounds of this connection being intelligible to us. Common understanding and the view of the world based upon it do not need these reasons; for both it is sufficient if they are able to predict one phenomenon from another by means of the empirical routine which they have made their own, and to draw the phenomenon into the circle of practical calculation. Philosophical insight, however, begins with a withdrawal from every such manner of consideration; it presupposes the "turning" of the soul to another ideal of knowledge. Not the phenomena in their bare succession, but the eternal and unchanging rational grounds, from which they proceed, are the unique object of knowledge. To grasp these rational grounds, this realm of λόγοι in the phenomena themselves was, indeed, according to Plato, denied to thought. Whoever, in the field of mathematics, had once grasped the nature of insight into the necessary, only turned back under compulsion to the consideration of a field in which, owing to the flowing and indefinite character of the objects, the same rigor of connection could never be reached. In this sense, empirical knowledge of the sequence of phenomena is not the completion and fulfillment of the pure knowledge of ideas, but only serves as a dark background, against which the clarity of purely conceptual investigation and knowledge stands out the more strongly.

The sceptical theory of knowledge. (*Protagoras, etc.*) Furthermore, it is highly probable that this opposition signifies no intellectual construction (of Plato's), but that it represents with radical sharpness a concrete historical opposition, which was already developed at the time of Plato.[12] In any case, the whole later development of scientific investigation in antiquity is dominated by this Platonic distinction. It is everywhere echoed in the controversy between the "empirical" and the "rational" physicians, that runs through Greek medicine. But the more investigation was applied to the discovery and establishment of individual facts, the more the value and order of knowledge was changed. Scientific empiricism expressed itself in the sceptical theory of knowledge, and affirmed as its positive significance and distinction the very feature, which Plato regarded as the lasting defect of all empirical knowledge. It is, indeed, not given to knowledge to comprehend the essence of things from a universal principle of reason. · What remains for us is only the observation of the customary sequence of phenomena, which enables us to use one phenomenon as the *sign* of another. The task of science is fulfilled in grouping and sifting such signs, each one of which awakens in us a certain memory, and thus directs our expectation of the future into fixed paths. The real causes of occurrences remain unknown to us; but we do not need to know them, as the real and final goal of all

---

[12] *Cf.* Natorp, *Forschungen zur Geschichte des Erkenntnisproblems im Altertum,* Berlin 1884, p. 146 ff.

theory lies in the practical consequences of our action. These consequences remain essentially the same, whether we logically comprehend how one event issues from another, or merely accept the fact of a certain empirical coexistence or succession, and rest in it.

The concepts of nature and purpose. (Plato.) We recognize, however, in the case of Plato, that the division he draws between rational and empirical knowledge,[13] produces no complete disjunction of the entire field of knowledge. Empirical knowledge, which is satisfied with the sequence of "shadows," is sharply characterized; but there remains an indefiniteness in the characterization of its ideal counterpart. This fact is the more significant since it has constantly recurred in the historical development of the problem. The real balance and division of the matter is rendered obscure as long as *one* member is exactly defined, while the other falls into two different meanings, between which opinion varies. At first, Plato opposes to the bare *sequence* of phenomena, insight into their *teleological* connection. We do not possess true knowledge of natural processes as long as we simply permit them to run off before us as before an indifferent spectator, but we first have true knowledge when we survey the total movement of the process as a purposively ordered whole. We must understand how one element demands another; how all the threads are mutually interwoven finally into one web, to form a single order of the phenomena of nature. The ethical idealism of Socrates lives in this view of nature. As little as the continuance of Socrates in prison can be explained by describing the position and relation of his muscles and cords, without considering the ethical reasons that determined him to obey the law,— just as little can an individual event be truly understood, as long as its place in the total plan of reality is not clearly distinguished. For example, if we attempt to explain the fact that the earth moves freely in the center of the universe, no sensuous connection, no mechanical vortex of bodies or any other cause of the same sort, can satisfy us; but "the good and right" alone is to be pointed out as the ultimate and decisive basis of the fact.[14] Sensuous being must be reduced to its ideal reasons; the conclusion of the world of ideas is the Idea of the Good, into which all concepts ultimately merge. Another view, however, is found in Plato opposed to this deduction

---

[13] *Cf.* especially *Republic*, 509 D ff.
[14] *Cf. Phaedo* 99 f., 109.

of natural phenomena out of purposes. It is rooted in Plato's interpretation of mathematics, which is for him the "mediator" between the ideas and the things of sense. The transformation of empirical connections into ideal ones cannot take place without this middle term. The first and necessary step throughout is to transform the sensuous indefinite, which as such cannot be grasped and enclosed in fixed limits, into something that is quantitatively definite, that can be mastered by measure and number. It is especially the later Platonic dialogues, as for example the *Philebus*, which most clearly developed this postulate. The chaos of sense perception must be confined in strict limits, by applying the pure concepts of quantity, before it can become an object of knowledge. We cannot rest with the indefinite "more" or "less," with the "stronger" or "weaker," which we think we discern in sensation, but we must strive throughout for exact measurement of being and process. In this measurement, being is grasped and explained.[15] Thus we stand before a new ideal of knowledge, one which Plato himself recognized as in immediate harmony with his teleological thought, and combining with it in a unified view. Being is a *cosmos*, a purposively ordered whole, only in so far as its structure is characterized by strict mathematical laws. The mathematical order is at once the condition and the basis of the existence of reality; it is the numerical determinateness of the universe, that secures its inner self-preservation.

Mathematics and teleology. (*Plato, Aristotle, Kepler.*) In Aristotle the two lines of thought have already separated, which were inseparably connected for Plato. The mathematical motive recedes into the background; and thus only teleology, the doctrine of final causes, remains as a conceptual foundation of physics. The outer process and its quantitative order according to law merely mirrors the dynamic process, by which the absolute substances maintain and develop themselves. The empirical-physical relation of bodies results ultimately from their essence, from the immanent purpose given them by their nature, and which they progressively strive to fulfill. Thus the elements are arranged in the cosmos according to the degree of their affinity, while those that agree with each other in any quality lie next each other; thus each body retains a tendency to its "natural place," prescribed to it by its property, even after it

---

[15] *Cf. Philebus* 16, 24 f.

has been forcibly deprived of this place. Here the true, inner causes of every physical connection are revealed; while the mathematical mode of consideration, on the other hand, does not attain to the causes but only to the quantities of being, and is limited to the "accidents" and their sphere. Thus a new opposition becomes henceforth effective in history. The unity of the teleological and the mathematical methods of consideration, which still existed in Plato's system of nature, is destroyed, and its place is taken by a relation of superordination and subordination. The line of division has moved; now not only is the sensuous observation of contingent, empirical uniformities excluded from the highest ideal knowledge of the supreme causes, but also the exact representation of the processes in pure concepts of magnitude is excluded from this ideal knowledge. Here the opposition between the empirical and the speculative views of nature is first sharply defined. In modern times, mathematical physics first seeks to prove its claims and independence by going back from the philosophy of Aristotle to that of Plato. Above all, it is Kepler of whom this reversion is characteristic.[16] With energy and clearness he repudiates a conception that would reduce the mathematician to a mere calculator and exclude him from the community of philosophers and the right to decide as to the total structure of the universe. Absolute substances and their inner forces are indeed unknown to the mathematical physicist and must remain so, in so far as he simply pursues his own task, free from all extraneous problems; but his abstraction from this problem in no way signifies his persistence in the ordinary empirical method, which is satisfied with the mere collection of individual facts. The mathematical *hypothesis* establishes an *ideal* connection among these facts; it creates a new unity to be tested and verified by thought, but which cannot be directly given by sensation. Thus the true hypothesis limits the field of mathematical physics to two different directions. It expands immediate experience into *theory* by filling up the gaps left by direct observation, and by substituting a continuous connection of intellectual consequences for isolated sense-data. On the other hand, it limits itself to representing this system of consequences merely as a system and dependence of *magnitudes*. At

---

[16] The more exact evidence for the following historical exposition is given in my work *Das Erkenntnisproblem in der Philosophie u. Wissenschaft der neuen Zeit*, I 258 ff., 308 ff., II 322 ff.

the same time, the mathematical *expression* of the hypothesis, its
algebraic-geometrical form, is the whole of its meaning. In defend-
ing the legitimacy of hypothesis, Kepler places its characteristic
function elsewhere than does the ordinary speculative philosophy
of nature. He is not concerned with a transition from the mathematically characterized phenomenon to its absolute causes, but with
a transition to a quantitative "understanding" of reality from the
first facts of perception (before they have been conceptually worked
over). The scientific physicist can simply leave alone the question as to the ultimate "forces," which shape being; yet all the more
he must seek to advance from a mere collection of observations to a
universal "statics of the universe," to a mastery of the all-inclusive
harmonious order prevailing in it. This order is not directly seized
and understood by the senses, but only by the mathematical intellect.
According to this view, the legitimate function of the concept does
not consist in revealing a path to a new non-sensuous reality; but it
plays its rôle in the conception of reality of mathematical empiricism
(*Empirie*), and gives it definite logical form.

The concept of hypothesis. (*Kepler and Newton.*) Physics, however, did not reach this conception of its problem without vacil-
lation and difficulty. The particular historical conditions, under
which modern natural science developed, forced the negative rather
than the positive part of the new task into the center of consideration. First of all, the theory had to ward off metaphysical claims;
and this warding off could only be accomplished by bringing to light
the *empirical* foundations of exact science. The logical factors, on
the other hand, remained in the background as long as all philosophical power was concentrated upon protecting pure experience from
the incursions of metaphysics. From this, one understands the
fundamental change in view between Kepler and Newton. Kepler,
although he strongly defends the claims of empirical investigation
against the metaphysics of substantial forms, nevertheless reverts
to the mathematical teleology of Plato in his final conception of the
world. The mathematical ideas are the eternal patterns and "arche-
types," according to which the divine architect ordered the cosmos.
Thus the more deeply we penetrate into the exact structure and exact
presuppositions of physics, the more danger there is that the strict
line of division between experience and speculation will again be
effaced. Newton's *regulae philosophandi* seek especially to meet

this danger. Induction is very definitely characterized as the only source of certainty in physical matters. Observation and scientific investigation teach us that those properties, which can neither be increased nor diminished and which are common to all bodies, constitute in their totality the *essence* of body. This expression can thus signify nothing more than an empirical generalization of certain facts of perception. In this sense, but only in this sense, we can conceive weight as an "essential" property of matter; *i.e.*, we can grasp it as such only in so far as we know of no experiment which might occasion us to doubt its universal empirical presence. The question as to the cause of the reciprocal attraction of cosmic masses cannot occupy the real physicist as such and lure him to speculative hypotheses; for attraction is to him nothing but a certain numerical value, which contains the measure of the acceleration which a body undergoes at each point of its path. The law of the change of this value from point to point contains the answer to all questions, that can be raised with scientific justification regarding the "nature" of weight. It is Newton's first disciples and pupils, who generalize these explanations and extend them to the whole field of natural science. The demand for a *physics without hypotheses* first appears distinctly with them, and the technical expression of "description of phenomena" is first formulated here. It is now recognized as a fundamental failure in method to attempt to form physical explanations on the model of logical definitions; we make such a failure if, instead of proceeding from the observation and collection of individual cases, we proceed from the hierarchy of concepts and species. Definitions, which claim to discover the ground and essence of any natural process, must be excluded from physics; they form no instrument of knowledge, but merely hinder the unprejudiced understanding of phenomena, on which depends the whole value of physics as a science.

The logical and ontological "hypotheses." Further development, however, even within the Newtonian school, showed very clearly what was problematical in this apparently final conclusion of methodology. If physics should be forbidden the use of hypothesis in every sense, then all elements would have to be removed, which had no immediate correlate in the field of perception. But the realization of this demand would mean nothing less than the destruction of the Newtonian mechanics and its systematic conception.

The concepts of *absolute space* and *absolute time*, which Newton takes as the starting-point of his deduction, lose every legitimate meaning when measured with the logical criteria, which Newton's methodology alone permits. And, nevertheless, it is precisely on these concepts that the possibility of distinguishing between *real* and *apparent* motion rests, and thus the very concept of physical reality itself. This antinomy is insoluble within the limits of the Newtonian system; its deeper ground lies in the indefiniteness with which the concept of the hypothesis is here interpreted. Aristotle and Descartes, the metaphysics of substantial causes and the first, although imperfect, plan of a complete mechanical explanation of the world are here alike condemned. There is a lack of certainty in distinguishing the assumption of some sort of "dark qualities" of things from the fundamental *theoretical* thought, which is assumed as a basis for defining and limiting the problem and empirical field of physics. And this same ambiguity remains to some extent in modern discussion, in spite of all attempts at sharper epistemological definition of the problem. The most striking expression of such ambiguity is in the concept of description itself. For this term serves to unite investigators, who merely agree with each other in opposing speculative metaphysics, but who entirely disagree in their positive interpretations of the logical structure of physics. Such an investigator as Duhem, for instance, forcibly and clearly develops the idea that every mere establishment of a physical fact involves certain theoretical presuppositions and thus a system of physical hypotheses, stands here directly on the side of an "empiricism," which rests upon a misapprehension of this fundamental double relation. Thus the difficulty involved in its historical development still persists undiminished in physics. The necessary and warranted struggle against *ontology* leads to an obscuring of the simple *logical* facts. Philosophical criticism must seek here for a strict separation of these two materially heterogeneous questions, long inseparably connected in history. The relation of physics and logic is still always described by distinguished scientific investigators as if we still stood in the midst of the conflict between Newton and Wolff, which stamped its form on the philosophy of the eighteenth century. This dispute, however, may be regarded as settled; for logic, in its new, critical form, has given up metaphysical claims. From this new standpoint, however, it is clear that the "phenomenalism" of a

Newton is not on the same plane intellectually with that developed and advocated by ancient scepticism. The problem arises to investigate the fundamental differences of these two views, which both agree in limiting physics to the field of "phenomena." The concept of the phenomenon itself differs, according as it is applied to an indefinite object of sense perception or to the theoretically constructed object of mathematical physics; and it is precisely the conditions of this construction which give rise ever anew to the epistemological question.

IV

Robert Mayer's methodology of natural science. The discoverer of the fundamental law of modern natural science agrees entirely in his methodological views with the series of great investigators that starts with the Renaissance. Robert Mayer begins with the same theoretical definition of the problem of physics as is found in Galileo and Newton in the most diverse applications. The logical continuity appears unbroken, in all the material remodeling of physics introduced with the principle of energy. "The most important, not to say the only, rule for the true investigation of nature is this: to remain persuaded that it is our task to learn to know the phenomena before we seek explanations or ask about higher causes. Once a fact is known on all sides, it is thereby explained and the work of science is ended. This assertion may be pronounced by some as trivial, or combated by others with ever so many reasons; yet certain it is, not only that this rule has been too often neglected down to the most recent times, but that all speculative attempts, even of the most brilliant intellects, to raise themselves above facts instead of taking possession of them, have up till now borne only barren fruit."[17] This is precisely such language as Kepler used against the alchemists and mystics of his time, or as Galileo used against the Peripatetic philosophy of the school. The question as to how heat arises from diminishing motion or how heat is again changed into motion, is declined by Robert Mayer, just as Galileo avoided the question as to the cause of weight. "I do not know what heat, electricity *etc.* are in their inner essence—just as little as I know the inner essence of a material substance or of any thing in general;

---

[17] Robert Mayer, *Bemerkungen über das mechanische Aequivalent der Wärme*, (*Mechanik der Wärme*, ed. by von Weyrauch, 3rd ed., Stuttgart 1893, p. 236).

I know this, however, that I see the connection of many phenomena much more clearly than has hitherto been seen, and that I can give a clear and good notion of what a force is." This, however, is all that can be required of an empirical investigation. "The sharp definition of the natural limits of human investigation is a task of practical value for science, while the attempt to penetrate the depths of the world-order by hypotheses is a counterpart of the efforts of adepts." In the light of this conception, only numbers, only the quantitative determinations of being and process ultimately remain as the firm possession of investigation. A fact is understood when it is measured: "a single number has more true and permanent value than a costly library of hypotheses."[18]

Hypotheses and natural laws. Here a new problem of permanent significance is indicated, along with the rejection of the false problem. A problem is held to be *explained*, when it is *known (bekannt)* perfectly and on all sides. This definition must, indeed, be accepted without limitation; but back of it arises the further question, as to under what *conditions* a phenomenon is to be taken as known in the sense of physics. The "knowledge" of a phenomenon, which exact science brings about, is obviously not the same thing as the bare sensuous cognizance of an isolated fact. A process is first known, when it is added to the totality of physical knowledge without contradiction; when its relation to cognate groups of phenomena is clearly established, and finally to the totality of facts of experience in general. Every assertorical affirmation of a reality, at the same time, implies an assertion concerning certain relations of law, *i.e.*, implies the validity of universal rules of connection. When the phenomenon is brought to a fixed *numerical expression*, this logical relativity becomes most evident. The constant numerical values, by which we characterize a physical object or a physical event, indicate nothing but its introduction into a universal serial connection. The individual constants mean nothing in themselves; their meaning is first established by comparison with and differentiation from other values. Thereby, however, reference is made to certain logical presuppositions which lie at the basis of all physical enumeration and measurement,— and these presuppositions form the real "hypotheses," that can no longer be contested by scientific phenomenalism. The "true hypoth-

[18] See Mayer's letter to Griesinger (*Kleinere Schriften u. Briefe*, ed. by Weyrauch, Stuttgart 1893, p. 180, 226 *etc.*)

esis" signifies nothing but a principle and means of measurement. It is not introduced *after* the phenomena are already known and ordered as magnitudes, in order to add a conjecture as to their absolute causes by way of supplement, but it serves to make possible this very order. It does not go beyond the realm of the factual, in order to reach a transcendent beyond, but it points the way by which we advance from the sensuous manifold of sensations to the intellectual manifold of measure and number.

The presuppositions of physical "measurement." Ostwald, in his polemic against the use of hypotheses, has laid great emphasis on the difference between the hypothesis as a *formula* and the hypothesis as a *picture*. Formulae contain merely algebraic expressions; they only express relations between magnitudes, which are capable of direct measurement and thereby of immediate verification by observation. In the case of physical pictures (*Bildern*), on the contrary, all such means of verification are lacking. Often, indeed, these pictures themselves appear in the guise of mathematical exposition, so that the given criterion of differentiation seems, at first glance, insufficient. But in every case there is a simple logical procedure, which always leads to a clear discrimination. "When every magnitude appearing in the formula is itself measurable, then we are concerned with a lasting formula or with a law of nature; . . . . if, on the contrary, magnitudes, which are not measurable, appear in the formula, then we are concerned with a hypothesis in mathematical form, and the worm is in the fruit."[19] While this postulate of measurability is justified, it is erroneous to regard measurement itself as a purely empirical procedure, which could be carried out by mere perception and its means. The answer given here signifies only the repetition of the real question; for the numbered and measured phenomenon is not a self-evident, immediately certain and given starting-point, but the result of certain conceptual operations, which must be traced in detail. In fact, it soon appears that the bare attempt to measure implies postulates that are never fulfilled in the field of sense-impressions. We never measure sensations as such, but only the *objects* to which we relate them. Even if we grant to psychophysics the measurability of sensation, this insight remains unaffected; for even granting this assumption, it is clear that the *physicist* at least never deals with colors or tones as sensuous

[19] Ostwald, *Vorlesungen über Naturphilosophie*, Leipzig 1902, p. 213 f.

experiences and contents, but solely with vibrations; that he has nothing to do with sensations of warmth or contact, but only with temperature and pressure. None of these concepts, however, can be understood as a simple copy of the facts of perception. If we consider the factors involved in the measurement of motion, the general solution is already given; for it is evident that the physical definition of motion cannot be established without substituting the geometrical body for the sensuous body, without substituting the "intelligible" continuous extension of the mathematician for sensuous extension. Before we can speak of motion and its exact measurement in the strict sense, we must go from the contents of perception to their conceptual limits. (*Cf.* above p. 119 ff.) It is no less a pure conceptual construction, when we ascribe a determinate velocity to a non-uniformly moving body at each point of its path; such a construction presupposes for its explanation nothing less than the whole logical theory of infinitesimal analysis. But even where we seem to stand closer to direct sensation, where we seem guided by no other interest than to arrange its differences as presented us, into a fixed scale, even here theoretical elements are requisite and clearly appear. It is a long way from the immediate sensation of heat to the exact concept of temperature. The indefinite stronger and weaker of the impression offers no foothold for gaining fixed numerical values. In order even to establish the schema of measurement, we are obliged to pass from the subjective perception to an objective functional correlation between heat and extension. If we give to a certain volume of mercury the value of 0 degrees, and to another volume of mercury the value of 100 degrees, then in order to divide the distance between the two points thus signified into further divisions and subdivisions, we must make the assumption that the differences of temperature are *directly proportional* to the volume of the mercury. This assumption is primarily nothing but an hypothesis suggested by empirical observation, but is in no way absolutely forced upon us by it alone. If we go from solid bodies to fluid, from the mercury thermometer to the water thermometer, then, for purposes of measurement, the simple formula of proportionality must be replaced by a more complex formula, according to which the correlation between temperature values and volume values is established.[20] In this example, we see how the simple

[20] *Cf.* the pertinent exposition of G. Milhaud, *Le Rationnel*, Paris 1898, p. 47 ff.

quantitative determination of a physical fact draws it into a network of theoretical presuppositions, outside of which the very question as to the measurability of the process could not be raised.

The physical "fact" and the physical "theory." This epistemological insight has been increasingly clarified by the philosophical work of the physical investigators themselves. It is Duhem above all, who has brought the mutual relation between physical fact and physical theory to its simplest and clearest expression. He gives a convincing and living portrayal of the contrast between the naïve sensuous observation, which remains merely in the field of the concrete facts of perception, and the scientifically guided and controlled experiment. Let us follow in thought the course of an experimental investigation; imagine ourselves, for instance, placed in the laboratory where Regnault carried out his well-known attempt to test the law of Mariotte; then, indeed, we see at first a sum of direct observations, which we can simply repeat. But the enumeration of these observations in no way constitutes the kernel and essential meaning of Regnault's results. What the physical investigator objectively sees before him are certain conditions and changes in his instruments of measurement. But the *judgments* he makes are not related to these instruments, but to the objects, which are measured by them. It is not the height of a certain column of mercury that is reported, but a value of "temperature" that is established; it is not a change which takes place in the manometer, but a variation in the pressure, under which the observed gas stands, that is noted. The peculiar and characteristic function of the scientific concept is found in this *transition* from what is directly offered in the perception of the individual element, to the form, which the elements gain finally in the physical statement. The value of the volume which a gas assumes, the value of the pressure it is under, and the degree of its temperature, are none of them concrete objects and properties, such as we could coördinate with colors and tones; but they are "abstract symbols," which merely connect the physical theory again with the actually observed facts. The apparatus, by which the volume of a gas is established, presupposes not only the principles of arithmetic and geometry, but also the abstract principles of general mechanics and celestial mechanics; the exact definition of pressure requires, for its complete understanding, insight into the deepest and most difficult theories of hydrostatics, of electricity, *etc.* Between the phenomena actually observed in the course of an experiment, and the final result of this experiment as the physicist formulates it, there lies extremely complex intellectual labor; and it is through this, that a report regarding a single instance of a process is made over into a judgment concerning a law of nature. This dependence of every practical measurement on certain fundamental assumptions, which are taken as universally valid, appears still more clearly when we consider, that the real outcome of an investigation never comes directly to light, but can only be ascertained through a critical discussion directed upon the exclusion of the "error of observation." In truth, no physicist experiments and measures with the particular instrument that he has sensibly before his eyes; but he substitutes for it an ideal instrument in thought, from which all accidental defects such as necessarily belong to the particular instrument, are excluded. For example, if we measure the intensity of an electric current by a tangent-compass, then the observations, which we make first with a concrete apparatus, must be related and carried over to a general geometrical model, before they are physically applicable. We substitute for a copper wire of a definite strength a strictly geometrical circle without breadth; in place of the steel of the magnetic needle, which has a certain magnitude and form, we substitute an infinitely small, horizontal magnetic axis, which can be moved without friction around a vertical axis; and it is the totality of these transformations, which permits us to carry the observed deflection of the magnetic needle into the general theoretical formula of the strength of the current, and thus to determine the value of the latter. The corrections, which we make and must necessarily make with the use of every physical instrument, are themselves a work of mathematical theory; to exclude these latter, is to deprive the observation itself of its meaning and value.[21]

Units of measurement. This connection appears from a new angle, when we realize that every concrete measurement requires the establishment of certain *units*, which it assumes as constant. This constancy, however, is never a property that belongs to the perceptible as such, but is first conferred upon the latter on the basis of

[21] *Cf.* the excellent exposition of Duhem, in which this connection is explained in its particulars and illuminated on all sides. (*La Théorie Physique, son objet et sa structure*, Paris 1906.)

intellectual postulates and definitions. The necessity of such postulates is seen especially in the fundamental physical problem of measurement, *viz.*, in the problem of the *measurement of time*. From the beginning, the measurement of time must forego all sensuous helps, such as seem to stand at the disposal of the measurement of space. We cannot move one stretch of time to the place of another, and compare them in direct intuition, for precisely the characteristic element of time is that two parts of it are never given at once. Thus there only remains a conceptual arrangement made possible by recourse to the phenomena of movement. For abstract mechanics, those times are said to be equal, in which a material point left to itself traverses equal distances. Here again we find the concept of the mass-point, and thus a purely ideal concept of a limit; and once more it is the hypothetical assumption of a universal *principle*, which first makes possible the unit of measure. The law of inertia enters as a conceptual element into the explanation of the unit of time. We might attempt to eliminate this dependence by going over from rational mechanics to its empirical applications, and seeking to establish a strictly uniform motion in the field of the concrete phenomena. The daily revolution of the earth, as it seems, offers the required uniformity in all the perfection that could ever be taken into account for purposes of measurement. The unit here would be directly given us by the interval, which lies between two successive culminations of the same star. More exact consideration, however, renders apparent the difference, which always remains between the ideal and empirical measure of time. The inequality of the stellar days is what is rather demanded now, on the basis of theoretical considerations, and is confirmed by empirical reasons. The friction, which arises from the continuous change of the ebb and flow of the tides, produces a gradual diminution of the velocity of the rotation of the earth, and thus a lengthening of the stellar day. The desired exact measure again eludes us, and we are forced to more remote intellectual assumptions. All of these gain their meaning only by relation to some physical law, which we tacitly assume with them. Thus the time, in which the emanation of radium loses its radioactivity, has recently been proposed as an exact unit of measurement; in this, the law of exponents, in accordance with which the diminution of effect takes place, serves as a foundation. It is analogous to this, that the principles and theories of optics are presupposed in order to

introduce the wave-lengths of certain rays of light as the foundation
of the *measurement of distance*. What guides us in the choice of
units is thus always the attempt to establish certain laws as universal.
We *assume* the empirically entirely "equal" stellar days to be unequal,
in order to maintain the principle of the conservation of energy.
The *real* constants are thus fundamentally, it has been justly urged,
not the material measuring-rods and units of measurement, but these
very laws, to which they are related and according to whose model
they are constructed.[22]

The verification of physical hypotheses. The naïve view, that
measurements inhere in physical things and processes like sensuous
properties, and only need to be read off from them, is more and more
superseded with the advance of theoretical physics. Nevertheless,
the relation of law and fact is thereby altered. For the explanation,
that we reach laws by comparing and measuring individual facts, is
now revealed as a logical circle. The law can only arise from
measurement, because we have assumed the law in hypothetical
form in measurement itself. Paradoxical as this reciprocal relation
may appear, it exactly expresses the central logical problem of
physics. The intellectual anticipation of the law is not contradictory,
because such anticipation does not occur in the form of a dogmatic
assertion, but merely as an initial intellectual assumption; because
it does not involve a final answer, but merely a question. The
value and correctness of this assumption are first shown when the
totality of experiences are connected into an unbroken unity on the
basis of it. The correctness of the assumption, on the other hand,
cannot indeed be assured by our verifying every hypothesis and
every theoretical construction directly in an individual experience, in
a particular sensuous impression. The validity of the physical
concept does not rest upon its content of real *elements of existence*,
such as can be directly pointed out, but upon the strictness of *connection*, which it makes possible. In this fundamental character,
it constitutes the extension and continuation of the mathematical
concept. (*Cf.* above p. 83 f.) Thus the individual concept can

---

[22] *Cf.* Henri Poincaré, *La mesure du temps, Revue de Métaphysique et de
Morale* VI, 1898. Concerning the theoretical presuppositions of the determination of units of measurement, *cf.* especially Lucien Poincaré, *La Physique
moderne, dtsch. v. Brahn,* Lpz. 1908; and also Wilbois, *L'Esprit positif, Revue de
Métaph.* IX (1901).

never be measured and confirmed by experience for itself alone, but it gains this confirmation always only as a member of a theoretical complex. Its "truth" is primarily revealed in the consequences it leads to; in the connection and systematic completeness of the explanations, which it makes possible. Here each element needs the other for its support and confirmation; no element can be separated from the total organism and be represented and proved in this isolation. We do not have physical concepts and physical facts in pure separation, so that we could select a member of the first sphere and enquire whether it possessed a copy in the second; but we possess the "facts" only by virtue of the totality of concepts, just as, on the other hand, we conceive the concepts only with reference to the totality of possible experience. It is the fundamental error of Baconian empiricism that it does not grasp this correlation; that it conceives the "facta" as isolated entities existing for themselves, which our thought has only to copy as faithfully as possible. Here the function of the concept only extends to the subsequent inclusion and representation of the empirical material; but not to the testing and proving of this material.[23] Although this conception has been obstinately maintained within the epistemology of natural science, nevertheless there are many signs that physics itself in its modern form has definitely overcome it. Those thinkers, also, who urge strongly that *experience* in its totality forms the highest and ultimate authority for all physical theory, repudiate the naïve Baconian thought of the *"experimentum crucis."* "Pure" experience, in the sense of a mere inductive collection of isolated observations, can never furnish the fundamental scaffolding of physics; for it is denied the power of giving mathematical form. The intellectual work of understanding, which connects the bare fact systematically with the totality of phenomena, only begins when the fact is represented and replaced by a mathematical symbol.[24]

The motive of serial construction. Yet if we conceive the final result of the analysis of physical theory in this fashion, there remains a paradox. Of what value is all the intellectual labor of physics, if we must ultimately recognize that all the complication of investigation and its methods simply removes us more and more from the concrete fact of intuition in its sensuous immediacy? Is all this expenditure of

---

[23] *Cf.* more particularly *Erkenntnisproblem* II, 125 ff.
[24] *Cf.* in particular Duhem, *La Théorie Physique*, p. 308 ff.

scientific means worth while, when the final outcome is and can be nothing but the transformation of facts into symbols? The reproach raised by modern physics, in its inception, against scholasticism, that scholasticism replaced the consideration of *facts* by that of *names*, now threatens to fall upon physics itself. Nothing but a new nomenclature seems to be gained, by which we are more and more alienated from the true reality of sensation. In fact, this consequence has occasionally been pointed out: the *necessity*, to which physical theory tends, has been opposed to the evidence and *truth*, which comes to consciousness in the experience of individual facts. This separation, however, rests upon a false abstraction; it is based in the attempt to isolate two moments from each other, which are inseparably connected together by the very presuppositions of the formation of concepts. It has been shown, in opposition to the traditional logical doctrine, that the course of the mathematical construction of concepts is defined by the procedure of the *construction of series*. We have not been concerned with separating out the common element from a plurality of similar impressions, but with establishing a principle by which their diversity should appear. The unity of the concept has not been found in a fixed group of properties, but in the rule, which represents the mere diversity as a sequence of elements according to law. (*Cf.* above 14 ff.) A consideration of the fundamental physical concepts has confirmed and broadened this view. All these concepts appear to be so many means of grasping the "given" in series, and of assigning it a fixed place within these series. Scientific investigation accomplishes this last definitely; but in order for it to be possible, the serial principles themselves, according to which the comparison and arrangement of the elements takes place, must be theoretically established. The individual thing is nothing for the physicist but a system of physical constants; outside of these constants, he possesses no means or possibility of characterizing the particularity of an object. In order to distinguish an object from other objects, and to subsume it under a fixed conceptual class, we must ascribe to it a definite volume and a definite mass, a definite specific gravity, a definite capacity for heat, a definite electricity, *etc.* The measurements, which are necessary for this, however, presuppose that the aspect, with respect to which the comparison is made, has been previously conceived with conceptual rigor and exactitude. This aspect is never given in the original impression, but has to be worked out theoretically, in order to be then applied to the manifold of perception. The physical analysis of the object into the totality of its numerical constants is thus in no sense the same as the breaking-up of a sensuous thing into the group of its sensuous properties; but new and specific categories of *judgment* must be introduced, in order to carry out this analysis. In this judgment, the concrete impression first changes into the physically determinate object. The sensuous quality of a thing becomes a physical object, when it is transformed into a serial determination. The "thing" now changes from a sum of properties into a mathematical system of values, which are established with reference to some scale of comparison. Each of the different physical concepts defines such a scale, and thereby renders possible an increasingly intimate connection and arrangement of the elements of the given. The chaos of impressions becomes a system of numbers; but these numbers first gain their denomination, and thus their specific meaning, from the system of concepts, which are theoretically established as universal standards of measurement. In this logical connection, we first see the "objective" value in the transformation of the impression into the mathematical "symbol." It is true that, in the symbolic designation, the particular property of the sensuous impression is lost; but all that distinguishes it as a *member of a system* is retained and brought out. The symbol possesses its adequate correlate in the *connection* according to law, that subsists between the individual members, and not in any constitutive part of the perception; yet it is this connection that gradually reveals itself to be the real kernel of the thought of empirical "reality."

The physical concepts of series. The relation, that is fundamental here, can be illumined from another standpoint by connecting it with the ordinary psychological theory of the concept. In the language of this theory, the problem of the concept resolves into the problem of "apperceptive connection." The newly appearing impression is first grasped as an individual, and first gains conceptual comprehension through the apperceptive interpretation and arrangement that it undergoes. If this reference of the individual to the totality of experience were lacking, the "unity of consciousness" itself would be destroyed,—the impression would no longer belong to "our" world of reality. In the sense of this conception, we can characterize the various physical concepts of measurement evolved by scientific theory as the real and necessary *apperceptive concepts* for all empirical knowledge in general. Without them, indeed, it has been shown there is no arrangement of the factual in series, and thus no thorough reciprocal determination among its individual members. We would only possess the fact as an individual *subject*, without being able to give any *predicate* for determining it more closely. It is only when we bring the given under some norm of measurement, that it gains fixed shape and form, and assumes clearly defined physical "properties." Even before its individual value has been empirically established within each of the possible comparative series, the fact is recognized, that it necessarily belongs to some of these series, and an anticipatory schema is therewith produced for its closer determination. The preliminary deductive work furnishes a survey of the possible kinds of exact correlation; while experience determines which of the possible types of connection is applicable to the case in hand. Scientific experiment always finds several ways before it, which theory has prepared, and between which a selection must be made. Thus no content of experience can ever appear as something absolutely strange; for even in making it a content of our thought, in setting it in spatial and temporal relations with other contents, we have thereby impressed it with the seal of our universal concepts of connection, in particular those of mathematical relations. The material of perception is not merely subsequently moulded into some conceptual form; but the thought of this form constitutes the necessary presupposition of being able to predicate any character of the matter itself, indeed, of being able to assert any concrete determination and predicates of it. Now it can no longer seem strange, that scientific physics, also, the further it seeks to penetrate into the "being" of its object only strikes new strata of numbers, as it were. It discovers no absolute metaphysical qualities; but it seeks to express the properties of the body or of the process it is investigating by taking up into its determination new "parameters." Such a parameter is the *mass*, which we ascribe to an individual body in order to render rationally intelligible the totality of its possible changes and its relations with respect to external impulses to motion, or the *amount of energy*, which we regard as characteristic of the momentary condition of a given physical system. The same holds of all the different magnitudes, by which physics and chemistry progressively determine the bodies of the real world.[25]

[25] *Cf.* the striking developments of G. F. Lipps, *Mythenbildung und Erkenntnis*, Lpz. 1907, p. 211 ff.

The more deeply we enter into this procedure, the more clear becomes the character of the scientific concept of the thing and its difference from the metaphysical concept of substance. Natural science in its development has everywhere used the *form* of this latter; yet in its progress it has filled this form with a new content, and raised it to a new level of confirmation.

V

The concept of substance in the Ionian philosophy of nature. The logical conception of substance stands at the pinnacle of the scientific view of the world in general; it is the concept of substance, which historically marks the line of distinction between investigation and myth. Philosophy has its own origin in this achievement. The attempt to deduce the multiplicity of sensuous reality from a single ultimate substance contains a universal postulate which, however imperfectly it may be fulfilled at first, nevertheless is the characteristic expression of a new mode of thought and a new question. For the first time, being is conceived as an ordered whole, not guided by alien caprice but containing within itself its principle of being. At first, however, this new conception seeks its confirmation only in the sphere of sensuous things, as they alone seem to constitute the fixed positive content of reality. The conceptual and critical work of investigation not having yet been conceived or begun, perception offered the only fixed limit separating reality from mythological and poetic fantasy. Thus it is some empirically given, particular material, which constitutes the meaning of "substance." But even within the Ionic philosophy of nature, tendencies begin to arise that transcend this conception. Anaximander's principle of the ἄπειρον already rises in logical freedom above the sphere of immediately perceptible reality. It contains indication of the thought that what constitutes the origin of sensuous being cannot have the same properties that it has. It cannot be clothed with any particular material quality, for all particular qualities must develop out of it. It thus becomes a being without definite, sensuous, differentiating properties, in whose homogeneous structure the oppositions of the warm and the cold, the wet and the dry still lie together unseparated. The realm of the material in general is not abandoned; it is rather precisely the pure abstraction of matter itself that first gains clear expression in the infinite and indeterminate substance of Anaximander.

The hypostatization of sensuous qualities. (*Anaxagoras.*) The problem of the particular qualities and properties, however, is not answered in this first attempted solution, but is rather raised for the first time. While the opposites are to evolve by "separating out" from the homogeneous ultimate principle, the manner and occasion of their differentiation remains at first wholly obscure. The question involved in this gives the motive to the further development of the speculative philosophy of nature. The unity, assumed by Anaximander in his principle of the infinite, represents merely a logical anticipation, which lacks exact foundation. In order to reach clarity on this point, thought must apparently take the opposite course. The true infinity of the ultimate substance is found not so much in its homogeneous and undifferentiated structure, as rather in the unlimited fullness and multiplicity of qualitative differences concealed within it. It is the cosmology of Anaxagoras, which draws out this tendency to its conclusion. With Anaxagoras, not only is a universal principle of motion first established, but also the physical explanation of the particular qualities enters on a new phase. It is vain to seek to deduce the particular from the universal, if the former is not already contained in the latter in some form. Thus the manifold, apparent properties of bodies, of whose existence and differences the senses inform us, point back to permanent and absolute properties of matter as their real origin. The wet and the dry, the bright and the dark, the warm and the cold, the thick and the thin are all fundamental properties of things; and on the manner and quantitative proportion in which these properties are mixed with each other, rest all the differences and contrasts between the composite sensuous substances, such as air and water, ether and earth. All the elementary properties enter in their totality into every composition, and are to be thought of as always contained in even the smallest material part, however far we pursue the analysis. What gives the particular substances their distinguishing form is not that they contain any one of the qualitative elements in isolation, but that one of them predominates in the composition, so that in ordinary popular consideration the remaining factors can be practically left out of account, although they are really never lacking. In this sense, "all is in all;" every particle, however small, even every physical point, represents a totality of infinitely many qualities, which interpenetrate in it. The detailed working-out of this theory is merely of historical interest;

but even aside from this, it contains an element of typical meaning, which has frequently come to light in the advance of physics. The analysis of Anaxagoras aims to go back of the concrete sensuous object, as it is first offered in intuition, to its conceptual principles; but it defines the content of these principles in expressions, which are throughout taken from sensuous perception. Properties and oppositions of sensation are immediately transformed into substantial causes, which exist in themselves and can operate for themselves, although only in connection with other causes of the same sort. The varied multiplicity of sensible qualities is thus retained; indeed, it is consciously increased to infinity. To each of these properties, which seem in appearance to go hither and thither, to appear and disappear, there corresponds in truth an unchanging and substantial being. For any sensuous property to arise newly in a subject or to disappear is a mere deception, with which we are misled by a superficial consideration of things; on the contrary, each of these properties *persists* and is only temporarily, as it were, concealed from view by others, which seem to take its place. Here we have, in fact, the characteristic attempt to construe the permanent being, which thought demands, without going beyond the sphere of the "given." It is no longer, as in the Ionic philosophy of nature, a particular empirical material like air or water, that represents the permanent existence of things; but this function is transferred to the totality of properties, from which each body results, and which can be discovered in it by perception. The hypostatization of these properties leaves their nature unchanged; it is true they gain thereby a changed *metaphysical* significance, but in principle they do not go beyond the character of the sensuous.

The hypostatization of sensuous qualities. (*Aristotle.*) Aristotelian physics also represents no intrinsic change in this respect. The fundamental qualities are here reduced again to a small number; instead of the infinitely numerous "germs" of things, there are merely the properties of warm and cold, wet and dry, from the combination of which are to arise the four elements: water and earth, air and fire. The nature of these elements determines the peculiarity of the motions they undergo, and thereby the total plan and order of the cosmos. Thus the structure of this physics also rests upon the same procedure of converting the relative properties of sensations into absolute properties of things. The view at the basis of this comes out with especial force and distinctness in its historical consequences.

The whole of natural science, especially the whole of chemistry and alchemy of the middle ages, is first intelligible, when considered in connection with the logical presuppositions of the Aristotelian system. The elevation of qualities into separate *essences*, which are different from the being of the body, and therewith at least in principle transferable from one body to another, is here the dominant view. The properties, which are common to a class of things, and which thus furnish the foundation for the construction of a certain *generic concept*, are separated off as physical parts and raised to independent existence. The solid bodies are distinguished from the fluid and volatile bodies by the presence of a certain absolute and separable property, that is immanent in them; transition to another state of aggregation means the loss of this quality and the assumption of a new substantial nature. Thus mercury, for example, can be changed into gold, if we successively remove from it the two "elements," on which its fluidity and volatility rest, and substitute other properties for them. In order to transform any body into another, in general it is sufficient to master the different "natures," in such a way that one is in a position to imprint them successively on matter. The transmutation of metals into each other is conceived and represented according to this fundamental view. We take from the particular body its particular properties, which are conceived as so many independent substances in it; for example, we separate from tin its creaking, its fusibility and its softness, in order to make it approach silver, from which it is at first separated by all these properties. The whole view, on which this conception of nature rests, appears clearly even in modern times in the physics of Bacon. Bacon's theory of forms goes back to the axiom, that what constitutes the generic common element of a group of bodies must be somehow present in them as a separable part. The form of heat exists as a peculiar somewhat, that is present in all warm things, and by its presence calls forth certain effects in them. The task of physics is exhausted in reducing the complex sensuous thing to a bundle of abstract and simple qualities, and explaining it from them. The hypothesis of a heat substance, like the assumption of a special electric or magnetic fluid, shows how slowly this conception has been replaced also in modern science.[26] Especially in the concepts of

---

[26] *Cf.* the excellent presentation by E. Meyerson, *Identité et Réalite*, Paris, 1908, p. 300 ff.; also Berthelot, *Les origines se l'Alchimie*, Paris, 1885, p. 206 ff., 279 ff., *etc.*

chemistry it reappears in diverse forms. Every element of the older chemistry is at once a bearer and a type of a certain striking property. Thus sulphur is the expression of the combustibility of bodies; salt, the expression of their solubility; while mercury comprehends and expresses the totality of metallic properties. It is always certain *reactions* according to law, for which a substantial substrate is directly assumed. The property of combustibility, which we seem to perceive sensuously in a number of bodies, is transformed by the assumption of phlogiston into a particular substance, that is mixed with bodies; and from this assumption, the whole structure of chemistry before Lavoisier follows with inner necessity.

Atomism and number. Along with the development we have traced here in its general features, there exists from the first another fundamental view of physical being and process. Ancient science gave a perfect expression to this view in the system of atomism. Atomism, through the mediation of the Eleatic system, goes back in its historical presuppositions to the fundamental concept of the Pythagorean doctrine. The concept of empty space, from which Democritus starts, is directly taken from the κενόν of the Pythagoreans. Here we face a change in the direction of thought. Being is no longer sought directly in the sensuous perceptible qualities, nor in what corresponds to them as an absolute correlate and counterpart; but being is resolved into the pure concept of number. On number, rests all the connection and inner harmony of things; precisely for that reason number is to be characterized as the substance of things; for it alone gives them a definite, knowable character. The mystical exaggeration, with which this thought was first grasped, gradually ceases with the advance of Greek science, till it finally yields to a purely methodological and rational explanation. This change is completed in the atomistic system; what was an abstract postulate for the Pythagorean, is here embodied in a concrete construction of mechanics. The sensuous properties of things are banished from the scientific picture of the universe; it is only according to "opinion," only according to the untested "subjective" view, that there is a sweet and a bitter, a colored and a colorless, a warm and a cold. For the representation of objective reality, on the contrary, all these properties are to be cast aside, for no one of them is capable of exact quantitative determination, and thus of a truly unambiguous definition. Thus only those properties of things, that are determinable in the sense of pure mathematics, remain as the "real" properties. The abstract number-schema of the Pythagoreans is now, however, supplemented by a new element, which enables it to develop to its full effect. In order to advance from number to material physical existence, it is necessary to have the mediation of the *concept of space*. Space, however, is here taken in a sense that makes it the pure image of number. It represents all the properties, and fulfills all the conditions of number. Its characteristic feature, accordingly, is the absolute uniformity of its parts; all inner differences are resolved into mere differences of position. The differences existing in the immediate space of perception are wholly removed, so that each particular point represents merely an equivalent starting-point for geometrical relations and constructions. Now if the real is determined from this point of view, nothing remains of it but what makes it a numerical order, a quantitative whole. Precisely in this is rooted the meaning and justification of the concept of the atom. The world of atoms is nothing but the abstract representation of physical reality, in so far as nothing is retained in this but pure determinations of magnitude. It was in this sense that Galileo, at the threshold of modern physics, understood and founded atomism. He explains, that in the concept of matter nothing else is fundamental save its conception as of this or that *form*, as in this or that *place*, as *large* or *small*, as in *motion* or *rest*. From all other properties, on the other hand, we can abstract without thereby destroying the thought of matter itself. No logical necessity forces us to conceive matter as either white or red, sweet or bitter, pleasant or evil-smelling: rather all these characterizations are mere *names* to which (since they cannot be reduced to exact numerical values) there corresponds no fixed, objective correlate. The substance of the physical body is exhausted in the totality of properties, which arithmetic and geometry and the pure theory of motion (which goes back to them both) can discover and establish in it.

The impact of atoms. However, with this acceptance of atomism, the problem is only raised in general terms, but is in no way solved completely. For the atom signifies no fixed physical fact, but a logical postulate; it is thus itself not unchanging, but rather a variable expression. It is interesting to trace how, in the transformations which the concept of the atom undergoes in the course of time, the intellectual motive, to which it owes its origin, continually works

out with greater clarity.   In the atom of Democritus, the dissolution
of sensuous determinations is not fully carried through.   If we take
the well-known report of Aristotle as our authority, the atoms are
distinguished, according to Democritus, not only in their position in
place, but also in their magnitude and form; they possess different
extension and different form, although no reason for this difference
is shown.   Yet to the extent that the dynamical interaction of the
atoms becomes a real problem, the logical necessity appears above
all of endowing each atom with an absolute *hardness*, by means of
which it excludes all others from its position in space.   The opposition
of the hard and the soft, like that of the light and the heavy, is thus
again taken up directly into the objective consideration of nature; a
residuum of the sensible properties of body is retained and put on
a level with the determinations discerned by mathematical thought.
The consequences of this dualism soon become clear in the development of the doctrine.   They concentrate into a real antinomy, when
we consider the relation that results between the physical concept of
being and the physical law of *process*.   This law, if we merely consider
its applications to mechanics, demands that the sum of energy remain
unchanged in every transference of motion from one body to another.
Yet if we attempt to apply this point of view to the impact of atoms,
a peculiar difficulty results.   If we consider the atoms as perfectly *hard*
bodies, then their properties and modes of action are to be determined according to the relations we can directly observe empirically
in inelastic masses; however, with every collision of wholly or partially inelastic bodies, there appears a certain loss of energy.   In
order to remove this contradiction of the law of the conservation of
energy, the theory must assume that a portion of the energy has
gone from the masses to their parts, that "molar" energy has been
changed into "molecular."   But this explanation is obviously not
available for the atoms themselves, since these, according to their
conception, are simple subjects of motion, and lack all possibility of
being further analysed into parts and subparts.

   *The postulate of continuity, and the "simple" atom of Boscovich and
Fechner*.   Kinetic atomism has sought in various ways to remove this
contradiction in its foundations, without ever having fully
succeeded.[27]   And a second problem no less difficult results, when

   [27] For criticism of the attempted solution of Secchi, according to which the
loss of energy, that must occur in the impact of absolutely hard bodies, is

we compare the mechanics of the atoms with the demands which result from the postulate of the *continuity* of process. The change of velocity, which two absolutely hard bodies undergo in the moment of their impact, can only consist in a sudden transition, in a jump from one magnitude to another, that is separated from it by a fixed, finite amount. If a slowly moving body is overtaken by a more rapidly moving one, and they both advance after the impact with a common velocity, which is determined by the law of the conservation of the algebraic sum of their motions, then this result can only be expressed by our ascribing to the one body an abrupt loss, to the other an abrupt increase in velocity. This assumption, however, leads to the discovery, that we can establish no definite value for the velocity of the two masses at the moment of impact, and that a gap remains in the mathematical determination of the whole process.[28] The advocates of extended atoms have occasionally replied to objections of this sort, that a false standard is applied here to the hypothetical structure on which mechanics should rest. The contradiction merely comes from the fact, that the atoms, which are meant to be nothing but rational constructions of thought, have certain properties ascribed to them, which properties are only deduced from analogy with the sensuous bodies of our world of perception. This very analogy, however, is to be abandoned from the standpoint of the theory of knowledge. The norm for the formation of the content of the atom, is not the behaviour of the empirical bodies of our environment, but the universal laws and principles of mechanics. Thus we are not directed for this norm to a mere vague comparison with directly observable phenomena, but we determine the conditions to be satisfied by the "real" "subject" of motion, on the basis of conceptual postulates. Thus we need not ask whether it is possible or impossible for absolutely rigid bodies in impact to satisfy the law of the conservation of energy, but conversely we assume the validity of this law as an axiom, to which we are bound in the theoretical construction of the atoms and their movements. The compati-

counterbalanced by a part of the rotary movement of the atoms being transformed into progressive movement, *cf.* Stallo, *The Concepts and Theories of Modern Physics*, German edition, Lpz. 1901, p. 34 ff.; For general criticism of the concept of the atom, *cf.* especially Otto Buek, *Die Atomistik und Faradays Begriff der Materie*, Berlin, 1905.

[28] *Cf.* more particularly, *Erkenntnisproblem*, II, 394 ff.

bility of this construction with the other assumptions of rational mechanics is the rule that alone can guide us, and not the similarity of atomic motions with any processes of known physical reality.[29] This reply is, in principle, thoroughly satisfactory; yet when we think it through, we find ourselves forced also from the logical side to that transformation of the concept of the atom, which natural science since Boscovich has carried out. In place of the extended but indivisible particle, there now appears the absolutely simple *point of force*. We see how the reduction of the sensible properties, which was already characteristic of Democritus, has here advanced another step. The *magnitude* and *form* of the atoms have now disappeared; what differentiates them is merely the position, that they mutually determine for each other in the system of dynamic actions and reactions. The negation of extension is joined to the negation of the sensuous qualities; therewith there is the negation of every determination of content in general, by which one empirical "thing" is distinguished from another. All independent, self-existent attributes are now completely effaced; what remains is merely the relation of a dynamic coexistence in the law of the reciprocal attraction and repulsion of the points of force. Boscovich urges energetically, and Fechner after him, that force itself, as it is here understood, resolves into the concept of law and that it is meant to be merely the expression of a functional dependence of magnitudes. The atom, which in its origin goes back to the pure concept of number, here reverts to its origin after manifold transformations; it signifies nothing but the member of a systematic manifold in general. All content, that can be ascribed to it, springs from the relations of which it is the intellectual center.

The concept of the atom and the application of differential equations. The scientific development, which the concept of the atom has undergone in the most recent modern physics, confirms this view throughout. In the conflict between atomism and energism, Boltzmann has sought to deduce the necessity of the atomic hypothesis from the fundamental method of theoretical natural science, from the procedure of applying differential equations. If we do not deceive ourselves concerning the meaning of a differential equation,— he explains,—we cannot doubt that the scheme of the world, that is assumed with it, is in essence and structure atomistic. "On closer

[29] See Lasswitz, *Geschichte der Atomistik*, II, 380 ff.

inspection, a differential equation is only the expression of the fact, that we have first to conceive a finite number; this is the first condition, then this number must increase until its further increase is no longer of influence. What use is it to suppress the demand for conceiving a great number of individuals, when in the explanation of the differential equation the value expressed by the same has been defined by that demand?" Thus he, who believes he can free himself from atomism by differential equations, does not see the wood for the trees.[30] This type of explanation is of high interest from the standpoint of the critique of knowledge; for the necessity of the atom is here meant to be deduced not from the *facts* of the empirical consideration of nature, but from the conditions of the *method* of exact physics. If, however, this is the case, then it is clear indeed that the "existence" assured to the atom in this way can be no other than belongs in general to the pure mathematical concepts. Thus Boltzmann expressly guards against the assumption that the absolute *existence* of the atoms is to be proved by his deduction; they are only to be understood and applied as images for the exact representation of phenomena.[31] Precisely in this assumption, however, the necessity finally appears of going from the *extensive* corpuscle to the *simple* mass-point, if the "image" is to gain its full sharpness and exactitude. The procedure of the infinitesimal calculus, to which Boltzmann appeals, itself urges this transition. If we start with the presentation of certain finite magnitudes, and then permit the latter to diminish continuously in order to gain a point of application for the differential equation, this process only reaches its mathematical conclusion if we permit the magnitudes under consideration to converge toward the limiting value null; while, according to atomism, a constant value could always be given, beyond which the ideal procedure could not go without involving itself in contradictions with the reality of phenomena. As long as we remain with magnitudes of a certain extension, we reach no definite logical determination, no matter how small we may choose these magnitudes. Along with whatever physical divisibility one may assume, there always remains

[30] Boltzmann, *Über die Unentbehrlichkeit der Atomistik, Annalen der Physik und Chemie* N. F. Vol. 60, p. 231 ff. (*Populäre Schriften*, Lpz. 1905, p. 141 ff.)

[31] *Cf.* Boltzmann, *Ein Wort der Mathematik an die Energetik* (*Pop. Schriften* p. 129 ff.)

the intellectual possibility of analysing the body further, and of ascribing different velocities to the many, distinguishable subgroups. It is only when we advance to the *material point*, that this indefiniteness is done away with, and a fixed subject of motion produced. (*Cf.* above p. 119 ff).

The changes in the concept of the atom. It has been urged from the side of energism against Boltzmann, that the concept of the material point, as it is assumed by mechanics, does not develop out of that of body by our abstracting from extension practically or entirely, but by our abstracting from rotary movement. "If we have to consider other than purely forward movements, we analyse the body into parts, which . . . . have absolutely nothing to do with atoms, into elements of volume, by which we can approach the material point that only moves forward, to any degree of approximation."[32] Here an important logical moment is indicated; the simplicity of the point is assumed for the sake of simplicity of movement. The assumption of the simple, not further reducible body is only a methodological device to advance to the abstraction of simple movement. In this sense the "atom," according to its fundamental physical meaning, is not defined and postulated as a part of matter, but as a subject of possible changes. It is only considered as an intellectual point of application for possible relations. We analyse complex movements into elementary processes, for which latter we then introduce the atoms as hypothetical substrata. Thus we are not primarily concerned with separating out the ultimate elements of things, but with the establishment of certain simple, fundamental processes, from which the variety of processes can be deduced. Thus we understand how the atom, in its modern physical application, loses the aspect of materiality more and more; how it is resolved into vortex movements in the ether, which, however, in accordance with their character fulfill the conditions of indestructibility and physical indivisibility. The postulate of identity, which is of course inevitable, is here satisfied not by any kind of material substratum, but by permanent *forms of motion*. It appears that, in general, as soon as any sort of physical process, which hitherto passed as simple, is regarded from a new point of view, by which it appears as the result of a plurality of conditions, the substratum, which we took as its

[32] *Cf.* Helm, *Die Energetik und ihre geschichtl. Entwicklung*, Leipzig, 1898, p. 215.

basis, at once divides. Thus when inertia no longer appears as an absolute property of bodies, but a way is offered to deduce it from the laws of electrodynamics, the hitherto material atom breaks up and resolves into a system of electrons. But the new unity gained in this manner can itself only be conceived as relative, and as thus in principle changeable. Sharper analysis of physical relations leads constantly to new determinations and differentiations within their subject. Thus we can say that the content of the concept of the atom may be regarded as changeable, while the *function*, which belongs to it, of defining the condition of knowledge at any given time and of bringing it to its most pregnant intellectual expression, remains unchanged. Only the point of application changes; but the procedure of postulating a unity remains constant. The "simplicity" of the atoms is fundamentally a purely logical predicate; it is determined with reference to the intellectual analysis of the phenomena of nature, and not by reference to our sensuous capacity of discrimination or with regard to technical-physical means of analysis. Every advance of analysis, every process of bringing a great field into a new connection, (which is especially possible in modern physics on the basis of phenomena of radio-activity) changes our view of the "constitution" of matter and of the elements out of which it is built up. The new unity, which we define, is always only the expression of the relatively highest and most comprehensive standpoint of judgment, by which we grasp the totality of physical things and processes in general.

The concept of the ether. An analogous development is presented, when we proceed from the concept of matter to the second principal concept of natural science, the concept of the ether. The difficulties, that arise here at first, spring from the fact that, in order to give this concept a definite content we have to utilize certain properties originally won through comparison with the objects of sense perception. Ether accordingly appears as a perfect fluid, which, however, possesses certain properties of perfectly elastic bodies. At first no fully unified picture results from the combination of these two aspects; the limiting case itself presents a different appearance, according as we approach it from one or the other direction, according as we seek to reach it through progressive idealization from different empirical starting-points. The conflict, that develops here, finds its solution in principle only when we resolve to abandon all direct sensuous representation of the ether, and to use it merely as a conceptual symbol of certain physical relations.[33] We discover a physical phenomenon, as for example, a certain effect of light at a certain point of space, while we are obliged to locate its "cause" at a point of space remote from it. In order to establish a continuous connection between these two conditions, we postulate a medium for them by conceiving the space between them continuously filled with certain qualities, which can be expressed by pure numerical values. The totality of such numerical determinations is really the fundamental conception which we express in the thought of the ether. The unitary and strictly homogeneous space is progressively differentiated through our inscribing upon it, as it were, a web of numbers. This gradation of the individual positions and their arrangement into different mathematical-physical series is what gives them a new content. "Empty" space, which only represents one principle of arrangement, is now in a certain sense covered over with a wealth of other determinations; these, however, are all held together by the fact that certain functional dependencies subsist between them. All that physics teaches of the "being" of the ether can, in fact, be ultimately reduced to judgments about such connections. When, according to the electro-magnetic theory of light, the identity is affirmed of the ether of light with that ether in which electro-magnetic effects are propagated, this is because the *equations*, to which we are led in the investigation of light vibrations, are identical in their form with those, which result for dielectric polarization, and because further the numerical constants, especially the constants for the velocity of propagation, mutually agree.[34] The assumption of the same substratum is here also only another expression of the thorough-going analogy of the mathematical relations: for the connections, that subsist between the values of the optical and electrical constants. The more inclusively and consciously physics makes use of the concept of the ether, the more clearly it appears that the object thus signified cannot be understood as an isolated, individual thing of perception, but only as a unification and concentration of objectively valid, measurable relations.

[33] *Cf.*, e.g., Pearson, *The Grammar of Science*, p. 178 ff., 262 ff.
[34] See Henri Poincaré, *Elektrizität und Optik*, German trans., Berlin, 1891, p. 159 ff.

The logical form of the concept of the physical object.   Now if we again survey the changes, which the scientific concept of substance has undergone from its first speculative beginnings, the unitary goal of its tendencies stands out clearly.   True, it must appear as a genuine impoverishment of reality that all existential qualities of the object are gradually stripped off; that the object loses not only its color, its taste, its smell, but gradually also its form and extension and shrinks to a mere "point."[35]   The "bit of wax," which Descartes took as the basis of his well-known analysis of the concept of the object, changed from a fixed, warm, bright, odorous thing into a mere geometrical figure of certain outlines and dimensions.   And even with this reduction, the intellectual process did not stop; it did not come to rest until extension itself had been dissolved into the mere appearance of simple and indivisible centres of force.   This progressive transformation must appear unintelligible, if we place the goal of natural science in gaining the most perfect possible *copy* of outer reality.   Every new theoretical conception, that science introduces, would then separate science further from its real task; owing to this peculiar method, empirical existence, which should be retained and verified unfalsified, would on the contrary threaten to vanish away.   Here, in fact, no reconciliation is possible; the exactitude and perfect rational intelligibility of scientific connections are only purchased with a loss of immediate thing-like reality.   This reciprocal relation between reality and the concepts of science, however, furnishes the real solution of the problem.   It is only owing to the fact that science abandons the attempt to give a direct, sensuous copy of reality, that science is able to represent this reality as a necessary connection of grounds and consequents.   It is only

[35] *Cf.*, e.g. the description of the "electron," thus of the fundamental element of "matter," by Lucien Poincaré, *La Physique moderne*, p. 249: "Thus the electron must be regarded as a simple electric charge without matter.   Our first investigations caused us to ascribe to it a mass one thousand times less than that of an atom of hydrogen; a more careful study now shows us that his mass was only a fiction; the electro-magnetic phenomena, which occur when we set the electron in motion, or permit its velocity to vary, resemble inertia to a certain extent, and this inertia depending on its charge, led us astray. The electron is thus simply a definite small volume at a point of the ether, which possesses special properties, and this point is propagated with a velocity, which cannot exceed the velocity of light."   *Cf.* also E. Meyerson, *Identité et Réalité*, p. 228 ff.

through going beyond the circle of the given, that science creates the intellectual means of representing the given according to laws. For the elements, at the basis of the order of perceptions according to law, are never found as constituent parts in the perceptions. If the significance of natural science consisted simply in reproducing the reality that is given in concrete sensations, then it would indeed be a vain and useless work; for what copy, however perfect, could equal the original in exactness and certainty? Knowledge has no need for such a duplication, which would still leave the logical form of the perceptions unchanged. Instead of imagining behind the world of perceptions a new existence built up out of the materials of sensation, it traces the universal intellectual schemata, in which the relations and connections of perceptions can be perfectly represented. Atom and ether, mass and force are nothing but examples of such schemata, and fulfill their purpose so much the better, the less they contain of direct perceptual content.

"Real" and "not real" elements in the concepts of the physical object. Thus we have two separate fields, two different dimensions as it were, of the concept; opposed to the concepts that represent an existence, are the concepts that merely express a possible form of connection. Yet there is no metaphysical dualism here; for although there is no direct similarity between the members of the two fields, they reciprocally refer to each other. The concepts of order of mathematical physics have no other meaning and function than to serve as a complete intellectual survey of the relations of empirical being. If this connection with empirical being is destroyed, a double antinomy arises. Behind the world of our experience arises a realm of absolute substances, which are themselves a kind of thing, yet which remain inaccessible to all the intellectual means for grasping the things of experience. The "genuinely real" of physics, the system of atoms and forces acting at a distance, remains in principle unintelligible. There presses upon us the inevitable idea of something existing eternally beyond presentation, something which, since we cannot reach into this "extra-phenomenal beyond," we can never attain. Thus the world of immediate experience pales into a shadow; while, on the other hand, that for which we exchange it remains before us as an eternally incomprehensible riddle. "The manifold forms of the absolute are not windows in our system of presentations, that afford a view into the extra-phenomenal world;

they only show how impenetrable are the walls of our intra-phenomenal prison." Physics itself, in its continuous and necessary progress, leads to a field permanently closed to research: to a *"terra nunc et in aeternum incognita."*[36] On the other side of the question, it is unintelligible how, with our physical concepts formed through transcending the "system of presentations," we could ever return to exactly this system, or how we could hope to master it on the basis of concepts produced in conscious contradiction to its real content. All these doubts are resolved, however, as soon as we consider the physical concepts no longer for themselves but, as it were, in their natural genealogy, in connection with the *mathematical* concepts. In fact, the physical concepts only carry forward the process that is begun in the mathematical concepts, and which here gains full clarity. The meaning of the mathematical concept cannot be comprehended, as long as we seek any sort of presentational correlate for it in the given; the meaning only appears when we recognize the concept as the expression of a *pure relation*, upon which rests the unity and continuous connection of the members of a manifold. The function of the physical concept also is first evident in this interpretation. The more it disclaims every independent perceptible content and everything pictorial, the more clearly its logical and systematic function is shown. (*Cf.* above p. 147 ff.) All that the "thing" of the popular view of the world loses in properties, it gains in relations; for it no longer remains isolated and dependent on itself alone, but is connected inseparably by logical threads with the totality of experience. Each particular concept is, as it were, one of these threads, on which we string real experiences and connect them with future possible experiences. The objects of physics: matter and force, atom and ether can no longer be misunderstood as so many new realities for investigation, and realities whose inner essence is to be penetrated,—when once they are recognized as instruments produced by thought for the purpose of comprehending the confusion of phenomena as an ordered and measurable whole. Thus there is only one reality, which is given to us, but it comes to consciousness in different ways; thus at one time we consider it in its sensuously intuitive character but in its sensuous isolation, while from the standpoint of science we merely retain those elements in it, which are at the basis of its intellectual connection and "harmony."

[36] *Cf.* P. du Bois-Reymond, *Über die Grundlagen der Erkenntnis in den exakten Wissenschaften*, p. 112 ff. (*Cf.* above p. 122 ff.)

The concept of non-being. In the history of physics, we can see how this characteristic interpenetration of the sensuous by the intellectual has been expressed with growing consciousness as a logical insight by the great empirical investigators also. Democritus, who first created the universal schema of the scientific conception of nature, also grasped the philosophical problem latent in it. Motion requires the *void* for its representation; empty space itself, however, is no sensuous "given," no thing-like reality. Thus it is impossible to relate scientific thought merely to being, as the Eleatic idealism had attempted; non-being is just as necessary and unavoidable a concept. Without this concept, intellectual mastery of empirical reality is not to be attained. The Eleatics in their denial of non-being not only robbed thought of one of its fundamental instruments, but they destroyed the phenomena themselves by giving up the possibility of understanding them in their multiplicity and mutability. The thought of non-being is thus no dialectical construction; but, on the contrary, it is taken as the sole means of protecting physics from the extravagances of a speculative idealism. Even when the *facts* themselves are regarded as the supreme standard for all intellectual conceptions, and when no other purpose of the concept is recognized than that of rendering the *fact* of motion, and thus of nature, intelligible, even so it has to be granted that in this fact a moment is involved, which is absent from direct intuition. Empty space is *necessary* for the phenomenon, even though it does not have the same sensuous form of existence as the concrete, particular phenomenon. In the conception of the real, this sensuous "nothing" has the same place and same inviolable validity as the "something:" μὴ μᾶλλον τὸ δὲν ἢ τὸ μηδέν.[37] The being, that belongs to the scientific principle in contrast to any concrete thing, gains clear definition here historically for the first time.[38] The physical concept is defined through a

---

[37] *Cf. Aristoteles de generatione et corruptione* A 8,325 a: ἐνίοις γὰρ τῶν ἀρχαίων ἔδοξε τὸ ὂν ἐξ ἀνάγκης ἓν εἶναι καὶ ἀκίνητον. τὸ μὲν γὰρ κενὸν οὐκ ὄν, κινηθῆναι δ'οὐκ ἂν δύνασθαι μὴ ὄντος κενοῦ κεχωρισμένου . . . ἐκ μὲν οὖν τούτων τῶν λόγων ὑπερβάντες τὴν αἴσθησιν καὶ παριδόντες αὐτὴν ὡς τῷ λόγῳ δέον ἀκολουθεῖν ἓν καὶ ἀκίνητον τὸ πᾶν εἶναί φασι καὶ ἄπειρον ἔνιοι . . . Λεύκιππος δ'ἔχειν ᾠήθη λόγους οἵτινες πρὸς τὴν αἴσθησιν ὁμολογούμενα λέγοντες οὐκ ἀναιρήσουσιν οὔτε γένεσιν οὔτε φθορὰν οὔτε κίνησιν καὶ τὸ πλῆθος τῶν ὄντων. ὁμολογήσας δὲ ταῦτα μὲν τοῖς φαινομένοις, τοῖς δὲ τὸ ἓν κατασκευάζουσιν ὡς οὐκ ἂν κίνησιν οὖσαν ἄνευ κενοῦ, τό τε κενὸν μὴ ὂν καὶ τοῦ ὄντος οὐθὲν μὴ ὄν φησιν εἶναι. τὸ γὰρ κυρίως ὂν παμπλῆρες ὄν.

[38] For the historical and systematic significance of the concept of the μὴ ὄν, see Cohen, *Platons Ideenlehre und die Mathematik*, Marburg 1879; *Logik der reinen Erkenntnis*, p. 70 etc.

double opposition: an opposition, on the one hand, to metaphysical speculation, and, on the other, to unmethodical sensuous perception. Geometrical space serves here as an example and type of a pure relational concept. Since it connects the atoms in a unity and renders possible motion and interaction among them, it can serve in general as a symbol of those principles, on which the connection of the real and given rests, without these principles themselves forming part of the intuitive reality. The senses are entangled in the "conventional" and subjective opposition of the warm and the cold, the sweet and the bitter; they do not exhaust the whole of objectivity. For this whole is completed only in the mathematical functional dependencies, which are inaccessible to the senses, since these are limited to the individual.

Matter and idea and Galileo's concept of inertia. The physics of modern times has retained these principles unchanged; thus Galileo, as an experimental investigator, directly joins on to Archimedes, while, in his philosophical view, he goes back to Democritus. Like Democritus, he describes and completes the conception of nature by the conception of necessity; only "the true and necessary things, which could not be otherwise," belong in the sphere of scientific investigation. The concept of truth, however, remains for him also distinct from the concept of reality. Just as the propositions of Archimedes concerning the spiral remain true, even in case there is no body in nature with a spiral movement, so in the same way in founding dynamics we can proceed from the presupposition of a uniformly accelerated movement towards a definite point, and conceptually deduce all the resulting consequences. If empirical observation agrees with these consequences, so that in the movement of a body with weight the same relations are found, which the theory evolved from hypothetical assumptions, then without danger of error we can regard the conditions, at first established purely intellectually, as realized in nature. But even if the latter were not the case, our propositions would lose nothing of their validity, since in and for themselves they contain no assertions about existence, but only connect certain ideal consequences to certain ideal premises. This general thought gains significant application in Galileo's exposition and defence of his supreme dynamic principle. For him the law of inertia has throughout the character of a mathematical principle, and even if its consequences are applicable to relations of outer reality, they nevertheless in no way signify a direct copy of any empirically given fact. The conditions, of which it speaks, are actually never realized; they are only gained by means of the "resolutive method." Thus, at one point in the "Dialogue concerning the two Systems of the World," when Simplicio is ready to grant the unlimited persistence of the motion of a body left to itself on a horizontal plane, in so far as the body itself is of sufficiently lasting material, Salviati-Galilei points out to him that this assumption is without significance for the real meaning of the principle of inertia; the material constitution of the particular body is merely an accidental and external circumstance, which is in no way used in the deduction and proof of the principle. Inertia is for Galileo,—as empty space was for Democritus,—a postulate that we cannot do without in the scientific exposition of phenomena, but which is not itself a concrete sensible process of external reality. It denotes an *idea*, conceived for the purpose of ordering the phenomena, yet not standing on the same plane methodologically with these phenomena. Hence this motion needs no real, but only a conceived substratum; the real subjects for the exact expression of the principle are the "material points" of mechanics, not the empirical bodies of our world of perception. We see how modern science has here retained the fundamental thought of Democritus only in order, in a certain sense, to pass beyond it; for what was there developed for the concept of the void, is here carried over to the concept of matter, to the $\pi\alpha\mu\pi\lambda\tilde{\eta}\rho\epsilon\sigma$ $\delta\nu$. Matter also in the sense of pure physics is no object of perception, but rather of construction. The fixed outlines and geometrical definiteness we have to give it are only possible because we go beyond the field of sensations to their ideal limits. The matter, with which exact science is alone concerned, never exists then as a "perception," but always only as a "conception." "When we consider space as objective and matter as that which occupies it," says a modern physicist of strictly "empiristic" tendency, "we are forming a construct largely based on geometrical symbols. We are projecting the form and volume of conception into perception, and so accustomed have we got to this conceptual element in the construct, that we confuse it with a reality of perception itself. It is the conceptual volume or form which occupies space, and it is this form, and not the sense-impressions, which we conceive to move."[39] Thus the con-

[39] Pearson, *The Grammar of Science*, p. 250 f.

cept of matter follows the same law that is universally characteristic
of the logical development of scientific principles. The sensuous
properties no ·longer constitute any essential part of its meaning.
Even the element of "weight," which at first seems an inseparable
part, is subordinated and excluded in the transition from the concept
of matter to the pure concept of mass. From mass, we attain to a
mere mass-point, which is distinguished by a certain numerical
value, by a certain coefficient. Matter itself becomes idea, being
increasingly limited to ideal conceptions, that are produced and
confirmed by mathematics.

VI

The concepts of space and time. The structure of pure mechanics
can logically be derived in different ways, according to the kind and
number of fundamental concepts from which one starts. While
the classical mechanics, that reaches a first conclusion in Newton's
Principia, is built upon the concepts of space and time, mass and
force, in modern expositions the concept of energy appears in place
of this last concept. The *Prinzipien der Mechanik* of Heinrich Hertz
have finally developed a new view; they rest merely on the establishment of three independent concepts: space, time and mass, and
from these it is sought to deduce the totality of the phenomena of
motion as an intelligible whole governed by law, through introducing
invisible masses along with the sensuously perceptible masses.
Even in this plurality of possible starting-points, it is evident that
the "picture" that we form of the reality of nature is not dependent
on the data of sense perception alone, but upon the intellectual
views and postulates that we bring to it. Among them, it is especially space and time, that uniformly recur in the different systems
and thus form the unchanging part, the real invariant, for every
theoretical founding of physics. Owing to this unchangeableness
of space and time, both concepts seem at first glance to be of sensuous
content; since sensation never appears outside these forms and,
conversely, since these forms are never given separated from sensation,
the psychological unity and interpenetration of the two moments
leads at once to their logical identification. The very beginning of
theoretical physics with Newton, however, destroys this apparent
unity. It is emphasized here that space and time are something
different when we grasp them after the fashion of immediate sensa-

tion, and when we grasp them after the fashion of mathematical concepts. And it is merely in the latter interpretation that their truth is affirmed. Absolute, motionless space and absolute, uniformly-flowing time are the true reality, while the relative space and the relative time offered us by outer and inner observation signify only sensuous, and thus inexact, measures for empirical movements. It is the task of physical investigation to advance from these sensuous measures, which are satisfactory for practical purposes, to the *realities* indicated and expressed through them. If there is objective knowledge of nature, it must present the time-space order of the cosmos not only as it appears to a sentient individual from his relative standpoint, but also as it is in itself in absolutely universal form. The pure concept alone gives this universality and necessity, because it abstracts from all the differences, which are based on the physiological constitution and the particular position of the individual subject.

Newton's concepts of absolute space and absolute time. From the epistemological point of view, the first scientific determination of the problem of *objectivity in general* is in the definition of space and time, and in the opposition of the sensuous and the mathematical meanings of the two concepts. This problem cannot yet be surveyed in its whole extent; but the decisive preliminary consideration is reached here.[40] It is clear that the philosophical oppositions in the fundamental conception of physics are expressed more strongly in this question than in any other. The struggle over principles has constantly referred back to the Newtonian theory of space and time in order to decide the general problem of the foundation of physics. What do the concepts of absolute space and absolute time mean, if experience can never give us true examples of these concepts? Can a conception claim any sort of physical significance, when on principle we have to refuse to apply it definitely to the reality accessible to us? It must appear as a barren intellectual game that we develop laws for absolute movements in pure mechanics, so long as we have no infallible mark for deciding regarding the absolute or relative character of an *actual* movement. The abstract rule in itself means nothing, if the conditions are not also known for its concrete application, by which we can subsume empirical cases under it. A contradiction remains here, however, in the Newtonian formulation. The laws of natural science, which are to be especially

[40] For the problem of "objectivity," see more particularly Chs. VI and VII.

understood as inductions from given facts, are ultimately related to objects, which (like absolute space and absolute time) belong to another world than that of experience, and are conceived as the eternal attributes of the infinite divine substance. This metaphysical determination recedes into the background in the further development of natural science; but the *logical* opposition, on which it is based, is not thereby removed. The question continually arises, whether in the foundation of mechanics we have to assume only such concepts as are directly borrowed from the empirical bodies and their perceptible relations, or whether we must transcend the sphere of empirical existence in any direction in order to conceive the laws of this existence as a perfect, closed unity.

The system of reference of pure mechanics. Henceforth the real difficulty is concentrated in this problem. The epistemological discussion of the mechanical concepts has failed to mark this difficulty with sufficient sharpness; following the course of history, it has placed the opposition of the "absolute" and "relative" in the center of consideration. But this opposition, which comes from the field of ontology, does not give adequate expression to the methodological questions pressing for solution. It is easy to see, of course, that "absolute" space and "absolute" time, if they are to be regarded with Newton as mathematical conceptions, do not exclude every sort of relation. Precisely this, in fact, is the essential character of all mathematical constructions, that no one of them means anything in itself alone, but that each individual is to be understood only in thorough-going connection with all the others. Thus it is nonsensical to seek to conceive a "place" without at the same time relating it to another distinguished from it, or to seek to establish a moment of time without regarding it as a point in an ordered manifold. The "here" gains its meaning only with reference to a "there," the "now" only with reference to an earlier or later contrasted with it. No physical determination, which we subsequently take up into our concepts of space and time, can impugn this fundamental logical character. They are and remain *systems of relations* in the sense that every particular construction in them denotes always only an individual position, that gains its full meaning only through its connections with the totality of serial members. Moreover, the conception of absolute motion only apparently contradicts this postulate. No physical thinker has ever interpreted this concept as excluding regard for any system of reference at all. The conflict is only with regard to the kind of system of reference, only with regard to whether it is to be taken as material or as immaterial, as empirically given or as an ideal construction. The postulate of absolute motion does not mean the exclusion of any correlate, but rather contains an assumption as to the nature of this correlate, which is here determined as "pure" space separated from all material content. Thereby the problem first loses its vague dialectical form and gains a definite physical meaning. The "relativity," that is inseparably connected with any scientific construction in general, can be left entirely out of account; for it constitutes the general and self-evident presupposition, which is without significance for the solution of any *special* question. But it is such special questions, which are here to be decided. It is above all necessary to become clear as to whether space and time, in the sense in which physics takes them, are only aggregates of sensuous impressions, or whether they are independent intellectual "forms;" as to whether the system, to which the fundamental equations of the Newtonian mechanics refer, can be pointed out as an empirical body, or whether it possesses an "intellectual" being. As soon as we decide in favor of the latter view, the further problem arises of mediating between the ideal beginnings of physics and its real results. The sensuous and intellectual elements stand at first in abstract opposition and require unification under a general point of view to determine their part in the unitary concept of objectivity.

The substitution of the fixed stars for absolute space. At first glance, it might seem that the answer to these questions required the mediation of no complicated, logical terms. The answer that empiricism has ready avoids all difficulties by resolving the problems into mere illusions. The principle of inertia would indeed be meaningless if we did not tacitly conceive it with reference to some system of coordinates, in connection with which the persistence of uniform motion in a straight line could be pointed out. But this unavoidable substratum we do not need to establish by wearisome conceptual deduction, since experience of itself definitely presses it upon us. The fixed stars offer us a system of reference, in connection with which the phenomenon of movement according to inertia can always be demonstrated with the exactitude of which empirical judgments in general are capable. It is a mistaken desire to ask more than this;

it is idle to seek to gain a notion as to what form the law of inertia would assume if we excluded reference to the fixed stars and replaced them by another system. What laws of motion would hold if the fixed stars did not exist, or if we lost the power of orientating our observations on them, we lack all possibility of judging, because we are here concerned with a case which is never realized in actual experience. The world is not given to us twice: once in reality and once in thought; but we must take it as it is offered in sensuous perception, without enquiring how it would appear to us under other conditions, which we logically imagine.[41] In this solution of the problem offered by Mach, the consequence of the empiristic view is drawn with great energy. According to this view, every scientifically valid judgment gains its meaning only as an assertion concerning a concrete, factually present existence. Thought can merely follow the indications of sensation, which reveal this being to us, but it can in no place go beyond them and take into account merely possible, hitherto not given cases. But as has already appeared on all sides, this inference, though unavoidable from the presupposition assumed, contradicts the known fact of scientific procedure itself. The fundamental theoretical laws of physics throughout speak of cases that are never given in experience nor can be given in it; for in the formula of the law the real object of perception is replaced by its ideal limit. (*Cf.* above p. 129 f.) The insight gained through them never issues from consideration of the real alone, but from the possible conditions and circumstances; it includes not only the actual, but also the "virtual" process. This is very clearly expressed in the principle of virtual velocities, which, since Lagrange, forms the real foundation of analytic mechanics. The movements of a material system, as here considered, do not need to be able to be actually carried out; their "possibility" means merely that we can intellectually formulate them without thereby being brought into contradiction with the conditions of the system. The further development of the principle in physics has allowed this methodological aspect to appear with increasing distinctness. In the development of modern thermodynamics, the principle of virtual changes is freed from its initial limitation to mechanical processes and is transformed into a universal principle, that includes all the fields of physics equally. By a virtual change of a system is now understood not only an infinitesimal

---

[41] *Cf.* Mach, *Die Mechanik in ihrer Entwicklung; Die Geschichte und die Wurzel des Satzes von der Erhaltung der Arbeit*, p. 47 ff.

spatial displacement of its individual parts, but also an infinitely small increase or decrease of temperature, an infinitely small change in the distribution of electricity on the surface of a conducting body;— in short, any elementary increase or decrease of one of the variable magnitudes that characterize the total state of the system, in so far as it is in accord with the general conditions which the system has to satisfy. Whether the transformation involved can be *physically* carried out is here of no importance; for the truth of our theoretical deductions is entirely independent of this possibility of an immediate realization of our intellectual operations. "If in the course of the deductions," Duhem remarks, "the magnitudes, to which the theory refers, are subjected to certain algebraic transformations, then we need not ask whether these calculations have a *physical meaning*, whether the particular methods of measurement can be directly translated into the language of concrete intuition, and correspond in this translation to real or possible facts. To raise any such question is rather to form a totally erroneous conception of the nature of a physical theory."[42] The discovery and first formulation of the principle of inertia thoroughly confirms this conception. Galileo, at least, leaves no doubt that the principle, in the sense that he takes it, has not arisen from the consideration of a particular class of empirically real movements. He certainly would have made the same answer to the objection, that the truth of the law of inertia presupposes the permanent existence of the fixed stars, that he made to Simplicio in a similar case: *viz.*, that the reality of the fixed stars belongs, like that of the moving body itself, only to the "accidental and external" conditions of the investigation on which its real theoretical decision does not depend. Into that *"mente concipio"* with which Galileo begins his general exposition, the existence of the fixed stars does not enter. The concept of uniform motion in a straight line is here introduced purely in an abstract phoronomic sense; it is not related to any material bodies, but merely to the ideal schemata offered by geometry and arithmetic. Whether the laws, which we deduce from such ideal conceptions, are applicable to the world of perception must be ultimately decided by experiment; the logical and mathematical meaning of the hypothetical laws is independent of this form of verification in the actually given.[43]

[42] *Cf.* Duhem, *L'évolution de la Mécanique*, Paris 1903, p. 211 ff.

[43] *Cf.* above p. 123 ff.; on the whole matter *cf.* now especially Natorp's exposition, *op. cit.* p. 356 ff.

The "intellectual experiment" and the law of inertia. In order to justify logically the deduction here actually made by Galileo, we need ultimately only to appeal to Mach himself. In Mach's development of the general methods of physics, the "intellectual experiment" takes an important place. He emphasizes that all really fruitful physical investigations have intellectual experiments as their necessary condition. We must anticipate (at least in its general features) the consequence which a definite arrangement of the experiment promises; we must compare the possible determining factors with each other and intellectually vary them, in order to give the observation itself a definite direction. It is this procedure of intellectual variation of the determining factors for a certain result which first gives us an entirely clear survey of the field of facts. Here the meaning of each individual moment is first made clear; here the perception is first analysed into an ordered complex, in which we clearly grasp the significance of each part in the structure of the whole. The essential features, on which its action according to law depends, are separated from the accidental, which can vary arbitrarily without thereby affecting our real physical deduction.[44] We need only apply all these considerations to the discovery and expression of the principle of inertia in order to recognize that the real validity of this principle is not bound to any definite material system of reference. Even if we had found the law at first verified with respect to the fixed stars, there would be nothing to hinder us from freeing it from this condition, by calling to mind that we can allow the original substratum to vary arbitrarily without the meaning and content of the law itself being thereby affected. For the assumption on which Mach's initial objection rests, *viz.*, that thought must never look beyond the circle of given facts, is now abandoned. The method of the "intellectual experiment" involves a characteristic activity of thought, by which we go from the real to the possible cases and also undertake to determine them. In fact, the logical meaning of the law of inertia would obviously remain unchanged even if, in the course of experience, reasons were found for us to ascribe certain movements to the fixed stars themselves. The principles of pure mechanics would lose none of their validity by this insight, but would be completely maintained in the new system of

[44] *Cf.* Mach, *Erkenntnis und Irrtum*, Leipzig 1905, "*Über Gedanken-experimente*," p. 180 ff.

orientation, which we would then have to seek. Such a transference would be impossible even in thought, if these principles merely reproduced the relations of moving bodies relative to a particular empirical system of reference. Mach himself must, according to his whole assumption, regard the fixed stars not as an element, which enters into the conceptual formulation of the law of inertia, but must conceive them as one of the *causal factors* on which the law of inertia is dependent.[45] In a formula that merely expresses the relation and interaction of definite physical objects, the one of the two factors can obviously not be replaced by another without the relation itself thus gaining an entirely new form. If the truth of the law of inertia depended on the fixed stars as these definite physical individuals, then it would be logically unintelligible that we could ever think of dropping this connection and going over to another system of reference. The principle of inertia would in this case not be so much a universal principle of the phenomena of motion in general, as rather an assertion concerning definite properties and "reactions" of a given empirical system of objects;—and how could we expect that the physical properties found in a concrete individual thing could be separated from their real "subject" and transferred to another? In any case, we see in this example that empiricism and empirical method (*Empirie*) are different. The only meaning that the principle of inertia could have, according to the empiristic assumptions, is one that corresponds in no way to the meaning and function it has actually fulfilled in scientific mechanics from the beginning. The logical principle of mechanics is not grasped and explained here, but is rather abandoned.

Streintz's concept of the "fundamental body." The same objection in principle can be made to every attempt to give the law of inertia a

[45] "A free body, affected by a momentary pair of forces, moves in such a fashion that its central ellipsoid, if its center is fixed, rotates without sliding on the tangential plane parallel to the plane of the pair of forces. This is a motion in consequence of inertia. In this, the body makes the strangest turns with reference to the heavenly bodies. Is it believed now that these bodies, without which the conceived motion cannot be described, are without influence on it? Does what one must name explicitly or tacitly, if one is to describe a phenomenon, not belong to the essential conditions, to the causal nexus of the same? The remote heavenly bodies in our example have no influence on acceleration, but have influence on velocity." (Mach, *Erhaltung der Arbeit*, p. 49.)

fixed basis by pointing out its system of reference as somehow present in the reality of things. Streintz has attempted to define as this system of reference any arbitrary, empirically given body in so far as it satisfies the double condition of not revolving and not being subjected to the influence of outer forces. The absence of rotation can be definitely indicated by certain instruments of measurement, designated by Streintz by the name of "gyroscopic compass;" every "absolute" rotation of a body brings about some physical effects, that can be directly perceived and measured. A similarly immediate and positive decision is never possible with regard to the second aspect, the absence of external forces: here we must simply be content with the fact that, as often as a variation from motion in a straight line or from uniform velocity has been observed in the movement of a point in a constant direction relative to a body, we have hitherto always succeeded in indicating some external bodies, which appear as the causes of this variation by virtue of their relative position to the moving point itself or to the assumed system of reference. If, now, we characterize the body defined by the two conditions indicated: by the absence of rotation and by perfect independence of all surrounding masses, as the fundamental body (FK), then we have in such a body a suitable system with reference to which the dynamic differential equations at the basis of physics are satisfied. These equations, which in the manner that they are ordinarily formulated are logically indefinite, thus gain a fixed and definite meaning. The principle of inertia, in particular, can now be expressed in the form, that every point left to itself moves in a straight line and with constant velocity with reference to this fundamental body.[46] This attempted deduction, it can be easily shown, rests however on a conversion of the real logical and historical relation. If Streintz's explanation were true, the mechanical principles would be merely inductions, which we have verified in particular bodies with definite physical properties, and which we have then taken as probably true for all bodies of the same sort. The claim to strict universality made by these principles would then be entirely unintelligible. It could not be understood by what right we opposed them to the observed facts as *postulates* predetermining the direction of our explanation, instead of transforming the principles (which have, indeed, only been gained by

[46] *Cf.* the more particular exposition by Streintz, *Die physikalischen Grundlagen der Mechanik*, Leipzig 1883, p. 13 ff., 22 ff.

definite observations), as soon as they failed to agree with the new experiences. But even if we abstract from this, the consideration would be decisive, that the fundamental body and the fundamental system of coördinates could never be discovered as empirical facts, if the meaning of both had not been previously established in ideal *construction*. The seemingly pure inductions, which Streintz makes the basis of his explanation, are already guided and dominated by the fundamental conceptions of analytic mechanics. It is only on the assumption of these conceptions that the meaning of the two aspects determining the fundamental body can be understood: the absence of rotation and the independence of external forces are the empirical criteria by which we recognize whether a definite given body satisfies the presuppositions of the theory, which we have previously developed independently. The *property* (*Merkmal*), by which we establish whether an individual case can be subsumed under a definite law, is logically strictly separated from the *conditions*, on which the validity of the law itself rests. The idea of inertia did not arise from observations of definite bodies, from which we were able, as it were, to read off sensuously the property of their being under no external influence; but it can only be explained conversely, on the basis of the idea that we *seek for* bodies of this sort and give them a special place in the structure of our empirical reality. Thus the attempt of Streintz, in so far as it is meant to be a true founding of mechanics, involves a circle; for in the experiments and empirical propositions, which form the basis of it, there is already a tacit recognition of the principles which are to be deduced. Analytic mechanics, as history shows, has come into existence without these experiments, while conversely the mere conception of these experiments could only arise on the basis of this mechanics.[47]

The theory of C. Neumann: the body alpha. Thus, if we still demand the connection of the law of inertia with some material system of reference, and if we would also explain the rational structure of mechanics, there remains finally only one escape, *viz.*, of assuming an unknown body not given in experience, and of explaining the fundamental dynamic equations with reference to it. The working out of this thought was first attempted by C. Neumann in his work on the principles of the Galileo-Newtonian theory, in which, along with the explanation of the fundamental physical problem, the methodological question also was clearly expressed. According to Neumann, the principle of Galileo can only be grasped in its conceptual meaning through the assumption of a definite existential background. Only in a world in which there exists at an unknown point of space an absolutely rigid body, unchanging in its form and dimensions to all time, are the propositions of our mechanics intelligible. "The saying of Galileo, that a material point left to itself moves in a straight line, appears as a proposition without content, as a proposition hanging in the air, which needs a definite background in order to be understood. Some special body in the cosmos must be given as a basis of judgment, as that object with reference to which all movements are to be estimated;—only then are we in a position to connect a definite content with those words. What body is it, to which we can give this special position? Unfortunately neither Galileo nor Newton have given us a definite answer to this question. But when we attentively examine the theoretical structure founded by them and expanded continuously up to the present time, its foundations can no longer be concealed from us. We easily recognize that all the motions present in the universe or conceivable are to be related to one and the same body. Where this body is found, what reasons are to be given for granting a single body so prominent, as it were sovereign, a position,—to this, at all events, we gain no answer."[48] We should not expect to find within physics the proof establishing the existence of this unique body, which is called by Neumann the "body alpha." For the proof is, in fact, of purely ontological nature; the postulate of a unitary logical point of reference is made into an assertion of an empirically unknowable existence. And to this existence, although it is to be of material nature, are ascribed all those predicates usually employed by the ontological argument: it is unchanging, eternal and indestructible. While, on the one hand, a being with absolute properties is deduced here from mere thought, on the other hand, the converse feature appears, *viz.*, that the conceivability of our ideal conceptions is made dependent on definite properties of being. If we conceive the body *alpha* annihilated by any force of nature, the propositions of mechanics would necessarily cease not only to be applicable, but even to be *intelligible*. The concept of the strict con-

---

[48] Carl Neumann, *Über die Prinzipien der Galilei-Newtonschen Theorie,* Leipzig 1870, p. 14 f.

stancy of direction, the concept of uniform motion of definite velocity, such as mathematical theory offers us, would at one stroke be deprived of all meaning. Thus not only definite physical consequences but also the most noteworthy logical consequences would be connected with a process in the outer world; thus it would depend on the being or not-being of an actual spatial thing whether our fundamental mathematical hypotheses had any meaning in them. But how could we ever make a rational judgment regarding a physical reality, if the meaning of these universal mathematical predicates were not already established? To all these questions, only one answer could ultimately be given by the advocates of the body *alpha*. It is not the *existence* of the body *alpha*, they could answer, but the *assumption* of this existence, on which the validity of our mechanical concepts depends. This assumption can never be taken from us; it is a pure postulate of our scientific thought, which herein obeys only its own norms and rules. Such an answer, however, places the problem on an entirely new basis. If it is in our power to deal with ideal assumptions, then we do not understand why this procedure should be limited to the assumption of *physical* things. Instead of the body *alpha*, we could then assume (in logically the only unobjectionable and intelligible manner) pure space itself, and grant it definite properties and relations. Here also we have moved in a circle; thought by its inner necessity has led us back to that very starting-point at which the first doubt and suspicion arose regarding the formulation of mechanical principles.

Space and time as mathematical ideals. We first escape the dilemma when we resolve to place our intellectual postulates in full clarity at the beginning, instead of somehow introducing them in a concealed form in the course of the deduction. The absolute space and the absolute time of mechanics involve the problem of existence just as little as does the pure number of arithmetic or the pure straight line of geometry. Absolute space and time arise in the sure and continuous development of these concepts; Galileo emphasized most sharply that the general theory of motion signified not a branch of *applied* but of *pure* mathematics. The phoronomical concepts of uniform and of uniformly accelerated motion contain nothing originally of the sensuous properties of material bodies, but merely define a certain relation between the spatial and temporal magnitudes that are generated and related to each other according to an ideal principle of construction.

Thus, in the expression of the principle of inertia, we can rely at first merely on a conceptual system of reference, to which we ascribe all the determinations required. By means of conceptual definitions, we create a spatial "inertial system" and an "inertial time-scale," and take both as a basis for all further consideration of the phenomena of motion and their reciprocal relations.[49] Thus there is no hypostatization of absolute space and absolute time into transcendent things; but at the same time both remain as pure *functions*, by means of which an exact knowledge of empirical reality is possible.[50] The fixity, that we must ascribe to the original and unitary system of reference, is not a sensuous but a logical property; it means that we have established it as a concept, in order to regard it as identical and unchanging through all the transformations of calculation. The ideal system of axes, to which we look, satisfies the fundamental postulate, which requires independence of all outer forces for the "fundamental system of coördinates:" for how could forces affect lines, pure geometrical forms? While considering these lines in our intellectual abstraction as absolutely constant, we evolve from them a general schema for possible spatial changes in general. Only experience can ultimately decide whether this schema is applicable to the reality of physical things and processes. Here also it is never possible to isolate the fundamental hypotheses and to point them out as valid individually in concrete perceptions; but we can always only justify them indirectly in the total system of connection that they effect among phenomena. (*Cf.* above p. 146 ff.) We develop the determination of the "inertial system" and the mathematical consequences connected with it purely in theory. In so far as any empirically given body seems to conform to these determinations, we ascribe "absolute" rest and absolute fixity to it also; *i.e.*, we affirm that a material point left to itself must move uniformly in a straight line with reference to that body. But at the same time, we know that this postulate is never exactly fulfilled in experience, but always only with a certain approximation. But just as there is

---

[49] *Cf.* more particularly regarding the mathematical construction of the "inertial system," Ludwig Lange, *Die geschichtliche Entwicklung des Bewegungsbegriffs*, Wundt's Philos. Studien III, (1886), p. 390 ff., 677 ff.

[50] See *Erkenntnisproblem* II, 344, 356 f., 559; *cf.* now especially the excellent treatment of Edm. König, *Kant und die Naturwissenschaft*, Braunschweig 1907, p. 129 ff.

no real straight line, that realizes all the properties of the pure geometrical concept, just so there is no real body, that conforms to the mechanical definition of the inertial system in all respects. Thus the possibility always remains open of establishing, by the choice of a new point of reference, a closer and more exact agreement between the system of observations and the system of deductions. This relativity is indeed unavoidable; for it lies in the very concept of the object of experience. It is the expression of the necessary difference, that remains between the exact conceptual laws we formulate and their empirical realization. That any system of given bodies is at rest (for example, the system of the fixed stars) does not signify a fact that can be directly established by perception or measurement, but means that a *paradigm* is found here in the world of bodies for certain principles of pure mechanics, in which they can be, as it were, visibly demonstrated and represented. The fixed stars stand in relations with the moving bodies of the real world, which relations are entirely according to the system of these propositions and find their complete expression in it. The individual material point of attachment, to which we connect our equations of motion, may change; yet the fundamental relation to a definite system of laws of mechanics and physics remains constant. Analogously, we substitute a more exact measure of time for the not wholly exact measure, which the stellar day affords, by taking our stand on the law of the conservation of energy and on the law of gravitation. That unit of time is taken as "absolutely" exact, whose application enables us, on the one hand, to avoid contradiction of the theoretical demands of the principle of energy, and on the other hand to avoid conflict between the secular acceleration of the moon as actually observed, and as calculated according to the Newtonian law.[51] Thus there is still a relation to the physical concepts of absolute space and absolute time. The meaning of these concepts does not consist in their stripping off every relation, but in their removing the necessarily assumed point of reference from the material into the ideal. The system, in which we seek our intellectual orientation, is no individual perceptible body, but a system of theoretical and empirical rules on which the concrete totality of phenomena is conceived to be dependent.

Hertz's system of mechanics. This meaning of the concepts of absolute space and absolute time was established in its general fea-

[51] Poincaré, *La mesure du temps* (*cf.* above, p. 145 f.).

tures by Leibniz. For him, both concepts were only another expression of the thorough-going spatial and temporal *determinateness*, which we demand for all being and process. This determinateness must be demanded, even if there is in the cosmos no strictly uniform course of any actual process of nature, or any fixed or immovable body. Theoretically it can always be gained; for we can always relate the non-uniform movements, whose law we know, to certain *conceived* uniform movements, and on the basis of this procedure calculate in advance the consequence of the connection of different movements.[52] The relation between theory and experience assumed here has found its clearest expression in modern times in the system of mechanics of Heinrich Hertz. Hertz in his exposition takes space and time at first merely in the sense in which they are offered to "inner intuition." The assertions, that are made of them, are "judgments *a priori* in the sense of Kant;" any appeal to the experience of sensuously perceptible bodies is foreign to them. It is only in the second book, where the transition is made from geometry and kinematics to the mechanics of material systems, that times, spaces and masses are conceived as signs of outer empirical objects; the properties of these empirical objects, however, cannot contradict the properties, which we have previously ascribed to these magnitudes as forms of our inner intuition or by definition. "Our assertions concerning the relations of times, spaces and masses should thus no longer satisfy merely the claims of our intellect, but at the same time they should correspond to all possible, especially to future, experiences. These assertions thus rest not only on the laws of our intuition and our thought, but on previous experience besides." By taking fixed units of measurement as a basis, especially within this field, and comparing the empirical spaces, times and masses with each other in accordance to them, we gain a general principle of coördination by means of which we allow certain mathematical symbols definitely to correspond to the concrete sensations and perceptions, and thereby translate the given impressions into the symbolic language of our inner intellectual images. The indefiniteness, that necessarily belongs to these ultimate units of measurement, is not indefiniteness of our images, and not of our laws of transformation and correlation, but it is the indefiniteness of the very outer experience to be copied. "We mean to say that by our senses we can

[52] Leibniz, *Nouveaux Essais*, Bk. II, Ch. 14.

determine no time more exactly than can be measured with the help of the best chronometer, no position more exactly than can be referred to the coördinate system of the remote fixed stars, and no mass more exactly than is done by the best scales."[53] Thus, while a perfect definiteness of all elements can be gained with the structures we generate from the laws of intuition and thought, in the field of empirical phenomena this is merely postulated. We measure the "reality" of our experiences by the "truth" of our abstract dynamic concepts and principles. The order of the world which we construe on the presupposition of the motionlessness of the fixed stars, is for us the true order of things, in so far as all actually observable motions with reference to this system have always hitherto corresponded with the greatest approximation to the axioms, by which mechanics characterizes the concept of "absolute movement." If this condition should no longer be fulfilled (and we must consider this in our calculations and assumptions as a possible case), still these axioms, the ideal according to which the construction has taken place, would be wholly unaffected in meaning; its empirical realization would only be shifted to another place.

Thus absolute space,—if we mean by it not the abstract space of mechanics but the definite order of the world of bodies,—is at all events never finally given, but is always only sought. But in this, there is no lessening of its objective meaning for our knowledge; for, as a sharper analysis shows, relative space also signifies no given fact in the sense of a dogmatic "positivism." Also when we consider any corporeal masses in their mutual positions and their relative *distances*, we have already gone beyond the limit of sensuous impressions. When we speak of "distance," we mean by this, strictly speaking, no relation between sensuous bodies, since these bodies might have very different distances from each other according as we took one point or another of their volume as a starting-point for measurement. In order to gain an exact geometrical meaning, we must substitute a relation between *points* for a relation between bodies, by considering the total mass of the body reduced to the center of gravity. We are thus obliged to have the direct empirical intuition transformed by means of the pure geometrical limiting concepts, in order to make a completely exact assertion regarding the relative position of two material systems. The positivistic scruples

[53] Heinrich Hertz, *Die Prinzipien der Mechanik*, p. 53 ff., p. 157 ff.

against the "pure" space and the "pure" time of mechanics thus prove nothing because they would prove too much; logically thought out, they would also have to forbid every representation of physically given bodies in a geometrical system in which there are fixed positions and distances. The physical space of bodies is no isolated essence, but is only possible by virtue of the geometrical space of lines and distances. This relation was also expressed by Leibniz in an especially striking and pregnant saying. It is indeed true, as he explains, that more is posited in the concept of body than in the concept of mere space; but it does not follow from this that the extension, which we perceive in bodies, differs from the ideal extension of geometry in any properties. Number also is something different from the totality of counted things, yet plurality as such has one and the same meaning, whether we define it in purely conceptual terms or represent it in some concrete example. "In the same sense, we can also say that we do not have to conceive two sorts of extension, the abstract of space and the concrete of bodies, for the concrete receives its properties only through the abstract."[54] We inscribe the data of experience in our constructive schema, and thus gain a picture of physical reality; but this picture always remains a *plan*, not a *copy*, and is thus always capable of change, although its main features remain constant in the concepts of geometry and phoronomy.

Construction and convention. It seems, indeed, that an element of arbitrariness is admitted into our scientific reflection, when we thus ground our assertion concerning reality in previous *constructions*. This conclusion is actually drawn, when the concepts of the "inertial system" and the "inertial time-scale" are described as mere *conventions*, which we introduce in order to survey the facts more easily, but which have no immediate objective correlate in empirical fact.[55] Poincaré, in an investigation of the conditions of time measurement, has deduced the general consequences very decisively. When we take any natural phenomenon as absolutely uniform, and measure all others by it, we are never absolutely determined in our choice from without; no measure of time is truer than any other, but the only proof we can give for any is merely that it is more convenient. The question raised here is open to no final answer so far as our

---

[54] See Leibniz, *Nouveaux Essais*, Bk. II, Ch. 4.

[55] *Cf.* Ludwig Lange, *op. cit.*; also *Das Inertialsystem vor dem Forum der Naturforschung, Philos. Studien*, Vol. II.

previous investigation is concerned; for it reaches over from the field of science into a methodologically alien field. Science has no higher criterion than truth; and can have no other than unity and completeness in the systematic construction of experience. Every other conception of the object lies outside its field; it must "transcend" itself even to be able to conceive the problem of another sort of objectivity. The distinction between an "absolute" truth of being and a "relative" truth of scientific knowledge, the separation between what is necessary from the standpoint of our concepts and what is necessary in itself from the nature of the facts, signifies a metaphysical assumption, that must be tested in its right and validity before it can be used as a standard. The characterization of the ideal conceptual creations as "conventions" has thus at first only one intelligible meaning; it involves the recognition that thought does not proceed merely receptively and imitatively in them, but develops a characteristic and original spontaneity (*Selbsttätigkeit*). Yet this spontaneity is not unlimited and unrestrained; it is connected, although not with the individual perception, with the *system* of perceptions in their order and connection. It is true this order is never to be established in a single system of concepts, which excludes any choice, but it always leaves room for different possibilities of exposition; in so far as our intellectual construction is extended and takes up new elements into itself, it appears that it does not proceed according to caprice, but follows a certain law of progress. This law is the ultimate criterion of "objectivity;" for it shows us that the world-system of physics more and more excludes all the accidents of judgment, such as seem unavoidable from the standpoint of the individual observer, and discovers in their place that necessity that is universally the kernel of the concept of the object.[56]

VII

The concept of energy. Necessary as are space and time in the construction of empirical reality, they are after all only the universal forms in which it is represented. They are the fundamental orders, in which the real is arranged, but they do not determine the concept of the real itself. A new principle is needed to fill these empty forms with concrete content. This principle has been conceived in different

[56] *Cf.* later Chs. VI and VII.

ways, from Democritus' concept of matter opposed to empty space as the $\pi\alpha\mu\pi\lambda\widehat{\eta}\rho\epsilon\varsigma$ ὄν, until its final logical definition in the modern conception of energy. Here, for the first time, we seem to have the ground of reality under our feet. Here we have a being that fulfills all the conditions of true and independent existence, since it is indestructible and eternal. Along with all physical reasons, energism claims an *epistemological* advantage. The *atom* and *matter*, which constitute the real type of objective reality for the older natural science, are reduced to mere abstractions through the closer analysis of the data and conditions of our knowledge. They are conceptual limits, to which we attach our impressions, but they can never be compared in real meaning with the immediate sensation itself. In energy, we grasp the real because it is the effective. Here no mere symbol comes between us and the physical thing; here we are no longer in the realm of mere thought, but in the realm of being. And in order to grasp this ultimate being, we need no circuitous route through complicated mathematical hypotheses, since it is directly revealed unsought in perception itself. What we sense is not the doubtful and in itself entirely indefinite *matter*, that we assume as the "bearer" of sensuous properties; but it is the concrete effect, which is worked on us by outer things. "What we see is nothing but radiating energy, which effects chemical changes in the retina of our eye, that are felt as light. When we touch a solid body, we feel the mechanical work that is involved in the compression of our finger-tips and also in the compression of the body touched. Smell and taste rest on chemical activities in the organs of the nose and mouth. Everywhere energies and activities are what inform us as to how the outer world is arranged and what properties it has; and, from this standpoint, the totality of nature appears to us as a division of spatially and temporally changeable energies in space and time, of which we gain knowledge to the extent that these energies go over to our body, in particular to the sense organs constituted for receiving definite energies."[57] The "thing" as a passive and indifferent substratum of properties is now set aside. The object *is* what it appears to be: a sum of actual and possible ways of acting. In this doctrine, a determination of purely philosophical reflection is admitted into the foundations of scientific thought; but the function of reflection is thereby limited and exhausted. Henceforth all

[57] Ostwald, *Vorles. über Naturphilosophie*, p. 159 f.

merely speculative points of view can be strictly excluded, and consideration limited to the reproduction of the empirical facts. In the accomplishment of this task, the primal reality itself without abstract or conceptual husk becomes increasingly clear.

Energy and the sense qualities. There is one doubt that must be felt with regard to this conception. Whatever physical advantages the concept of energy may have over that of matter and the atom, logically at least, both stand on the same plane and belong to the same sphere of consideration. This is shown negatively in the similar difference of both from the sensuously given. The notion that "energies" can be seen or heard is obviously no less naïve than the notion that the "matter" of theoretical physics can be directly touched and grasped with the hands. What is given us are qualitative differences of sensation: of warm and cold, light and dark, sweet and bitter, but not numerical differences of quantities of work. When we refer sensations to such magnitudes and their mutual relations, we have brought about just such a translation into a foreign language as energism criticizes in the mechanical view of the world. To *measure* a perception means to change it into another form of being and to approach it with definite theoretical assumptions of judgment. (*Cf.* p. 141 ff.) The advantage, which energism might claim over mechanism, could never consist in that it did without these presuppositions, but rather in that it understood them more clearly in their logical character. It could not be a matter of entirely excluding "hypotheses," but only of not making them into absolute properties of things as is done by dogmatic materialism.

Energy and the concept of number. If we conceive the problem in this way, it is seen that energism contains a motive from the beginning, which protects it more than any other physical view from the danger of an immediate hypostatization of abstract principles. Its fundamental thought, from an epistemological point of view, does not go back primarily to the concept of space, but to that of number. It is to numerical values and relations that the theoretical and experimental inquiry are alike directed, and in them consists the real kernel of the fundamental law. Number, however, can not be misunderstood as substance, unless we are to revert to the mysticism of Pythagoreanism, but it signifies merely a general point of view, by which we make the sensuous manifold unitary and uniform in conception. The evolution of the conception of energy offers a concrete physical example for this general process of knowledge. We saw that the first step in the mathematical objectification of the given was to conceive it under certain serial concepts. The given only becomes an object of knowledge when it is "established" in this sense, by ascribing to it a definite place in a manifold, ordered and graded according to some point of view. But the real task of knowledge of nature is not thereby exhausted; in fact, in principle it is not yet begun. The insertion of the sensuous manifold into series of purely mathematical structure remains inadequate, as long as these series are separated from each other. As long as this is the case, the "thing" of popular experience is not yet wholly grasped in its logical meaning. It is not enough for us to express the individual physical and chemical properties in a pure numerical value, and to represent the object as a totality of such "parameters." For the object means more than a mere sum of properties; it means the *unity* of the properties, and thus their reciprocal dependency. If this postulate is to find its adequate expression in science, we must seek a principle, which enables us to connect the different series, in which we have first arranged the content of the given, among themselves by a unitary law. Heat, motion, electricity, chemical attraction mean at first only certain abstract types, to which we relate the whole of our perceptions. In order to advance from them to a representation of the real process, a thorough mediation is needed, so that all these different fields again becomes members of an inclusive *system*.

The concept of the measure of work. From this point, the general meaning of the conception of energism can be surveyed. The structure of mathematical physics is in principle complete when we have arranged the members of the individual series according to an exact numerical scale, and when we discover a constant numerical relation governing the transition from one series to the others. Only when this is done is the way determined from one member to any other, and prescribed by fixed rules of deduction, no matter what the series. Only then it becomes clear how all the threads of the mathematical system of phenomena are connected on all sides, so that no element remains without connection. This relation was first established empirically in the case of the equivalence of motion and heat; but, once discovered, it was extended beyond this starting-point. The conception was soon extended as a universal postulate to the totality of possible physical manifolds in general. The law of energy directs us to coördinate every member of a manifold with one and only one member of any other manifold, in so far as to any *quantum* of motion there corresponds one *quantum* of heat, to any *quantum* of electricity, one *quantum* of chemical attraction, *etc.* In the concept of *work*, all these determinations of magnitude are related to a common denominator. If such a connection is once established, then every numerical difference that we find within one series can be completely expressed and reproduced in the appropriate values of any other series. The unit of comparison, which we take as a basis, can arbitrarily vary without the result being affected. If two elements of any field are equal when the same amount of work corresponds to them in *any* series of physical qualities, then this equality must be maintained, even when we go over to any other series for the purpose of their numerical comparison. In this postulate, the essential content of the principle of conservation is already exhausted; for any quantity of work, which arose "from nothing," would violate the principle of the mutual one to one coördination of all series. If we wish to represent the system schematically, we have here a number of series A, B, C, of which the members $a_1\, a_2\, a_3 \ldots a_n$, $b_1\, b_2\, b_3 \ldots b_n$, $c_1\, c_2\, c_3 \ldots c_n$, stand in a definite physical relation of exchangeability, such that any member of A can be replaced by a definite member of B or C without the capacity of work of the physical system in which this substitution is assumed being thereby changed. We briefly represent this relation of possible substitution not by always coördinating each individual member with the multitude of corresponding equivalents, but by ascribing to it once for all a certain value of energy, which draws all these coördinations into a single pregnant expression. We do not compare the different systems with each other directly, but create for this purpose a *common series*, to which they are all equally related. It is chiefly owing to technical circumstances that we traditionally choose mechanical work as this common series, since the transformation of the different "types of energy" into this form is relatively easy and exactly measurable. In itself, however, any arbitrary series could be taken as the basis for expressing the totality of possible relations. In any case, it appears that energy in this form of deduction is never a new *thing*, but is a unitary *system of reference* on which we base measurement. All that can be said of it on scientific grounds is exhausted in the quantitative relations of equivalence, that prevail between the different fields of physics.

Energy does not appear as a new objective somewhat, alongside of the already known physical contents, such as light and heat, electricity and magnetism; but it signifies merely an objective correlation according to law, in which all these contents stand. Its real meaning and function consists in the *equation* it permits us to establish between the diverse groups of processes. If we clothed the principle itself, which demands the definite quantitative correlation of the totality of phenomena, in the form of a particular thing, even in the form of "the" thing, the all-inclusive substance, we should create the same dogmatic confusion, that energism charges against materialism. *Science* at least knows nothing of such a transformation into substance, and cannot understand it. The *identity* science seeks, and to which it connects the chaotic individual phenomena, has always the form of a supreme mathematical law; not, however, of an all-inclusive and thus ultimately propertyless and indeterminate object. Conceived as a particular thing, energy would be a somewhat, which was at once motion and heat, magnetism and electricity, and yet also nothing of all these. As a principle, it signifies nothing but an intellectual point of view, from which all these phenomena can be measured, and thus brought into one system in spite of all sensuous diversity.

The formal presuppositions of energism. At this point, in the midst of the controversial questions of contemporary philosophy of nature, we are led to make a general logical remark. Paradoxical as it may appear, here where consideration seems wholly given over to the *facts*, the effect of general logical theories is evident. Whether we conceive energy as a *substance*, or as the expression of a *causal relation*, depends finally on our general idea of the nature of the scientific construction of concepts in general. However frankly the physical investigator intends to stand face to face with nature itself, it can nevertheless be shown that in the construction of energism motives have been at work, which have their real origin in definite "formal" convictions. At this point, we recognize anew how deeply the problems of "form" penetrate into those of "matter," and how lasting is their influence. Two different views were opposed to each other in the problem of the concept. The one, that has remained dominant in traditional logic, bases the concept on the procedure of abstraction, *i.e.*, on the separating out of an identical or similar part from a plurality of similar perceptions. The content thus gained is, strictly speaking, of the same property and nature as the object from which it is abstracted; it represents a property that in general is not isolated, yet which can always be pointed out as a constitutive part in these objects, and therefore possesses a concrete existence. The concept is accordingly the "presentation of the common;" it is the unification of those individual features, that belong uniformly to definite classes of objects. Opposed to this conception, however, is another that is founded above all on the analysis of the mathematical concepts. In this concept, we do not attempt to separate the given by comparison into classes whose individual examples all agree in certain properties, but to construct the given from a postulate of unity by a process according to law. Here it is not so much the particular *parts* of the given, that are isolated, but rather the connections and relations on which its organization rests; these are investigated in their characteristic relational structure. (*Cf.* above esp. p. 14 ff. and 81 ff.) The meaning of this opposition in views of the concept now appears from a new angle; it is this opposition, that is noticeable in modern discussion as to the formulation of the principle of energy. Rankine, who first created the name and concept of a general "energetics," proceeds from purely methodological considerations in his treatise devoted to the first establishment of the new conception. Physics, as he explains, is characteristically separated from the purely abstract sciences, such as geometry, by the fact that the definitions fundamental in the development of an abstract science do not necessarily correspond to any existing things, and the theorems deduced from them need not necessarily be laws of *real* processes and phenomena, while the true scientific concept is to be nothing but the designation of certain properties common to a *class of real objects*. There is, in general, a two-fold way to separate out such properties. We can, by a pure "abstractive" method, separate from a group of given things or phenomena that group of determinations which is common to all members of the class, and which belongs to them directly in their sensuous appearance; or we can go behind the phenomena to certain *hypotheses* for the explanation of the field of physical facts in question. Only the first procedure strictly corresponds to the demands of scientific and philosophic criticism. For only here are we sure that we do not falsify the observations by arbitrary interpretation; only here do we remain purely in the field of the facts themselves, for while we divide the facts into definite classes, we add no foreign feature to them. It is an advantage in principle of the new science of energetics, that from the beginning it uses merely this abstractive procedure. It does not refer the phenomenon of heat to molecular movements, the phenomenon of magnetism to any hypothetical fluid, but it grasps both merely in the simple form in which they are offered to perception. "Instead of compounding the different classes of physical processes in some obscure way out of motions and forces, we would merely separate out those properties, that these classes have in common, and in this way define more inclusive classes, which we represent by appropriate terms. In this way, we finally reach a group of principles applicable to all physical phenomena in general, and which, since they are merely inductively derived from the facts themselves, are free from that uncertainty, which always attaches even to such mechanical hypotheses whose consequences seem completely confirmed by experience."

Rankine's deduction of energetics. The first result gained from this mode of investigation is the general concept of energy. It signifies nothing else than the capacity to bring forth changes; and this capacity is the most universal determination that we can distinguish in the bodies of our world of perception, and without it they would cease to be physical phenomena for us. If we discover definite universal laws concerning this property, then, allowing for particular circumstances, they must be applicable to every branch of physics in general, and must represent a system of rules, which every natural process as such obeys.[58] The way in which Rankine establishes and proves these rules concerns merely the historical development of physics;[59] but the logical form he chooses for his thoughts is of the most general philosophical interest. The laws of energy, we see, owe their universality to the circumstance that the *property* of things, which we have called energy, is spread throughout the physical universe, and somehow attaches to every body as such. No part of reality can escape these laws, because each part is known as real only by this distinguishing property. This form of deduction

---

[58] Rankine, *Outlines of the Science of Energetics, Proceedings of the Philosophical Society of Glasgow.* Vol. III, London and Glasgow, 1855, p. 381 ff.

[59] *Cf.* especially on Rankine, Helm, *Die Energetik*, p. 110 ff., also A. Rey, *La Théorie de la Physique chez les Physiciens contemporains.* Paris 1907, p. 49 ff.

already determines the general intellectual *category*, under which energy is conceived here. It stands in principle on the same plane with the perceptible things, whose essential being it constitutes. Energy is, as it were, concrete substantiality itself, the one indestructible and eternal being.

Criticism of the method of physical "abstraction." From the standpoint of epistemology, the gap to be pointed out is not so much in Rankine's physics as in his theory of method. According to Rankine, the most general property by which the objects of physical reality are distinguished, is the capacity to produce and to receive effects. Things first gain their real objective character, when they are conceived as members of actual or possible *causal relations*. The unprejudiced, "abstractive" analysis, which Rankine regards as the ideal of true science, shows with certainty that causality is no property that can be pointed out as an immediate part of perceptions. Both rationalistic and empiristic criticism agree at least in one conclusion, that there are no direct *impressions* corresponding to the concepts cause and effect. Thus, if abstraction, as here understood, is only a division and grouping of the material of perception, it is clear that precisely that aspect must escape it, on which the concept of energy is founded. And even if we grant the "power of producing effects" to be a quality inhering in bodies just as does any other sensuous property, such as their color or odor, even so the real problem would not yet be solved. In the construction of energetics, we are not concerned as to whether this power of producing effects can in general be shown to exist, but with the fact that it can be exactly *measured*. But as soon as we enquire as to the methods by which this *numerical* determination is made possible, we are referred to a system of intellectual conceptions and conditions, which has no sufficient basis in the purely abstractive procedure, as has been shown on all hands. The mathematical foundation of energetics already involves all those methods of "construction of series," which can never be adequately grounded from the ordinary standpoint of abstraction.

The problem of abstraction in modern logic. Modern logic, at any rate, has substituted for the old principle of abstraction a new one, which may be introduced here. In this new principle of abstraction, the procedure is not from *things* and their common properties, but from *relations* between concepts. If we define a symmetrical and transitive relation R for a number of members a, b, c, . . . (so that from the relations aRb and bRc, the relations bRa, cRb and aRc follow),[60] then the connection produced in this way can also always be expressed by introducing a new identity x, which stands in a definite relation R′ to every member of our original series. Instead of comparing the members directly with each other in their possible relations, the possible relations between the serial members can now be represented by establishing the relation of each to this x, and thus forming the relations aR′x, bR′x, cR′x. The relation R′ is here an asymmetrical, many-one relation, so that the members a, b, c can stand in the relation mentioned to no other term than x, while x, on the contrary, can stand in the corresponding relation R′ with several members.[61] We have an example of this procedure in the relation between series, which we call their "similarity." Two series s and $s′$ are said to be similar to each other in the ordinal sense, when there subsists between them a definite, reciprocal relation of such a sort that to every member of s one member of $s′$ corresponds (and conversely), and when, if in the series s, a member x precedes a member y, the correlate of x in $s′(x′)$ precedes the correlate of $y(y′)$. Here we have a symmetrical and transitive relation, by which a plurality of series s, s′, s″ . . . s[n] etc. can be connected. On the basis of this relation, we can now produce, by the principle of abstraction, a new concept which we call the common "type of order" of all these series. To all the series bound together in this manner, we ascribe one and the same conceptual property. We replace the system of coördinations by the assumption of an identical property, which belongs to all the series uniformly. It is clear, however, that we do not claim to have discovered a new self-existent *thing;* our claim is only that a common *ideal point of reference* is thereby produced, with reference to which we can make our assertions regarding the relations of the given series more pregnant, and bring them to a single, concentrated judgment. If we now apply this result to the *physical* construction of concepts, an essential feature of the modern concept of energy is clearly revealed. Here also we start from the establishment of certain dependencies between empirically physical series. We dis-

---

[60] For the concept of the transitive and symmetrical relation see above, Ch. II.

[61] *Cf.* more particularly Russell, *Principles of Mathematics,* p. 166, 219 ff. etc.

cover that the manifolds, which at first seem to stand side by side isolated and independent, are connected with each other by a relation of "equivalence," by virtue of which there corresponds to a value in one series one and only one value of the other. We extend this connection by taking more and more fields of physical process into account, until finally on the ground of observation and general deductive reasons we draw the conclusion, that whenever any arbitrary groups of physical phenomena are given, definite relations of equivalence must prevail between them. Here is given a thoroughgoing transitive and symmetrical relation between physical contents;[62] and it is the validity of this universally applicable relation, which leads us to introduce a new being by coördinating a certain *work-value*, a certain quantity of *energy*, to every individual member of the compared series. This being, however, would obviously lose all meaning, if we wished to separate it from the whole system of judgments, in which it has arisen. The being, posited in it, is not that of an isolated sensuous property to be perceived for itself, but it is the "being" of certain laws of connection. At this point, we recognize anew what deep, actual oppositions may be concealed behind the differences regarding the logical schema. If we follow the traditional doctrine of abstraction, then we are almost necessarily forced to a substantial interpretation of energy, as the example of Rankine shows; while the functional theory of the concept finds its natural correlate in a functional determination of the supreme physical "reality." In the one case, consideration ends in the assumption of a *property* common to all bodies, and, in the other case in the creation of a highest common *standard of measurement* for all changes in general.

Energy as a relational concept. Some of the representatives of energism have already made the latter logical interpretation. Here we must remember above all Robert Mayer, who also determined the general theoretical position of the new concept, which he introduced. The transformation of force into motion, of motion into heat, meant for him, as he emphasizes, nothing but the establishment of the fact that certain quantitative relations are found between two different groups of phenomena. "How heat arises from disappearing motion

---

[62] If we denote the relation of equivalence by A, then from aAb, bAa evidently follows, as on the other hand the validity of aAb and bAc also involves that of aAc; thus the condition of symmetry and transitivity is fulfilled.

or, in my manner of speaking, how motion is transformed into heat: to demand an answer to this is to demand too much of the human mind. How disappearing O and H give water, why a material of other properties does not result, would trouble no chemist. But whether he does not come closer to the laws, to which his objects, the materials, are subjected, when he understands that the resulting amount of water can be precisely determined from the disappearing amounts of oxygen and hydrogen, than when he is conscious of no such connection,—this cannot be questioned."[63] "In the sense of its founder," Helm justly remarks here, "energetics is a pure 'system of relations,' and is not to establish a new absolute in the world. *If* changes occur, then this definite mathematical relation subsists between them,—that is the formula of energetics; and certainly it is also the only formula of all true knowledge of nature." "As often as the spirit of investigation has rested in the bed of sloth of an absolute, it has been undone. It may be a comfortable dream that our questions can find their answer in the atoms, but yet it is a dream. And it would be no less a dream, if we saw an absolute in energy, and not rather the temporarily most adequate expression of quantitative relations among the phenomena of nature."[64] Thus energy, like the atom, is more and more divested of all sensuous meaning with the advance of knowledge. (*Cf.* above p. 217 ff.) This development appears most clearly in the concept of potential energy, which even in its general name points to a peculiar logical problem. As Heinrich Hertz has emphasized, there is a peculiar difficulty in the assumption, that the alleged substantial energy should exist in such diverse forms of existence as the kinetic and the potential form. Potential energy, as it is ordinarily conceived, contradicts every definition that ascribes to it the properties of a substance; for the quantity of a substance must necessarily be a positive magnitude, while the totality of potential energy in a system is under some circumstances to be expressed by a negative value.[65] Such a relation can, in fact, according to Gauss' theory of negatives, only be explained

---

[63] Mayer to Griesinger (*Kleinere Schriften und Briefe*, p. 187).

[64] Helm, *Die Energetik*, p. 20, 562. The same definition of energy as a mere "causal standard of measurement" (*Kausalmass*) is given by H. Driesch, *Naturbegriffe und Natururteile.* Berlin, 1904 (*Abh. zur Didaktik u. Philos. der Naturwiss.*, Heft 2).

[65] See H. Hertz, *Die Prinzipien der Mechanik*, p. 26.

where that which is counted has an opposite, *i.e.*, "where not substances (objects conceivable in themselves) but relations between two objects are counted." (*Cf.* above p. 56.)

Energetics and mechanics. Even where energy is at first introduced as a unitary and indestructible object, as in the case of Robert Mayer, this very category of object gradually assumes a new meaning; this new meaning is for the purpose of doing justice to the new content, which appears in a double form of existence. "The elevation of a kilogram to a height of 5 meters," says Robert Mayer, "and the movement of such a weight with a velocity of 10 meters per second, are one and the same object; such a motion can be transformed again into the elevation of a weight, but then naturally ceases to be motion, just as the elevation of a weight is no longer elevation of a weight, when it is transformed into motion."[66] If mere elevation above a certain level (thus a mere state) is here assumed to be *identical* with the fall over a certain distance (thus with a temporal process), then it is clearly evident that no immediate substantial standard is applied to both, and that they are not compared with each other according to any similarity of factual property, but merely as abstract measuring values. The two are the "same" not because they share any objective property, but because they can occur as members of the same causal equation, and thus can be substituted for each other from the standpoint of pure magnitude. We begin with the discovery of an exact numerical relation, and *posit* that new "object" we call energy as an expression of this relation. Here a radically new turn is taken, in opposition to atomism. The real advantage of energism over the "mechanical" hypotheses, as ordinarily understood by its adherents, is that it keeps closer to the given facts of perception, in so far as it permits us to relate two qualitatively different fields of natural phenomena, without previously having reduced them to processes of movement, and thus having divested them of their specific character. The processes remain unaffected in their specific character, as all our assertions are merely directed to their causal connection. But, on the other hand, precisely this exclusive reference to the numerical rule of relation involves a new intellectual moment. The atom, even while its purely conceptual meaning gradually becomes more pronounced, always appears as the analogue, as it were, the reduced model of the empirical, sensuous

---

[66] Mayer, *Kleinere Schriften und Briefe*, p. 178.

body, while energy belongs to another field in its origin. Energy is able to institute an order among the totality of phenomena, because it itself is on the same plane with no one of them; because, lacking all concrete existence, energy only expresses a pure relation of mutual dependency.

From the epistemological standpoint, the claim, which energism makes to understand the different groups of physical processes in their specific character (instead of transforming them into mechanical processes and thereby extinguishing their individual features) is one that now seems limited although justified within a certain sphere. In fact, the general logical possibility appears here of our shaping nature into a *system*, without our being obliged to require representation of this system in a unitary, intuitive *picture*, such as is offered by mechanism. But it is an error to see in this tendency to a "qualitative" physics a reversion to the general Aristotelian view of the world. "We are forced," says a prominent, modern advocate of energism, "to take up into our physics other features than the purely quantitative elements, of which the geometrician treats, and thus to agree that matter has *qualities;* we must face the danger that we shall be accused of a reversion to the occult faculties of scholasticism, of recognizing the quality, by which a body is warm or bright, electrical or magnetic, as an original and not further reducible property in it; in other words, we must abandon all the attempts that have continually been made since the time of Descartes, and connect our theories again with the essential concepts of the peripatetic physics." But the further working out of the thought destroys the appearance of a deeper connection. The qualities of Aristotle are something entirely different from the qualities of modern physics; for while the former signify only hypostasized sensuous properties, the latter have already passed through the whole *conceptual system of mathematics*, and have thereby received a new logical form and character. What energism abandons is only the "explanation" of the particular qualities out of certain mechanical motions; what it retains, on the other hand, and what is the condition of its existence, is the expression of quality in a definite *number* which fully represents and replaces it in our consideration. The question whether heat *is* motion can remain in the background, as long as the indefinite sensations of warmer and colder are replaced at the same time, by the concept of an exact *degree of temperature* and objectified in it. What is here retained of quality is not its sensuous property, but merely the peculiarity of its mathematical serial form. Duhem, whose judgment regarding the connection between the energistic and the peripatetic physics was just cited,[67] says we can develop a theory of heat and can define the expression "quantity of heat" without borrowing anything from the specific *perceptions* of cold and warm.[68] In the schema of theoretical physics, the definite empirical system under investigation is replaced by a system of numerical values, which express its various quantitative elements. (*Cf.* above p. 150 f.) Energism shows that this form of numerical order is not necessarily connected without analyzing the things and processes into their ultimate intuitive parts, and recompounding them from the latter. The general problem of mathematical determination can be worked out without any necessity for this sort of concrete composition of a whole out of its parts.

Physics as a science of qualities. In this conception, however, physics only completes and applies a thought already recognized in the general doctrine of the principles of mathematics. There is, and can be, a "physics of qualities," because and in so far as there is a mathematics of qualities. The gradual development of this latter can already be traced in its general features. There is a continuous development beginning with Leibniz, who first saw the essence of mathematics in a doctrine of the possible forms of deductive connection in general, and who therefore demanded the completion of ordinary algebra (as the science of quantity) by a general science of quality (*Scientia generalis de qualitate*), up to modern projective geometry and the theory of groups. In this whole development, it clearly appears that there are wide and fruitful fields perfectly accessible to mathematical determination without their objects being extensive magnitudes which have arisen by the repeated additive positing of one and the same unity. The projective theory of distances shows how it is possible to place the elements of a spatial manifold in exact correlation to fixed numerical values, and to imprint a definite order upon them by virtue of this correlation, without applying the ordinary metrical concept of distance. (*Cf.* above p. 84 ff.) This thought is carried over by universal energetics to the

<hr />

[67] Duhem, *Lévolution de la Mécanique*, p. 197 f., as also H. Driesch, *Naturbegriffe und Natururteile*, p. 51 ff.

[68] Duhem, *op. cit.*, p. 233 f.

totality of physical manifolds. It is sufficient for the numerical formulation of processes if a definite scale of comparison is provided for the individual qualities, and if, further, the values within these different scales are reciprocally coördinated with each other by an objective law. This connection, however, can be established and retained independently of any mechanical interpretation of the particular groups of phenomena. The objection frequently raised against energism, that it destroys the homogeneity of process since in it nature falls into separate classes of phenomena, is not conclusive. If we take the *mathematical* general concept as the starting-point and standard of judgment, not only such contents are "like," which share some intuitive property that can be given for itself, but all structures are "like," which can be deduced from each other by a fixed conceptual rule. (*Cf.* above especially p. 81 ff.) Here, however, this criterion is satisfied; the connection of concepts produced by the equivalence between the different series gives a no less definite logical connection than reduction to a common mechanical model. The intellectual postulate of homogeneity is just as effective in the energistic as in the mechanistic conception of the natural processes; the difference only consists in that, in the one case, its realization is based purely on the concept of number, while, in the other case, it also requires the concept of space. The conflict between these two conceptions can ultimately only be decided by the history of physics itself; for only history can show which of the two views can finally be most adequate to the concrete tasks and problems. Abstracting from this, however, energism is in any case of preëminent epistemological interest in so far as the attempt is made to establish the minimum of conditions, under which we can still speak of a "measurability" of phenomena in general.[69]   Only those princi-

[69] It has been objected occasionally to the logical possibility of the goal that general energetics sets itself, that every measurement of things and processes involves the presupposition that they are compounded out of homogenous parts, and can thus be represented by repeated addition of the same unit. Every measure would thus necessarily be a determination of extension; reference to a unit of measurement would thus already contain the transformation of all qualitative differences into extensive distances, and thus reduction to a spatial and mechanical scheme. (*Cf.* Rey, *La Théorie de la Physique chez les Physiciens contemporains*, p. 264, 286, etc.) Here, however, the concept of "measure" is obviously taken too narrowly. If we understand under the "measurement" of a manifold only its mathematical determination in general,

ples and rules are truly general, on which rests the numerical determination of any particular process whatever and its numerical comparison with any other process. The comparison itself, however, does not presuppose that we have already discovered any unity of "essence,"—for example, between heat and motion; but, on the contrary, mathematical physics *begins* by establishing an exact numerical relation, on the basis of which it also maintains the homogeneity of such processes as can in no way be sensuously reduced to each other. That the different forms of energy "in themselves" are of kinetic nature is a proposition that the theory of knowledge cannot defend, for the latter is merely directed on the fundamental aspects of knowing, not on those of absolute being. The demands of the theory of knowledge are rather satisfied when a way is shown for relating every physical process to values of mechanical work, and thus for producing a complex of *coördinations*, in which each individual process has its definite place. A representation of the processes of nature absolutely without hypotheses cannot, indeed, be gained in this way; for translation into the language of the abstract numerical concepts, no less than translation into the language of spatial concepts, involves a theoretical transformation of the empirical material of perception. But it is of logical value to separate the general presuppositions strictly from the particular assumptions; and to separate the "metaphysical," because mathematical, principles of knowledge of nature from those special hypotheses, which serve only in the treatment of a particular field.

VIII

The problem of the construction of concepts in chemistry. The exposition of the conceptual construction of exact natural science is incomplete on the logical side as long as it does not take into consideration the fundamental concepts of chemistry. The epistemological interest of these fundamental concepts rests above all on the intermediate position which they occupy. Chemistry seems to begin with the purely empirical description of the particular substances and their composition; but the further it advances, the more it also tends

---

i.e., the correlation of its elements with the particular members of the series of numbers, then mathematics itself shows that such a correlation is also possible where the objects of the group in question are not compounded out of spatial parts.

toward *constructive* concepts. In physical chemistry, this goal in fact is reached; a leading representative of this discipline is able to give as the connecting feature of physics and chemistry, that both *create* the systems, which they investigate, on the basis of the empirical data.[70] In so far as chemistry has reached this modern form, it stands on no other ground logically than physics itself. Its fundamental laws, as for example the law of phases (*Phasenregel*) of Gibbs or the law of chemical masses (*Massenwirkung*), belong to the same purely mathematical type as the propositions of theoretical physics. It is, however, of special interest to trace the way in which the ideal, which was realized in theoretical physics from its first beginnings in Galileo and Newton, has only been gradually reached in chemistry. The limits of purely empirical and of rational knowledge stand out very clearly in the constant shifting, which they undergo in the progress of chemical knowledge. The intermediate terms and conditions of exact understanding are brought into sharp relief. The power of scientific moulding is especially striking with the more stubborn material, with which chemistry works. Ultimately physics is only apparently concerned with thing-concepts; for its goal and real field is that of pure law concepts. Chemistry places the problem of the individual thing decisively in the foreground. Here the particular materials of empirical reality and their particular properties are the object of the enquiry. But the "concept," in the specific meaning it has in mathematics and physics, is not available for this new problem. For it is only the symbol of a certain form of connection, that has more and more lost all material content; it only signifies a type of possible arrangement, not the "what" of the elements arranged. Are we here concerned with a gap that is to be filled out by new determinations belonging to the same logical direction of thought, or must we recognize and introduce at this point a form of knowledge different in principle?

The chemistry of sensuous qualities and Richter's law of definite proportions. This question can only be answered if we follow the concrete historical development of chemical doctrines themselves,— not grasping the vast wealth of their content in detail, but laying bare the great logical lines of direction of their advance. In fact, a few general characteristics soon appear of themselves, according to which the development can be divided and surveyed in all its diversity.

[70] Nernst, *Die Ziele der physikalischen Chemie.* Göttingen 1896.

The older form of the chemical doctrine of elements, which pre-
dominated up to the time of Lavoisier, and which found its last
characteristic expression in the theory of phlogiston, conceives the
element as a generic property belonging to all the members of a
definite group and determining their perceptible type. Here the
elements are only the hypostatizations of the especially striking
sensuous qualities. Thus sulphur by its presence confers on any
body the property of combustibility, salt the property of solubility,
while mercury is the bearer of the metallic properties, which are
found empirically in any material.[71] This conception is only transcended in principle when along with the task of dividing bodies into
classes according to their generic properties is added the other task
of gaining exact *quantitative* proportions concerning their mutual
relation. The demand for strict numerical determination here also
forms the decisive turning point. The law of the definite proportions,
in which the different elements are connected with each other, constitutes the starting-point of modern chemical theory. It is interesting that this law is at first conceived entirely independently of any
conception of the constitution of matter, and in particular independently of the atomic hypothesis. In the original, still incomplete
form, in which it was first asserted by J. D. Richter, it signifies
primarily nothing but the validity of certain harmonic relations, that
prevail between different series of bodies. If we consider a series of
acids $A_1 A_2 A_3 \ldots$ and a series of bases $B_1 B_2 B_3$, there prevails
between the two a certain relation, which is expressed by saying, that
we coördinate with each member of the first series a definite number
$m_1 m_2 m_3 \ldots$, while we let correspond to the members of the second
series other constant numerical values $n_1 n_2 n_3$, to be gained by
observation. The manner, in which an element of the first series
combines with an element of the second, is definitely determined by
these numbers; the two weights, according to which an acid A_p
combines with any base B_q, are related according to the corresponding numerical values m_p and n_q. Richter seeks to prove in detail
that the series of weights of the bases forms an arithmetical series,
and that of the acids a geometrical series; and that a law is thus
found here, which is assumed to be analogous to that of the distances

[71] See Ostwald, *Leitlinien der Chemie*, p. 4 ff; *cf.* also Meyerson, *Identité et
Réalité*, p. 213 ff.

of the planets from the sun.[72] This conception has not proved to be satisfactory empirically; but it is nevertheless characteristic and significant in its general tendency. It is, as we see, the general Pythagorean doctrine of the "harmony" of the cosmos, which here is present at the cradle of modern chemistry, as it is also at the cradle of modern physics. In this connection, Richter is to be compared with Kepler, that is, if we consider not his whole achievements but merely his intellectual tendency, for with Kepler he shares the conception of the thorough-going numerical arrangement of the universe, which is continued in all particular fields of phenomena.

Dalton's law of multiple proportions. The interpretation of the law of constant combination-numbers by its real scientific founder adds a new concrete feature to this general view. Here it is actually at first only asserted, that there is a characteristic equivalence-number for each element, and that, when two or more elements enter into a combination, their masses are related as whole multiples of these numbers. But this rule of "multiple proportion" is combined by Dalton with a certain interpretation, and enters only in this form into the system of chemical doctrines. The concept of combination-weight is transformed into that of atomic weight. The law of multiple proportions means, that the atoms of different simple bodies are different in their masses, while within the same chemical genus the atom is always unchangeably one and the same constant mass, and thus the mass suffices to characterize a given simple material in its specific character. In place of the empirically gained proportion-numbers of the individual bodies, assertions appear regarding an essential property of their ultimate constitutive parts. Since, however, all our knowledge only concerns the *relations*, according to which the elements enter into combinations, no definite determination is possible of the *absolute* values of the atomic weights. If we take the atomic weight of hydrogen as a unit of comparison, then we can, without contradicting the known facts of composition, determine that of oxygen by the value $O = 8$ instead of $O = 16$, whereby we would

[72] For the following data from the history of chemistry, *cf.*, besides the well-known general historical works, especially Wurtz, *La Théorie atomique*, Paris 1879; Duhem, *Le Mixte et la combinaison chimique*, Paris 1902; Lothar Meyer, *Die modernen Theorien der Chemie*, 5th ed., Breslau 384; Ladenburg, *Vorträge über die Entwicklungsgeschichte der Chemie*, 3rd. ed., Braunschweig 1902.— On Richter, see espec. Duhem, p. 69 ff., Ladenburg, p. 53 ff.

have to double the number of atoms of oxygen in all our formulae; we could successively take the values S = 8, 16, 32 ... as the atomic weight of sulphur, in so far as we formed the chemical formulae in agreement with one of these assumptions, and thus, for example, characterize sulphide of hydrogen according to our choice by the expression HS_2 or HS or H_2S. The decision between all these possible determinations is made on the basis of several criteria, which are only gradually worked out in the history of chemistry. One of the most important criteria is the rule of Avogadro, according to which similar quantities of molecules of different combinations occupy the same volume as perfect gases under the same conditions of pressure and temperature. Along with the determination of atomic weights from the density of vapors, which is hereby made possible, there is their determination from heat capacity, which rests on the law of Dulong-Petit; and also the determination on the basis of isomorphism, resting on the law of Mitscherlich, that the same crystal form having different combinations, indicates an equal number of atoms connected in the same way. It is only the totality of all these different points of view, mutually confirming and correcting each other, that finally after many experiments gives a unitary table of atomic weights, and thus lays the basis of a definite system of chemical formulae.[73]

The atom as a relational concept. The development here completed offers a general logical problem, if we abstract from all details. If we asked merely the individual investigators who coöperated, it would seem to possess only *one* perfectly definite and clear meaning for them all. The objective existence of the different types of atoms is presupposed; it is only necessary to discover their *properties*, and to define them more exactly. The further we advance and the more diverse groups of phenomena we consider, so much the more definitely the wealth of these properties appears. The substantial "inwardness" of the atom is revealed and takes on fixed and tangible form for us. We trace, especially in the involved chemical constitution-formula, how the atoms are situated relatively to each other, and how they are mutually connected in the unified structure of the molecule. We see how in their combination they generate, by their number and relative position, a certain structural outline, such as is

[73] *Cf.* more particularly esp. Lothar Meyer, Bk. I, Pt. II–IV; Ostwald, *Grundriss der allgemeinen Chemie,* 4th ed., Lpz. 1909.

expressed, *e.g.*, in the forms of crystals. If we look into the closer
empirical grounding of these assertions, however, the general picture
soon changes. It becomes clear that the atom is never the given
starting-point, but always only the goal of our scientific statements.
The wealth of content it gains in the progress of scientific investigation never belongs to *it* fundamentally, but is related to another
kind of empirical "subject." As we apparently investigate the
atom itself in its manifold determinations, we at the same time
place these different groups of circumstances in a new relation to
each other. We speak of the number of atoms contained in a
definite volume of a gaseous substance, and thereby express a relation,
that, according to the law of Gay Lussac, subsists between the
numerical value of the density of the gas and the value of its combination-weight. We ascribe to the atom of all simple bodies the
same heat-capacity and thereby express the fact that, if we arrange
combination-weights of the chemical elements in a series a a' a''..a_n,
and the values of their specific heats in another series b b' b'' ..b_n,
then there is a definite correlation between these two series, in
so far as the products ab, a' b', a'' b'' *etc.* possess the same constant
value. The characteristic logical function of the concept of the
atom appears clearly in these examples;—we may abstract from all
metaphysical assertions regarding the existence of atoms. The atom
functions here as the conceived unitary center of a system of coördinates, in which we conceive all assertions concerning the various
groups of chemical properties arranged. The diverse and originally
heterogeneous manifolds of determinations gain a fixed connection
when we relate them to this common center. The particular
property is only apparently connected with the atom as its absolute
"bearer," in order that the system of relations can be perfected. In
truth, we are concerned not so much with relating the diverse series to
the atom, as rather with *relating them reciprocally* to each other through
the mediation of the concept of the atom. Here again appears the
same intellectual process, that we previously met; the complicated
relations between certain systems are not expressed by our comparing
each system individually with all the others, but by putting them
all in relation to one and the same identical term. (*Cf.* above p.
196 ff.) The attempt to determine exactly the atomic weight of the
individual elements compels us to appeal constantly to new fields
of chemico-physical phenomena as criteria. To the extent that this

determination advances, the circle of empirical relations is extended. If we conceive this progress completed, then in the "absolute" atomic weights all possible relations would be expressed, that the particular series could enter into among themselves. The real positive outcome of chemical knowledge here is in the systematic analysis of these relations. The originally confused factual material is organized; it is no longer unrelated, but is arranged around a fixed central point. When we ascribe to one and the same subject the observations on vapor density, on heat capacity, on isomorphism *etc.*, they thereby enter into true conceptual relation. But, indeed, it is not the only logical value of this "subject," that it describes and unifies previous observations. This unification is rather directly productive; it produces a general schema for future observations, and indicates a definite direction for these. The progress of science would be slow, its exposition unwieldy and tiresome, if, on approaching a new field of facts, it had every time explicitly to repeat the wealth of previously gained material and to present it in all details. While the concept of the atom concentrates all these features, it retains their essential content; on the other hand, it leaves all the forces of thought free to grasp the new empirical content. The totality of what is empirically known is condensed, as it were, to a single point, and from this point issue the different lines of direction, in accordance with which our knowledge advances into the unknown. Those manifolds already discovered and defined according to law function as a fixed logical unity in opposition to those manifolds newly to be discovered; and it is this unity of the fundamental point of connection, which renders possible our assumption of an ultimate identical subject for the totality of possible properties.

The "regulative" use of the concept of the atom. The meaning, belonging to the general concept of substance within the actual processes of experience, is clearly evident in this example. Empirical knowledge cannot avoid the concept of substance, although genuine philosophical progress in such knowledge is in understanding and evaluating it *as a concept*. True, the direct, living work of investigation has another standpoint from the beginning, and grasps the problem as if from another side than that of epistemological reflection. Investigation finds its interest in the new fields of facts to be mastered, while it can take known facts as a given condition needing no further analysis. The totality of the "factual" in this sense is fixed; it is the permanent substratum, which gives the fundamental mould for all further observations. What has already been reached at any point, what has been won, must be taken by the investigation as something assured and given; for only thus is it possible to move the field of the problematical to another point, and to push it further on, so that new questions are always coming in for consideration. Thus the passive *fixity*, established by science at certain points, is an element in its own activity. In fact, it is justified and unavoidable, that science should condense a wealth of empirical relations into a single expression, into the assumption of a particular thing-like "bearer." The critical self-characterization of thought, however, must analyze this product once more into its particular factors, although it conceives this product as *necessary* for certain purposes of knowledge. This is done because critical thought is not directed forwards on the gaining of new objective experiences, but backwards on the origin and foundations of knowledge. The two tendencies of thought here referred to can never be directly united; the conditions of scientific *production* are different from those of critical *reflection*. We cannot use functions in the construction of empirical reality and at the same time consider and describe them. Nevertheless, the two standpoints, and therewith the constant alteration of standpoints, are desirable, in order to judge knowledge as a whole in the motives of its advance, and in the permanent logical conditions of its existence. The peculiar character of knowledge rests on the tension and opposition remaining between these standpoints. In the light of this, it can be understood that the chemical concept of the atom also shows a different form, according to the way we approach it. To the first naïve consideration, the atom appears as a fixed substantial kernel, from which different properties can be successively distinguished and separated out, while, conversely, from the standpoint of the critique of knowledge, precisely those "properties" and their mutual relations form the real empirical data, for which the concept of the atom is created. The given factual material is united in a single focus with that which is conceptually anticipated, and by virtue of a natural illusion this focus appears as a real, unitary object, instead of a mere "virtual" point. Thus the atom of chemistry is an "Idea," in the strict meaning Kant gave this term,—in so far as it possesses "a most admirable and indispensably necessary regulative use, in directing the understanding to a certain aim, towards which all the lines of its rules converge and which, though it is an idea only (*focus imaginarius*), that is, a point from which, as lying completely outside the limits of possible experience, the concepts of the understanding do not in reality proceed, serves nevertheless to impart to them the greatest unity and the greatest extension."[74] This function remains as a permanent characteristic of the concept of the atom, although its *content* may completely change; thus *e.g.*, the atom of matter becomes the atom of electricity, the electron. Precisely this sort of change shows that what is essential in the concept does not consist in any material properties, but that it is a formal concept, that can be filled with manifold concrete content according to the state of our experience.

* * * * *

The concept of valency and the theory of types. The second important step in the construction of chemical concepts, after the conception of the atom, and after the value of the atomic weights of the individual elements has in general been established, is to connect according to conceptual standpoints the various, originally separate determinations thus gained, and to collect them into classes of definite character. The empirical facts leading to such relative distinctions and combinations within the total system are given in the relations of chemical *substitution*. If we trace how the atoms of various simple materials replace each other in combinations, and how they can be reciprocally substituted for each other, certain fundamental rules result governing this form of relation. The form of substitution can be determined once for all and expressed by certain *numerical values*, which we attach to each element along with the numerical value of its combination weight. If we take the atom of hydrogen as unity, it appears, for example, that an atom of chlorine can in certain combinations replace an atom of hydrogen, while an atom of oxygen always replaces two, an atom of nitrogen three, an atom of carbon four atoms of hydrogen. Thus a new point of view is achieved for the correlation of the individual elements, and a new characteristic constant for each simple material. The "valency" of the elements is the expression of a definite property in them, that belongs to them independently of their chemical affinity. Now if we arrange the chemical combinations according to this new principle,

---

[74] *Kritik der reinen Vernunft*, 2nd. ed., p. 672. Müller's trans. p. 518.

they are divided into various *types*, in which the members belonging to the same type are characterized by the fact that they can all be produced from a certain form by progressive substitutions, which take place according to the rules of the valency of the particular atoms.

Logical aspects of the concept of type. The concept of the "type" here comes in for consideration not in its significance for the special problems of chemistry, but as a paradigm of certain logical relations. In fact, it displays most distinctly a characteristic feature already established in general by the analysis of the exact scientific concept. The chemical concept of type is also not formed on the pattern of the generic concept, but on that of the serial concept. The different combinations, belonging under a type, are not so conceived because of the external similarity of their sensuous properties, or because of the direct agreement in their chemical functions. They belong together in so far as they can be changed into each other by means of the relation, which subsists between the valency of the individual atoms, while the remote members of the series need no further analogy than is established by this law of derivation itself. In the history of chemistry, the concept of type was only gradually separated from the concept of chemical analogy.[75] The first step in this separation is found in the relation of substitution itself, since here elements, which seem entirely different from each other in their nature and properties, can replace each other. The conception of substitution, as formulated by Dumas, was at first rejected by Berzelius as paradoxical and inconsistent from this point of view; chlorine cannot take the place of hydrogen in any combination, as the former (according to the theory of electro-chemical dualism, which Berzelius advocated,) is negatively electric, while hydrogen possesses positive electricity. The more, however, the theory of substitution made its way, the more the converse view gained acceptance, *viz.*, that entirely dissimilar materials can replace each other in certain combinations, without altering the nature of the combination. The consequences of this view appear even more sharply, when not only the elements, which can be substituted for each other, are contrasted individually, but the whole group of materials, that can issue from repeated substitutions, is considered. Here also the demand for

[75] *Cf.* more particularly esp. Duhem, *Le Mixte*, p. 97 ff., Wurtz, *op. cit.*, p. 189 ff.

analogy was at first upheld, until further investigations showed that the series, which arise in this way, can contain members completely different from each other in all their perceptible properties and in their essential chemical determinations. To the "chemical type" of Dumas, for which he demanded *similar fundamental properties* in all members, there is now opposed the "molecular type" of Regnault, which includes materials of very different properties, and considers these materials as issuing from each other by substitution. The conditions, on which the *unity* of the type rests, thus correspond throughout to those that we found realized in the field of the construction of mathematical concepts. Then there were geometrical systems and groups of systems, whose elements were not connected by any intuitive feature common to them, but merely by the definite rule of relation, which prevailed from member to member; and the same is true here. The "valency" of the particular elements establishes among them such a relation as produces in its continued application definite systems of characteristic serial types. The variation according to law of this "parameter" generates and founds the form of the concept, which accordingly does not rest on a similarity in the *content* of what is connected, but on the type of connection.

The chemical concept as a relational concept. The chemical concept is indeed distinguished from the mathematical, in that the relation by which we proceed from one member to another is established by mathematics purely *constructively*, while the relation of equivalence, on the other hand, is discovered as an *empirical* relation between the various elements. However, if we abstract from this difference of origin, we recognize that, once the decisive property for comparison is gained, the further conceptual construction on both sides takes exactly the same direction. Here, once a general principle of coordination has been defined, our concern is with carrying this principle through the whole manifold of materials given by observation, and thus with shaping the latter aggregate into a system, within which we grasp the reciprocal action and interdependence of the particular members according to fixed rules. In this regard, the theory of types constitutes the first approach to chemical *deduction*, since it shows us how, from certain starting-points, to construct the manifold of bodies by adherence to a few general principles, and how to group them around fixed central points. The sensuously heterogeneous now becomes homogeneous, when we organize it into certain numerical relations. It is the numerical and relational aspect, which is here again decisive; for it constitutes the genuinely characteristic property of the scientific interpretation of the chemical concepts. The "valency," ascribed to the particular atoms, must at first appear as a real *qualitas occulta,* if it is conceived as a substantial quality in them. We do not know the peculiar property of the atom of chlorine, by which it can only be combined with *one* atom of hydrogen; we do not know by what force the atom of oxygen combines with two atoms of hydrogen, and the atom of carbon with four of hydrogen. And this riddle is not solved when, in explanation of the different relative valencies, reference is made to the *states of motion* of the individual atoms,—which states are supposed to agree or to be opposed to each other in such a way that they can only combine with each other in an entirely definite relation.[76] For here something absolutely unknown and empirically undemonstrable is substituted for the relations of substitution, which are alone known. What distinguished the concept of valency from all scholastic qualities is the intellectual renunciation, which it implies. It does not seek to penetrate into the substantial nature of the connection of atom and atom, but merely to represent the facts of this connection according to universal quantitative principles of order. The chemical constitution-formula at first seems to offer a direct intuitive picture of the serial order and position of the atoms among themselves; but what it finally achieves is not such a knowledge of the ultimate, absolute elements of reality, as rather a general analysis of the bodies and materials of experience. The formula of a definite compound does not teach us to know it merely in its composition, but inserts it into various typical series, and thus refers to the totality of such structures, that can arise by substitution out of a given combination. The individual member becomes the representative of the whole group to which it belongs, and it can issue from the group by variation according to law of certain fundamental parts. Since the constitution-formula represents this connection, this formula is indeed the real scientific expression of the empirical reality of the body; for it means nothing else than the thorough-going objective connection, in which an individual "thing" or particular event stands with the totality of real and possible experiences. (*Cf.* here esp. Ch. VI.)

[76] *Cf.* Wurtz, *La Théorie atomique,* p. 175.

The concept of the "radical" and the theories of "composite radicals."
The conception of substitution becomes especially significant, when it is applied not merely to the individual atom but to *whole groups of atoms.* The theory of the "composite radical" now arises and becomes the real foundation of organic chemistry. Here, according to the definition of Liebig, the radical is considered as a constant part in a series of compounds, in so far as it can be replaced in the latter by other simple bodies, or in so far as in its combination with a simple body this body can be excluded and replaced by equivalents of other simple bodies. Regarding the manner in which the radicals "exist" in compounds, there is disagreement at first. In the "kernel theory" of Laurent, the relation is at first conceived and described in a thoroughly realistic sense. The kernels are as such present in a plurality of bodies, which arise from them by their combination with other atoms; they *preëxist* with respect to the more complex structures. In the further development of the theory, this view is more and more superseded. When Gerhardt, in particular, shows that it is possible to assume *two* radicals in a compound, then the conception of the real existence of isolated groups is destroyed. Since the formulae of chemistry are only meant to express by equations certain relations of structure and reaction, since they are not intended to represent what bodies are in and for themselves, but only what they *were* or *could become*, there is nothing to prevent us, it is now urged, from erecting several rational formulae for one and the same body, according as we wish to express its connection with one or another group of compounds. The conflict concerning the nature and absolute properties of the radical is thus resolved; for the radicals now appear as the result of certain ideal analyses, which we make, and which give different results, according to the standpoint of comparison taken as a basis. The radical now possesses no independent reality, but is meant to express what Gerhardt calls "the relations, by which elements or groups of atoms replace each other."[77] We thus stand at the beginning of a conception, which in general abandons the questions as to whether and how the elements continue to exist in the compounds into which they enter, in order instead to discover and represent, according to general rules, merely the measurable relations, that subsist between the initial and final conditions of a

[77] Gerhardt, *Traité de Chimie organique;* cited by Ladenburg, *op. cit.* p. 235. (*Cf.* p. 194 ff.)

process of chemical transformation. As soon as this phase is reached, however, chemistry takes its place in the general plan of energetics,[78] and thus passes from the circle of empirical descriptive sciences into that of mathematical science.

* * * * *

The reconstruction of the systematic form of chemistry. However, before this subordination of chemistry to a more general scientific problem occurs, certain standpoints and tendencies appear within it, which point to this reconstruction in systematic form. The first phase of the determination of the material manifold is marked by the fact that every element is characterized by its atomic weight. Thus every simple body gains, in the expression of Leibniz, a definite "characteristic number;" and this number is taken implicitly as what brings the wealth of its empirical properties conceptually to complete expression. This representation of the material manifold in a manifold of numbers already indicates a new problem. As the real methodological advantage of the field of number is that each member in it is deduced and constructively developed from an initial structure according to unified rules, this demand henceforth is extended to all physical and chemical determinations, such as are known to be dependent on certain numerical values. They must no longer be conceived as a lawless aggregate; but it must be possible to represent them in their sequence and gradual transformation by an exact law.

The periodic system of the elements. This general demand has its first fulfillment in the establishment of the periodic system of the elements. The various properties of simple bodies, their hardness and malleability, their fusibility and volatility, their conductibility for heat and electricity, *etc.*, now appear as periodic functions of their atomic weights. If we conceive all the elements arranged in a series, we find that, in advancing in the series, the properties of the various elements change from member to member, but that after going over a certain period, the same properties recur. The place of an element in this fundamental systematic series determines in detail its physico-chemical "nature." One of the founders of the periodic

---

[78] On the "energetic" conception and treatment of chemistry, see esp. Ostwald, *Elemente und Verbindungen, Faraday-Vorlesung*, Leipzig, 1904; also Duhem, *Le Mixte*, Chs. IX and X.

system, Lothar Meyer, has clearly characterized the new principle involved in it. "Matter" is here removed from the field of scientific constants into the field of variables. "Up till now there have been introduced into the calculations of physics as variable magnitudes, on which phenomena depend, place and time in particular, and, under certain circumstances, heat, temperature, electricity, and a few other magnitudes; matter appeared, as expressed in magnitude and number, only as mass in the equations; its quality was only considered in that the constants for each kind of matter had a different value in the differential equations. To treat these magnitudes, which are dependent on the material nature of the substances, as variables was hitherto not customary; but this advance has now been made. The influence of the nature of matter had previously been considered in physical phenomena by determining the physical constants for the most various substances. But this material nature always remained something qualitative; the possibility was lacking of introducing this fundamental variable expressed in number and measure into the calculations. An approach to such an introduction, although a very primitive one, has now been made by proving that the numerical value of the atomic weight is the variable, by which the substantial nature and the properties dependent on it are determined."[79] The qualitative nature of the particular material is made mathematically conceivable by discovering a point of view from which the material can be arranged in series with a definite law of progression. The significance of this point of view appears especially in that now members of the manifold, that were hitherto unknown empirically, can be demanded and predicted on the basis of the general systematic principle, and that advancing experience confirms this demand.

Chemistry and mathematics. The deductive element, that enters into chemistry, can be understood in its peculiar quality most clearly, when we compare it with the ideal of deduction developed, on the one hand, in the speculative and metaphysical view of nature, and on the other hand, in mathematical physics. If we abstract from the part played by the problem of matter in the philosophy of nature, the problem of matter has repeatedly had an important epistemological rôle in the history of philosophy. Thus Locke, for instance, develops his whole view of the problems and limits of scientific

[79] Lothar Meyer, *op. cit.*, p. 176.

investigation in connection with the example of chemical knowledge of the fundamental elements and their properties. For him, genuine knowledge is only attainable where it is possible to gain universal insights into necessary connections. We can speak of genuine knowledge, in the strict sense, only where all the properties of the object are certain and perfectly intelligible from its original nature, thus where it is possible from acquaintance with an object to conclude directly and to determine *a priori* all its properties. This postulate, however, which is fulfilled in all our "intuitive" judgments regarding mathematical relations, is not satisfied in our scientific knowledge of nature. Here, where we are merely concerned with the collection and description of various facts of perception, it remains forever impossible to establish that *dependence* of the individual members on each other, by which alone they could become a rationally connected whole. No matter how many properties of a substance we may discover by observation and investigation, the question as to their inner connection is not advanced a step. If we collect ever so many properties of gold, its malleability, its hardness, its non-combustibility, *etc.*, still we cannot discover from them one single *new* determination, and we can never understand the form of connection, by which definite properties of one kind always correspond to other definite properties of another kind. Such an insight, as would make our knowledge of nature a genuine science like mathematics, would only be possible if, instead of merely collecting observations concerning the empírical coexistence or empirical incompatibility of properties, we could grasp the problem "at the other end;" if we could start from some sort of determination of the essence of gold, to deduce from it the totality of secondary properties.[80] Modern science has in part fulfilled the ideal, which Locke here abandons; but it had first to give this ideal a new meaning. Modern science agrees with Locke, that the deduction of the particular properties of a material from its "substantial essence" transcends the problems of exact and empirical knowledge; but it does not thereby disclaim all *conceptual connection of the empirical data themselves*. Science today collects the plurality of elements into a fundamental series, whose members succeed each other according to a definite principle, and then determine the individual properties of bodies as functions of their position in this series. *How* from the assumed fundamental property,

---

[80] Locke, *Essay on Human Understanding*, Bk. IV, Ch. 6.

the further properties follow, how, from a definite atomic weight, there results a definite malleability and hardness, fusibility and volatility, remains indeed unanswered; nevertheless, the fact of this dependence itself is used in the attempt to calculate and predict certain special properties on the basis of certain special data. The functional connection thus established contains, indeed, less than metaphysical insight into ultimate essences, but at the same time it offers more than a mere empirical collection of disconnected particulars. The order of elements, that now arises, offers at least an *analogue of mathematics*, and thereby an analogue of exact and "intuitive" knowledge. We penetrate no deeper into the absolute being of bodies by this means; but we grasp the rules of their systematic connection more definitely. (*Cf.* above p. 207 ff.) Yet at the same time, this solution leads to a new problem. There arises the problem of allowing the atomic weights, that were first introduced as *discrete* values, to proceed from each other by *continuous* transformation, and of determining the law according to which, in such a variation, the dependent properties must change. If we regard this problem as solved, then we would have logically entered a new form of conceptual construction; instead of a number of rules concerning the concomitant appearance of properties, we would have henceforth a unitary, mathematically representable law of causal dependence between the variations of different magnitudes. The atomic weights, by which we express the special character of the elements, would no longer stand next to each other as rigid, given values, but could be traced in their origin out of each other. The chemical concept would have become the physical concept. The latest phase of natural science, which has resulted from consideration of the phenomena of radioactivity, seems to attest such a change directly; for here science assumes a continuous transformation of the elements into each other, and for it the particular material with its sensuous definiteness is only a transition-point in a dynamic process. When the chemical atom is resolved into a system of electrons, it loses the absolute fixity and immutability previously ascribed to it, and appears as a mere relative resting-point—as a cross-section, which thought makes in the continuous flow of process. However we may judge of the positive truth of such assumptions, at any rate, they show very distinctly the way in which the scientific concept advances. Chemical research begins with a plurality of actual observations, which at first stand unconnected side by side; and these it defines in fixed determinations of number and measure. These numerical values gained through observation are soon arranged into series; the series proceed according to a rule, and the later members can be determined from the preceding. As, however, empirical manifolds are transformed into rational in this way, there arises the problem of reducing the laws of *structural relations* to a deeper-lying causal law of *process*, and of completely grounding the former in the latter. In this progressive mastery of the empirical material, the peculiarity of the logical process is revealed; through it, the concept, while obeying the facts, at the same time gains intellectual dominance over the facts.

IX

The concept of natural science and "reality." The real methodological interest of the construction of the concepts of chemistry lies in the fact that in it the relation of the universal to the particular is set in a new light. The consideration of the *physical* concepts and methods allows only one side of this fundamental relation to appear clearly. The goal of theoretical physics is and remains the universal laws of process. The particular cases, in so far as they are taken into account, serve only as paradigms, in which these laws are represented and illustrated. The further this scientific problem is followed, the sharper the separation becomes between the system of our *concepts* and the system of the *real*. For all "reality" is offered to us in individual shape and form, and thus in a vast manifold of particular features, while all *conception*, according to its function, turns aside from this concrete totality of particular features. Here is again revealed the antinomy that found its first striking expression in the system of Aristotle. All knowledge seeks to be knowledge of the universal, and is only fulfilled in this goal; while true and original being does not belong to the universal, but to the individual substances in the dynamic succession of their realization. The historical struggles, that took place regarding the Aristotelian system during the middle ages and far down into modern times, are for the most part to be explained from this point of view. The conflict of "nominalism" and "realism" represents only a further development of the problem, already latent in the first beginnings of the Aristotelian metaphysics and theory of knowledge.

Rickert's theory of the scientific construction of concepts. In contemporary philosophy, the opposition here referred to finds its sharpest formulation in Rickert's theory of the scientific construction of concepts. The direction of thought upon the "concept," and its direction upon the real, mutually exclude each other. For to the extent that the concept progressively fulfills its task, the field of perceptible facts recedes. The simplification, which conception undertakes with regard to the intensive and extensive manifold of things, means a continuous impoverishment of its significance for reality. The final goal of the material sciences and of all other natural sciences is to remove empirical intuition from the content of their concepts. Science does not *bridge* the gap between "thoughts" and "facts," but it is science, which first creates this gap and constantly increases it. "Whatever the content of the concept may be, it stands in the most decided opposition to the empirical world of the intuitive. . . . . The individual in the strict sense disappears even in the most primitive construction of concepts, and natural science finally comes to the view, that all reality is always and everywhere the same, and thus contains absolutely nothing individual. . . . . But this is universally not the case; and as soon as we only consider that every bit of reality in its intuitive form is different from every other, and further that the particular, the intuitive and the individual, is the only reality that we know, then the significance of the fact, that all conceptual construction annuls the individuality of reality, must come to mind. If nothing individual and intuitive enters into the content of scientific concepts, it follows that nothing real enters into them. The gap between the concepts and the individuals, which is produced by natural science, is thus a gap between concepts and reality in general."[81]

Criticism of Rickert's theory. If this logical consequence is justified, then scientific investigation has hitherto been entangled in a strange self-deception regarding its goal. For all the great exact and empirical investigators believed, and still believe, that the task of their science is to permeate the real more and more with knowledge, and to raise it to more *definite* intuition. In place of an accidental and fragmentary consideration of things, which is different for each individual observer, a more perfect survey of them should be gained;

[81] Rickert, *Die Grenzen der naturwissenschaftlichen Begriffsbildung*, Tübingen und Leipzig 1902, p. 235 ff.

and in place of the narrow naïve picture of the world, a more comprehensive insight should be achieved, such as to reveal to us the finer structural relations of the real, while permitting us to trace them in detail. But how can this demand be satisfied, if the logical instrument of investigation, the scientific *concept*, is in direct conflict with it? We must now recognize, that what ought to sharpen our apprehension of the details of empirical intuition, dulls it; that what seems to confirm and extend our knowledge of facts, rather separates us further and further from the real kernel of the "factual." The conceptual understanding of reality amounts to the annihilation of its characteristic import. Peculiar as this result may seem, it follows necessarily from the premises of Rickert's theory. If the concept is, what the dominant logical doctrine holds it to be, nothing but a "presentation of what is common," then it is incapable of grasping the particular as particular. Its function then is not essentially different from that of the *word*, with which it is placed on a level by Rickert, who in this follows Sigwart. As Sigwart explains, all that is presented either exists individually or in abstraction from the conditions of its individual existence, and in this case it is called *universal*, in so far as what is presented, as inwardly present, can be thought of as existing in any group of individual things or cases. The expression for this inwardly present meaning of what is presented is the word as such. For instance, there is no entirely definite intuitive content corresponding to the word "bird," but rather only a certain vague outline of form along with a vague presentation of wing movement, so that a child may call a flying beetle or butterfly a bird; the same is originally true of all our universal presentations. They are only possible because we have, along with the concrete and complete sense perceptions, also less perfect and definite contents of consciousness. The indefiniteness of the memory-images of our actual sensations involves that, along with the vivid and immediately present, sensuous intuitions in the real process of consciousness, pale residua of them are always found, which retain only one or another feature of them; and it is these latter, which contain the real psychological material for the construction of the universal presentation. From this indefiniteness results the capacity of the presentation to be applied to what is different, not merely in space and time but in content: "the more indefinite, the easier the application." The apparent variety of the conceptual function, its ability to introduce constantly new and more remote elements for comparison, thus rests rather on the poverty of the psychological substratum. The scientific concept arises in the same way and under the same conditions. It differs from the naïve concepts of language and the popular view of the world merely in that the procedure, that is there unconsciously effective, is here practised with critical awareness. The methods of natural abstraction, when left to themselves, are very involved, and never attain a complete, *definite* result; the achievement of science is to remove this ambiguity by establishing by universal definition certain rules for the selection of the perceptual material. The various abstract forms thus gain an exact delimitation with respect to each other, since each of them comprehends a single group of properties. The essence of the concept is found in this constancy and all-round differentiation of the presentational content, signified by a definite word.[82] But remoteness from the living intuition of the particular facts is even greater than before. For in the case of the word-meanings, the concrete presentation of the content they are to signify still stands in the background of consciousness, although it need not be explicitly clear; while in the case of the scientific concept, the purer it is, the more it frees itself from this final residuum of intuition. The scientific concept thus becomes a whole that can be completely surveyed and mastered by thought, but, on the other hand, it must fail to grasp and reproduce reality, which is always present only in individual form.

Word-meanings and mathematical concepts. What first strikes us in this deduction is that it separates the scientific concept from the connection, in which it logically arises, and from which it continues to draw its real force. The exact scientific concepts only continue an intellectual process already effective in pure mathematical knowledge. Criticism of popular word-meanings does not affect these concepts, for from the beginning they stand on other ground and are rooted in entirely different presuppositions. The theoretical concepts of natural science are in no sense merely purified and idealized word-meanings; all of them have an element totally foreign to the word as such. As we have seen, they always contain reference to an exact serial principle, that enables us to connect the manifold of intuition in a definite way, and to run through it according to a prescribed law.

[82] See Sigwart, *Logik*,[2] I, 45 ff., I, 325, *etc.* *Cf.* Rickert, *op. cit.*, p. 32 ff., 47 ff. (See also Ch. I.)

For the "concept" in *this* sense, the antinomy on which Rickert founds his argument does not arise. Here no insuperable gap can arise between the "universal" and the "particular," for the universal itself has no other meaning and purpose than to represent and to render possible the *connection and order of the particular itself*. If we regard the particular as a *serial member* and the universal as a *serial principle*, it is at once clear that the two moments, without going over into each other and in any way being confused, still refer throughout in their function to each other. It is not evident that any concrete content must lose its particularity and intuitive character as soon as it is placed with other similar contents in various serial connections, and is in so far "conceptually" shaped. Rather the opposite is the case; the further this shaping proceeds, and the more systems of relations the particular enters into, the more clearly its peculiar character is revealed. Every new standpoint (and the concept is nothing but such a standpoint) permits a new aspect, a new specific property, to become manifest. Here logic again unites with the view of concrete science. In fact, every true scientific concept proves its fruitfulness just by pointing the way to hitherto unknown fields of "facts." In turning aside from the particular material of intuition, the scientific concept does not lose sight of it completely, but always shows us a direction which, if followed further, teaches us new peculiarities in the manifold of intuition. Thus if the chemical "concept" of a certain body is given by its constitution-formula, in which it is grasped as a particular material in its characteristic structure, it is at the same time brought under the various chemical "types," and is thus set in a definite relation to the totality of remaining bodies. The ordinary chemical formula only gives us the composition in general, but not the type of construction of the individual elements; here it is enriched by a wealth of new relations. The general rule now in our possession enables us to trace how and according to what law the given material is transformed into another; the rule involves not only the form of the existence of the material at a definite moment, but the totality of its possible spatial and temporal phases. The further the chemical construction of concepts proceeds, the more sharply the particulars can be distinguished. Materials, which were called similar, because "isomeric," from the standpoint of the undeveloped concept, are clearly separated and distinctly defined in character, from the standpoint of the developed concept. Thus we never meet here the vague "universality" of popular word-meanings. The particular serial member, whose place in the system is to be determined, can be retained throughout; nevertheless, its relation with other members of a group possesses a sharply defined meaning, by which it is distinguished from other forms of relation. The individual in its peculiarity is threatened only by the universality of the blurred generic image, while the universality of a definite law of relation confirms this peculiarity and makes it known on all sides.

Rickert's confusion of "meanings" and "presentations." Thus Rickert's criticism ultimately concerns only a form of the concept that he himself recognizes as unsatisfactory. This form of conceptual construction comes under the "theory of subsumption,"[83] which is nevertheless rejected as far as the *foundation* of the exact concepts is concerned. If we recognize, with Rickert, that all the *thing-concepts* of natural science have a tendency to be transformed more and more into *relation-concepts*, it is thereby implicitly admitted that the real logical value of the concept is not connected with the form of abstract "universality." "A science can only be valuable with regard to knowledge of the whole of the corporeal world," Rickert says, "if it has before it in the first beginnings of the construction of its concepts the final goal of all natural science, insight into the necessity of things according to natural law. If a science has this goal before it, it will always seek to abandon the purely classificatory construction of concepts as soon as possible; *i.e.*, it will never be satisfied with concepts, that are mere complexes of properties, but any collection of such elements into a concept occurs only on the assumption, that the elements either stand directly in a necessary connection according to natural law (*i.e.*, unconditionally universal connection), or at least represent the preliminary stages of such concepts, in which a necessary connection according to natural law is expressed. The relation of the world of meanings to the world of perceptions assuredly constitutes our knowledge, as least in so far as we are concerned with knowledge in the sense of the natural sciences; but for that very reason the meanings cannot be presentations, but must be judgments in their logical value, and must either

[83] *Cf.* the pertinent critical remarks of M. Frischeisen-Köhler, *Die Grenzen der naturwissenshaftlichen Begriffsbildung*, Arch. f. system. Philosophie, XII, (1906), p. 225 ff.

contain laws or prepare the way for them."[84]   In connection with this clear explanation, the critical point in Rickert's theory can soon be pointed out; the center of the problem is falsely shifted from the necessity of conceptual "meanings" to the universality of the generic presentations.   Only of "presentations" can it be said, that the more general they become the more they lose their intuitive sharpness and clarity, until they are finally reduced to mere schemas without significance for reality.   Judgments, on the contrary, determine the individual the more exactly the wider the sphere of comparison and correlation to which they relate it.   Increase of extension is here parallel with determination of the content.   (See above Ch. I.) The universality of a judgment does not signify the quantity of the judgment, but the quality of the judgmental connection, so that judgments concerning individuals can be completely universal. The proposition S is P, in this case, does not signify that the property P is uniformly contained in a plurality of subjects, but that it belongs to this particular subject unconditionally and with objective necessity. When we conceive the given of sensation as made necessary by scientific laws, we thereby change nothing in its material content, but merely represent it from a new standpoint.   A whole "individual" thing does not pass over into a whole "universal" thing; but a relatively loose aggregate of empirical determinations is united into a system of objectively valid connections.   A peculiar kind of object is not produced, but the very same empirical reality is given a new categorical form.   The transition to "universality" is thus a secondary aspect, which does not concern the real tendency of the construction of concepts.   In so far as it enters, the transition to "universality" is only a symptom and expression of that transition to necessity, which is posited and demanded by the problem of scientific knowledge itself.[85]

[84] *Op. cit.*, p. 71, 73.

[85] I find an indirect confirmation of this view in the latest exposition of the theory of Rickert, which is contained in the writing of Sergius Hessen, *"Individuelle Kausalität"* (*Ergänzungshefte der Kantstudien*, Nr. 15, Berlin 1909). In order to reveal clearly the opposition between the scientific and the historical construction of concepts, Hessen distinguishes two different forms of causality.   The causality, which natural science affirms and makes the basis of its explanations, can be reduced to the idea of universal lawfulness.   According to this view, to conceive an event causally means to subsume it under general laws; what is known in this way is thus never grasped in its absolutely

The concept as the expression of individual relations.   True, in one respect the separation between the scientific concept and the "reality" given us in sensuous impressions remains.   None of the fundamental concepts of natural science can be pointed out as *parts* of sensuous perceptions, and thus verified by an immediately corresponding impression.   It has become increasingly evident that, the more scientific thought extends its dominion, the more it is forced to intellectual conceptions that possess no analogue in the field of

unique and unrepeatable character, but only as an example of a general concept.   The content of the idea of causality in general is, however, not exhausted by this one-sided scientific schema.   For causality ultimately signifies nothing else than the "necessity in the temporal sequence of the parts of reality;" we must postulate such a necessity also where we are concerned with the succession of purely *individual* events, which can thus never recur in precisely the same way.   The specifically "historical causality" is founded on the application of this point of view; its concept arises, as soon as we insert the idea of necessity and determinateness into a unique, temporally determined process, without attempting to conceive it as a special case of universal laws.   (*Cf.* Hessen *op. cit.* esp. p. 32 ff., p. 73 ff., *etc.*)   Here it appears that there is an inclusive *unity* for the scientific and the historical "concept," from which both are deduced; and this unity is constituted by the idea of necessity.   Hessen himself ascribed this necessity at first to the "objective reality," which as such is to be conceived as free in principle from every form of conceptual interpretation, whether in the direction of the scientific or the historical concepts.   A more exact epistemological analysis shows, however, that this reality is not taken in the sense of an absolute metaphysical existence, but as a *regulative idea*, which guides our diverse, methodically separate conceptions toward a common goal.   (*Cf.* esp. p. 88 ff.)   In other words, it thus appears that the methodological distinction of the "universal" concepts of natural science from the "individual" concepts of history does not exclude a connection between the two, but rather requires it; what is logically distinct from the standpoint of "universality" tends to coincide, when we exchange this standpoint with that of necessity.

If we hold to the latter thought as the truly original and decisive idea, it is further clear that the distinction in degree of "universality" can never become an unconditional opposition.   In so far as we apply the idea of necessity to a particular temporal occurrence, thus in so far as we assert that this individual A necessarily demands and draws after it this individual B, we implicitly assume an element of universality even in this establishment of a *unique* state of affairs.   For in this judgment, the case is excluded of the total complex A ever recurring in precisely the same character; but at the same time it is asserted that, *if* A were repeated in this fashion, then B and only B would be demanded as real.   Whoever sees more in history than a mere positivistic "description" of the sequence of various events, whoever grants it a particular form of concrete sensations. Not only the hypothetical concepts, like the atom and the ether, but purely empirical concepts, like matter and motion, show that scientific investigation along with the "given" elements of perception cannot do without the purely ideal *limiting concepts* not given in direct experience; along with the "real," scientific investigation must have the "not-real." (See above p. 120 ff.) Nevertheless it would be a mistake to assume that exact science, owing to this characteristic feature of its concepts, withdraws

causal judgment, has already recognized in history this form of the "universal." The universality does not belong to the categorical but to the hypothetical part of the assertion; the form of connection of A and B is *ideally* projected into universality, although the particular elements possess only a unique reality. The historical concept, which seeks to grasp this reality, indirectly refers to the universal form of necessity, just as, on the other hand, the exact concept of natural science, which is meant in the first place to be an expression of a universal connection, seeks verification and application in the temporally limited particular case. Only, in the two, the direction of the reference of the "particular" to the "universal" is different, while the correlation of the two moments is seen to be necessary in both cases.

Thus we are not concerned here with an opposition between the "concept" and absolute "reality," but with a distinction wholly within the system of concepts. Hessen himself emphasizes this fact, and therewith the conceptual character also belonging to history. "The opposite opinion, which makes history a perceptual science and connects it with reality, is guilty of an historical concept-realism as dangerous as that of natural science." The historical concepts are "in general products of a more or less intense abstraction," and are thus as little perceptual as the concepts of the natural sciences. "As an individualizing science of civilization, history implies a removal from reality; it stands in principle as close to reality as the natural sciences; it also works with concepts, and indeed—with individual concepts. This must be especially emphasized against historical concept-realism" (p. 27 ff.). Here it appears from another angle, that the separation of the concepts of natural science from those of history, presupposes a certain connection between the two. The conceptual function as such must be understood and derived in its unitary form before the differentiation into various types of concepts can begin. This fundamental form, however, is not found in the generic concept, but in the serial concept, which is unavoidable for any sort of "shaping" of the perceptually given. An essential task of the historical concepts is the insertion of the individual into an inclusive systematic connection, such as has constantly established itself more distinctly as the real goal of the scientific construction of concepts. This "insertion" can occur under different points of view and according to different motives; nevertheless it has common logical features, which can be defined and isolated as the essence of "the" concept.

more and more from the tasks offered by concrete empirical existence. Precisely in this apparent turning away from the reality of things, science is directed upon them in a new way. For those very concepts, that have no direct intuitive content, have a necessary function in the shaping and construction of intuitive reality. The determinations expressed by the scientific concepts are not perceptible properties of the empirical objects, like their color or taste; but, on the other hand, they are *relations* of these empirical objects. The judgments thus formed, although they cannot be resolved in their content into mere aggregates of sense impressions, are nevertheless related in their use to the totality of these impressions, to which they seek to give systematic form. The methodological opposition thus never becomes metaphysical; for thought only separates itself from intuition in order to return to it with new independent instruments, thereby to enrich it in itself. Every relation, which theory has discovered and given mathematical form, indicates a new way from the given to the not yet given, from real to "possible" experience. It is thus true indeed that the relational concepts of natural science have no immediate copy in the individual things; but what they lack is not so much the element of individuality as rather the thing-like character. They render insight into relations possible, and guarantee it, although they themselves can never be perceived after the fashion of isolated objects. Thus energy, for example, does not signify a homogeneous thing, in which all inner differences of the different types of energy are cancelled, but it is a unitary principle of connection, that as such can only be verified in the qualitatively different. Identity of serial form is concealed behind every assumption of identical objects in natural science; and such identity is only found in the manifold of serial members, which must be retained as such. There is thus no contradiction between the universal validity of principles and the particular existence of things, because there is ultimately no rivalry between the two. They belong to different logical dimensions; thus neither can seek directly to take the place of the other.

The problem of the constants of natural science. The problem gains sharper conception, when we bring it back to the field of mathematics.[86] Reference has justly been made, in opposition to Rickert's

---

[86] The "concrete universality" of the mathematical concepts (*cf.* above Ch. I.) has also incidentally been recognized and emphasized from the

theory, to the significant *rôle* due to the establishment of certain facts of magnitude, to the establishment of certain numerical constants in the structure of natural science.[87] It is only when the values of these constants are inserted in the formulae of these general laws, that the manifold of experiences gains that fixed and definite structure, that makes it "nature." The scientific construction of reality is only completed when there are found, along with the general causal equations, definite, empirically established, quantitative values for particular groups of processes: as, for example, when the general principle of the conservation of energy is supplemented by giving the fixed equivalence-numbers, in accordance with which the exchange of energy takes place between two different fields. As Robert Mayer has said, these numbers are the desired foundation of an exact investigation of nature.[88] The definite number breaks through the traditional logical schema, which recognizes the concept only as a generic concept and as comprehending a plurality of examples under it. The "two" or the "four" does not exist as a

standpoint of Rickert. "The gap for conceptual knowledge between the universal and the particular," says Lask in his work, *Fichtes Idealismus und die Geschichte*, "and the consequent irrationality is bridged in the mathematical view through the possibility of construction. The individual cases realizing the mathematical concept can be generated by the concept itself. From the concept of the circle, we can attain by construction the mathematical individuality of the particular circle, and thus go from the universal to the individual in its individuality. . . . . In mathematics, also, the intuitive object is an individual, concrete and given object; but it is given *a priori*, not *a posteriori* like the material of sensation it is a logical unique! something individual, but at the same time capable of being construed *a priori*" (p. 40 f.). We see here also that Rickert's criticism would have taken another form if he had conceived the concepts of natural science decisively and from the beginning as products of constructive mathematical procedure, rather than as results of "abstractive" procedure. The insight once gained for mathematics would have had to be transferred to physics; for precisely here lies the real problem—that mathematics is no "logical unique," but that it progressively provides the "special" natural sciences with its own characteristic form of concept. The form of mathematical "deduction" is already contained in the form of physical "induction," by which we grasp the empirically real, and thus the same methodic mastery of the particular by the universal is achieved. (*Cf.* esp. Ch. 5.)

[87] *Cf.* Riehl, *Logik und Erkenntnistheorie* (*Die Kultur der Gegenwart*, I, 6, p. 101 f.); *cf.* esp. Frischeisen- Köhler, *op. cit.* p. 255.

[88] R. Mayer, *Bemerkungen über das mechanische Äquivalent der Wärme* (1851) (*Die Mechanik der Wärme*, p. 237).

genus, that is realized in all concrete twos or fours of objects, but it is a fixed member in the series of unities and occurs only once, although there can be no doubt that it possesses no sensuous, but a purely conceptual "being." (See Ch. II.) On the one hand, it appears that the scientific concept is in no way denied the establishment of the individual; although, on the other hand, it never grasps the individual as isolated, but only as a particular element in an ordered manifold. Instead of ascending to abstract and empty genera of being and process, investigation seeks to connect the empirical constants, which it discovers and which are represented by definite number-individualities, into series according to necessary laws.[89] The essential object of scientific consideration is the "structural relations," along with the laws of causal dependence. These structural relations are finally reduced to definite numbers, as the example of chemical knowledge shows, and the attempt is made to understand these numbers as an ordered sequence. Theory considers and defines the possible forms of serial connection in general, while experience shows the definite place, taken by an empirical "real" being or an empirically real process in this connection. In the developed scientific conception of the world, the two elements are inseparably united. The universality of the functional rule is only represented in the particular numerical constants, and the particularity of the numerical constants is only represented in the universality of a law mutually connecting them. This reciprocal relation is also repeated and confirmed within the special sciences. No natural science renounces the establishment of particular facts; nor can it establish them without the decisive concurrence of the idea of law. Even those who start from the opposition of the historical concept of the individual and the scientific generic concept have to agree expressly, that this intellectual division corresponds to no real separation in the sciences themselves. Everywhere the two motives interpenetrate; and it is only by the dominance of one or the other that the position of a particular science in the general system of knowledge can be ascertained. But if this is the case, then it is questionable by what right we can characterize a type of problem and the treatment of it by the name of one science, when the problem

[89] *Cf.* here especially A. Görland, *Aristoteles und Kant bezüglich der Idee der theoretischen Erkenntnis untersucht.* Giessen, 1909. (Philos. Arbeiten hrg. von Cohen u. Natorp II, 2), p. 433 ff.

is shared by the most various disciplines. If we collect under the generic concept of the "historical" all those scientific procedures, which are directed on the gaining of pure "facts," even so it is by no means shown that the concept thus produced represents a true methodological unity. For the establishment of facts occurs in the different special sciences under very different conditions. The general theory of the special discipline is always necessarily presupposed, and also gives the judgment of fact its definite form. Thus every astronomical "factum" involves in its formulation the whole conceptual apparatus of celestial mechanics, and further the fundamental doctrines of optics, in fact all the essential parts of theoretical physics. (See above p. 142 ff.) The "historical" part of each science is methodologically connected with its "theoretical" part by a genuine inner dependence, while between the descriptive parts of two different disciplines, on the other hand, there subsists only a loose connection. The unity here is not one of principle, but is merely classificatory. The procedure, by which astronomy gains its facts, is conceptually connected with the procedure, by which it constructs its general theoretical conceptions; but it is sharply and definitely distinguished from the way in which, for example, biology determines and selects its empirical material. Here also it proves to be impossible to divide our knowledge in such a way that, on the one side, there should be purely the universal, and on the other side, purely the particular. Only the relations of the two moments, only the function fulfilled by the universal in connection with the particular, gives a true ground of division.

Magnitudes and other forms of relations. That this function is completed in none of its activities, that beyond every solution a new problem arises, is indeed indubitable. Here, in fact, the "individual" reality shows its fundamental character of inexhaustibility. But it is the characteristic merit of the true scientific relational concepts, that they attempt this task in spite of the impossibility in principle of its completion. Each new construction, since it is connected with the preceding, forms a new step in the determination of being and process. The individual, as an infinitely distant point, determines the direction of knowledge. This last and highest goal points indeed beyond the circle of scientific concepts and methods. The "individual" of natural science includes and exhausts neither the individual of aesthetic consideration nor the ethical personalities, which are the subjects of history.   For all particularity in natural science reduces to the discovery of definite magnitudes and relations of magnitudes, while the peculiarity and value of the object of artistic consideration and of ethical judgment lies outside its field of vision.   But this delimitation of the different methods of judging produces no dualistic opposition between them.   The concept of natural science does not deny the object of ethics and aesthetics, although it cannot construct this object with the means at its disposal; it does not falsify intuition, although it consciously regards intuition from one dominant standpoint, and selects one particular form of determination.   The other types of consideration, which go beyond natural science, are not in contradiction with it, but in a relation of intellectual supplementation.   They also do not approach the individual as a separated and isolated element, but they produce new and significant points of connection.   It is a new teleological order of the real, that is added to the mere quantitative order, and in which the individual first gains its full meaning.   Logically speaking, the individual is taken up and shaped by different forms of relation.   The conflict of the "universal" and the "particular" resolves into a system of complementary conditions, such as only in their totality and co-operation can grasp the problem of the real.

PART II

THE SYSTEM OF RELATIONAL CONCEPTS AND THE PROBLEM OF REALITY

CHAPTER V

ON THE PROBLEM OF INDUCTION

1

The metaphysical tendency in induction and deduction. The real result of the methodological analysis of scientific knowledge is to deprive the opposition of the universal and the particular of its metaphysical character. The law and the fact appear no longer as two eternally sundered poles of knowledge; but they stand in living, functional connection, related to each other as means and end. There is no empirical law, which is not concerned with the connection of the given and with inferring not-given groups of facts; as, on the other hand, each "fact" is established with reference to a hypothetical law, and receives its definite character through this reference. Empirical natural science, since it first entered "the sure course of a science," has itself taken no considerable part in the struggle of the philosophical parties concerning the rights of "induction" and "deduction." When empirical science examines its own procedure, it has to recognize that there is in this struggle a false and technical separation of ways of knowing that are alike indispensable to its very existence. The motive peculiar to all *metaphysics of knowledge* is here revealed. What appears and acts in the process of knowledge as an inseparable unity of conditions is hypostatized on the metaphysical view into a conflict of things. Permanence and change, being and becoming, unity and plurality, all of which signify only partial aspects of certain fundamental ways of knowing, appear in unconditional opposition. Thus in the philosophy of nature, there is a metaphysics of the particular along with the metaphysics of the universal. While in this latter, concepts expressing the necessary connection of experiences are raised to independent realities, in the former, the simple sensation in its individual character is made the bearer and content of true reality. The real content of existence, which resists every analysis, is sought only in the isolated impressions and their qualitative properties. Advancing intellectual insight serves only to bring out more clearly this fundamental existence, and to resolve all assertions regarding being into it more completely.

The empirical theory of judgment. If this demand is to be satisfied, the motive of particularization must be sharply and clearly carried through. All our judgments can mean nothing else than the establishment of a state of fact, given here and now, and conceived merely in its spatial and temporal particularity. An assertion reaching beyond this would fall into the field of mere fiction. The validity claimed by any true judgment must be strictly limited to the moment of making the judgment; for as the perception, as a real process, does not go beyond this instant of time, so the concept must recognize these natural limits, if it is not to lose its definite character. Present and past sensations constitute the kernel of all our judgments, both the rational judgments and judgments of fact. Indeed the element of past sensations already threatens to break through the general schema; for the past does not "exist" for consciousness in the same sense as that in which the concept of reality is here taken. When we compare a temporally present impression with others, that have occupied consciousness at an earlier point of time, we have already taken the first step from the "given" into the "not-given." This step may be taken without danger, in so far as it is assumed, that the remembered perception is like the actual one in all essential parts. The past is conceived as *present*, in spite of its temporal remoteness, and with all the definiteness of the immediate impression. Judgment rests solely on the comparison of actual and reproduced contents of perception.

Mach's "thought-experiment." Consistent "empiricism" is obliged to extend this consequence to all fields of knowledge. Mathematics and physics, physics and biology are, from this point of view, equivalent; for it is not the analysis of the object, but the psychological analysis of the act of judgment, which has led to this explanation. The form of judgment must be the same everywhere, because the presentational material, on which its form exclusively rests, is always the same for the different disciplines of knowledge. The method of observation and of investigation is independent of whether we experiment with things themselves, or with our presentations and memories of things. We may cite an instance from Mach. If, for example, the geometrical problem is proposed of inscribing a square in a right-angled triangle, having the two sides a and b, and the hypotenuse c, while one corner of the square is to coincide with the right angle, and the three other corners are to lie on the sides a, b

and c; then thought must submit the given conditions to an investigation, in order to find a solution of the problem. Now if we conceive an arbitrary distance to be measured off from the vertex of the right angle on one of the two legs, and the corresponding square to be constructed, then the corner of this square will not in general fall on the hypotenuse, but to the right or left of it, outside or inside the surface of the triangle. Between these two possible cases, as is further shown, there is a continuous transition, since by continually increasing the originally chosen distance, we can move its end-point from within the triangle to a position outside the triangle. Experience directly shows that this movement cannot take place except by crossing the hypotenuse once, as the line which divides the two parts of the plane of the triangle, and thus marking a point on it, which represents the point demanded by the problem. "Such a tentative (*tatonnierende*) division of the field of presentations, in which we have to seek the solution of the problem, naturally precedes its perfect solution. Popular thought may be satisfied with a practically sufficient, approximate solution. Science demands the most general, shortest and most comprehensible solution. We obtain this, when we remember that all inscribed squares have the line which bisects the angle at the intersection of a and b in common as a diagonal. Consequently if we draw this bisecting line from this known point, we can complete the required square from its point of intersection with c. Simple as this purposely chosen . . . . example is, it brings out clearly what is essential in all solution of problems, experimentation with thoughts, with memories. . . . ."[1]

Criticism of Mach's theory. However, even this example reveals a latent presupposition at the basis of the whole argument. "Memory" in the strict psychological sense can produce no new content; it can only repeat what has been offered by sensuous presentation. It can thus recall into consciousness those cases, that have been intuitively presented to us, but it is incomprehensible how it can venture any assertion regarding a *whole group* of forms, without having run through the particular examples individually. Yet this latter possibility is excluded by the nature of the problem in the case mentioned; the number of possible squares is infinite, and is thus absolutely inexhaustible for concrete sensuous imagination. The judgment of memory as such can never survey an infinite group of

[1] Mach, *Erkenntnis und Irrtum*, p. 39 f.

possible cases, but only a limited number of actual cases. However many points of the line bisecting the angle we may have investigated, we can never decide whether the next point that we select will show the same characteristics as those previously observed, if we commit ourselves solely to the described method of experimentation with presentations and memories. From this point of view, there is nothing to hinder the assumption that in further advance there might be found points of the line bisecting the angle, which do not satisfy the assumed condition; or that conversely, there are points that fulfill the condition without belonging to this line. The solution first gains the character of necessity and determinateness, when we go back from the particular example to the process of *construction*, in which the bisecting line arises and it gains its mathematical properties. In becoming aware of this unitary rule of construction, we thereby grasp the totality of determinations of the complete structure; for these determinations arise only by virtue of the generating law and can be deduced from it in all strictness. We do not proceed here from a plurality of particular cases to the connecting law, but from the unity of the geometrical procedure to the particularities of application. Only in this way is a relation posited, that affirms not only the present presentation, as it is found in consciousness, but a permanent ideal connection. A presupposition is established which is meant to hold not of this or that individual triangle with its particular properties, but of "the triangle" absolutely. It is no matter whether this claim can be finally justified; certainly as a mere psychological phenomenon, it breaks through the scheme of knowledge of the consistent sensationalistic view.

Thus those thinkers, who proclaim the postulate of "radical empiricism" most decidedly within psychology, are obliged even from this standpoint to recognize the logical and methodological difference here. The unprejudiced verdict of "pure experience" is opposed on this point to the dogmatic deductions of sensationalism. Unprejudiced analysis of the *facts* of knowledge shows clearly that to reduce mathematical and logical relations to assertions regarding the frequent empirical coexistence of particular presentations is a vain endeavor. Mathematical and logical relations do not report whether and how often certain empirical contents have been found in existence in space and time, but rather establish a necessary connection between ideal structures, the validity of which is to remain unaffected by all changes in the world of existing sensuous objects. To interpret a logical or mathematical proposition as a mere reproduction of particular, actual "impressions" and their empirical relations, is to falsify the real *meaning* of the proposition in the attempt to discover its *origin;* such an interpretation ascribes to the proposition a sense, which, by the nature of the subject to which it refers, it neither does nor can possess. No metaphysical construction can set aside the psychological and logical *phenomenon* of this difference; "relations between ideas" are separated in principle from purely factual determinations of the coexistence and sequence of particular empirical properties.[2]

Locke's theory of empirical judgment. The more sharply this separation is carried through, the more strongly, on the other hand, the character of the empirical judgment is revealed. The character of this latter seems to consist in nothing else than a conscious limitation of the validity of the judgmental connection to the temporal moment of making the judgment. In this sense, the relation of the two kinds of truths was already grasped by Locke. According to him, the validity of mathematical knowledge rests on the principle of the immutability of the same relations between the same intellectual objects. What is proved of one triangle can be carried over without further mediation to all triangles; for a particular intuitive presentation of the triangle does not stand for itself in the proof, but is only meant to be an accidentally selected sensuous image for a universal and permanent relation. This insight is denied in all the judgments, which go beyond the field of our intellectual presentations to the existence of things. External things manifest themselves to consciousness in no other way than in the sensuous impressions they arouse in us; their certainty can thus be of no other kind than that of these impressions themselves. The existence of sensation reaches no further than its immediate presence. Once this is removed we lose our only criterion for the existence of things and the basis of all assertions regarding the more exact properties of this existence is taken from us. Consequently, judgments concerning the existence of things have only relative and limited truth; for, however

[2] Genuine psychological "empiricism" maintains this separation throughout; this is brought out with especial clarity by James in his polemic against Spencer and Mill on this point. (*The Principles of Psychology*, London 1901, esp. Vol II, 645, 654, 661, *etc.*)

convincing and evident they may seem as long as the direct sensation
is given, we have no assurance that the momentary witness of
sensation will ever be repeated in exactly the same way. Accordingly
there is necessary knowledge only of such objects as, like the objects
of pure mathematics, renounce all concrete reality; while at the very
moment this reality is taken into account, the character of knowledge
is completely changed.

The "element of eternity" in all empirical judgment. Although this
distinction is illuminating from the abstract standpoint of theory of
knowledge, it offers a difficult problem when we consider it in connection with the concrete procedure of natural science. Locke's
account, which has been frequently repeated since with slight variations, might appear as a correct account of what the purely empirical
and inductive propositions of natural science *ought* to be; but it
certainly does not touch what they are in reality. No judgment of
natural science is limited to establishing what sensuous impressions
are found in the consciousness of an individual observer at a definite,
strictly limited point of time. If there are judgments that speak of
this, they are the narrative judgments of psychology, and not the
theoretical and descriptive judgments of general natural science.
As the mathematician, who treats of the relations between geometrical forms or between pure numbers, permits nothing in his statements regarding the properties of the particular presentations, in
which he sensuously represents these relations, so also the investigator, who gives out the results of an experimental research, constantly
goes beyond a simple report of his particular perceptual experiences.
What he establishes is not the sequence and play of certain sense
impressions, that have occurred in him and again disappeared into
nothing, but the constant "properties" of constant things and processes. In this advance from the mere process of sensation to definite
"objective" assertions, the metaphysical concept of "transcendence"
is indeed wholly absent. The change, that makes possible the judgment of natural science, only transforms the data of sense into a new
mode of being in so far as it imprints upon sense-data a new *form
of knowledge*. This element of knowledge can be separated out and
retained independently of all further metaphysical assertions that
may be added. It is, first of all, a new sort of temporal validity,
which is now ascribed to the judgment. Even the simplest judgment
concerning any empirical matter of fact ascribes to the latter an

existence and a permanence, which the fleeting sense-experience as such cannot establish. The proposition, that sulphur melts at a definite temperature, that water freezes at a definite temperature, means even in its simplest terms, and abstracting from the various theoretical assumptions contained in the mere concept of "temperature," something that is to be limited to no isolated temporal moment. It asserts that, as often as the conditions embraced by the subject concept are realized, the consequences expressed in the predicate concept will be always and necessarily connected with them. For thought, the moment of immediate perception is extended to the whole course of time, which is surveyed in its totality at one glance. It is this logical function, which gives each experiment its peculiar significance as proof. Each scientific conclusion based on an experiment rests on the latent assumption, that what is found to be valid here and now, remains valid for all places and all times, in so far as the conditions of investigation are unchanged. Only through this principle are the "subjective" facts of sense-perception changed into the "objective" facts of scientific judgment. From a new side, it is seen how far, as Goethe said, all that is factual is already theory; for it is only the thought of the necessary determinateness of phenomena, that leads us to arrest a single transitory observation, and establish it as a fact.

The postulate of necessary determinateness. Even investigators, who think they stand exclusively on the ground of empirical "facts" and who repudiate all independence of the intellect with respect to the data of immediate perception, have expressly attested to this intellectual function. In spite of all supposed scepticism this one conviction is occasionally expressed. "The relations between different phenomena," says Ostwald, "which have once been known, remain as indestructible parts of all future science. It can, and indeed often does, happen that the form in which such relations were first expressed is seen to be imperfect; it is seen that the relations are not to be maintained as wholly universal, but are subject to other influences, which alter them, of which no thought could be taken at their discovery and first formulation, since they were then unknown. But however science may be transformed, a definite indestructible residuum of that first knowledge remains, and a truth once gained by science has an eternal life in this sense, *i.e.*, it exists as long as human science will exist."[3] This element of "eternity" is

---

[3] Ostwald, *Grundriss der Naturphilosophie*, p. 15.

also characteristic of empirical judgments concerning facts. No connection between observations, once it has been established objectively, can be absolutely annulled in the further course of investigation. The new facts, which we discover, do not displace the earlier experiences in every sense, but only add to them a definite conceptual determination. And this change really affects not so much the judgmental connection, as rather the *subject* to which it refers. For example, if we conceive a certain material as determined by the scheme of its physical and chemical properties, then no matter what contrary instances may arise in the course of advancing observation, or what change may appear in its action, the previously affirmed connection in the determinations of the material is in no way set aside. If the empirical judgment were bound to a single moment of time, then a simple relation of annihilation and re-creation would prevail here; the later moment would set aside the earlier, and with it all the "truths," which were indeed only established and asserted of the earlier. Just as, in the real course of phenomena, the later moment replaces the earlier, so also it would involve an inner change in the empirical laws of the thing. But in truth, each body has an identical structure and character, which we ascribe to it once for all. We never express divergent results by assuming that one and the same body has changed in its fundamental properties, but by calling the identity of the observed object into question. What we now see standing before us is not the *same* empirical object, as was previously offered; but we consider it as modified by certain conditions, that are to be discovered and established. Thus the truth of the earlier judgment "*S is P*" is not invalidated by its opposite "*S is not P*," but, while maintaining the first proposition, we trace the transformation, which the judgment must undergo, when S passes into S'. The advance of observation thus involves a continuous advance in analysis; it distinguishes with increasing exactness cases which on first vague consideration appear wholly similar and reveals the characteristic differences of each individual case. If we conceive this work of analysis as complete and a wholly determinate subject therewith gained, then the determinateness of this subject will also involve in it the determinateness and necessity of the judgmental connection. The element of uncertainty, contained by the empirical judgment in contrast to the rational, thus concerns only the subsumption of the given under an ideal case. The question is not whether to a strictly defined content a the predicate b belongs or not, but whether a given content satisfies all the conditions of the concept a, or is to be determined by a different concept a'. The problem is not whether a is truly b, but whether x, which is offered by mere perception, is truly a. Here lies the real superiority of the mathematical construction of concepts; for the objects of mathematical construction are only what our ideal construction makes them, while every empirical content conceals unknown determinations in itself, and therefore we can never decide with full certainty under which of our previously developed hypothetical concepts it is to be brought.

Judgments of perception and judgments of experience. Locke's analysis of the empirical judgment is thus seen to be intrinsically inadequate; for it conceals that element of necessary connection, which is also characteristic of assertions concerning facts, and which gives them their real meaning. For Kant, this necessity was the real fundamental problem; yet in first introducing his critical question, he still appears dependent in one point upon Locke. The distinction between judgments of perception and judgments of experience, to which Kant appeals, has not so much a direct, systematic, as a didactic meaning; it joins on to the sensualistic conception of judgment, so as to gain from it a new meaning and a deeper interpretation. Empirical judgments, in so far as they have objective validity, are to be called judgments of experience; while those, that are only subjectively valid, are to be called judgments of perception. The latter concept thus includes everything that dogmatic empiricism regards as the genuine mark and character of experience. The "judgment of perception" at least is nothing else, and is intended to be nothing else, than a report concerning a momentary and individual experience; it does not connect subject and predicate according to any standpoint of intellectual dependence and coherence, but only takes them as they are accidentally found together in an individual consciousness, according to the "subjective" rules of association. In the judgment of perception, we only establish the coexistence of two contents, without setting them in any relation of mutual dependence. The further the Kantian distinction proceeds, however, the more it appears that the judgment of perception is only meant to be a methodologically constructed limiting-case, to throw light upon the newly gained concept of scientific objectivity by force of contrast; but that the distinction carries with it no real separation of judgments into two heterogeneous classes. Every judgment claims a certain measure of objectivity within its self-chosen narrower sphere, no matter how limited its subject-concept. It is never satisfied with establishing a mere coexistence of presentations, but it erects a functional coördination between them, so that whenever the one content is given, the other is taken as required. The "is" of the copula is the expression of this connection, and thus enters as a necessary factor into every assertion regarding an individual, empirical object. The proposition, that body is heavy, does not mean that, as often as I have hitherto lifted a body, certain touch and pressure sensations have been felt, but it is meant to establish a connection, based in the object and independent of the condition of this or that sensing individual. Even the individual "*a posteriori*" judgment always contains an "*a priori*" element in the necessity of the connection, which it affirms.[4] In the final conception of the system of experience, the instrumental concept of the mere judgment of perception is transcended and excluded. Indeed the individual as individual can be the object of a scientific statement; hence a state of being, which is given here and now, constitutes the content of judgment. But in this case also, we do not step outside the field of objective necessity into that of mere "contingency;" on the contrary, we seek to conceive the particular itself as necessary, by giving it a fixed place within the causal process ruled by exact laws. The sphere of the necessary is narrowed until it is adequate for closer determination of the apparently "contingent." For example, we determine the astronomical position of the heavenly bodies for a given point of time in this sense, by taking as a basis the universal relations offered by the principles of mechanics, as well as by the law of gravitation. The real goal of "induction" is not the absolutely isolated, temporal fact (*Setzung*) as such, but the subordination of this fact to the whole process of nature.

Experience as aggregate and as system. The "secret of induction," often referred to, does not begin where we draw a conclusion from several observations regarding *all* the cases, but is already fully contained in the establishment of any individual case. The solution of the problem of induction can only be sought in this broadening of its import. In fact, it cannot be understood how the mere repetition

---

[4] *Cf. Kritik der reinen Vernunft*, 2nd ed., p. 141 f.

and arrangement of particular observations should lend the particular a new logical value. The mere accumulation of elements cannot entirely change their conceptual meaning; it can only render more distinct the determinations already contained in the elements. There must be a factor concealed in the individual case, that raises it out of its limitation and isolation. The function, that constitutes the real kernel of inductive procedure, is that by means of which we trace an empirical content beyond its given temporal limits and retain it in its determinate character for all points of the time series. The relation, which only manifests itself at first for a single, indivisible moment, grows until in some way it determines the totality of future points of time. Thus each individual judgment involves an element of infinity, in so far as its content is transferred to the totality of times, and is continued through this totality as if by constant, identical re-creation. The permanent empirical object along with its constant empirical properties is, in mathematical language, always the integral of the momentary properties, of which the individual enquiry gives evidence. But the logical process of integration would not be possible, if there were not a reference to the whole already in the element, *i.e.*, if the varying content of experience, no matter how scattered and detached it may appear, did not always involve reference to its permanent laws (*gesetzliche Form*). It is through this reference that our limited, spatio-temporal circle of experience, which is all we have, becomes the test and image of the system of reality in general. It is only by conceiving all phenomena as connected by necessary relations, that we can use any individual phase as a representation and symbol of the total process and of its universal rules. It is this symbolic meaning, which every inductive inference claims for itself; the particular determination offered by the sensuous impression becomes a norm, that has to be retained as a permanent feature in the intellectual structure of empirical reality. Each particular experience, that has been established according to the objective methods and criteria of science, claims to be absolute; what methodically tested experiment has once shown, can never be entirely logically annulled. The task of induction is to unify these various assertions, which frequently seem to cross and contradict each other, by indicating a definite sphere of validity for each of them. What, to the ordinary sensuous view, is an identical group of conditions connected now with one, now with another circum stance, is here separated into definitely distinguished, particular cases; these cases vary in some theoretically-discoverable circumstance, and their variation is thus conceived as necessary.

Discrete and continuous "wholes." In the relation of the inductive individual case to the totality of scientific experience, a characteristic recurs, which can always be established where the problem is to define a "whole," that is not merely the sum of its parts, but is a systematic totality arising from their relations. Logic traditionally distinguishes between "discrete" and "continuous" wholes. In the first, the parts precede the whole, and, independently of the connection into which they subsequently enter, are possible and distinguishable as independent pieces. The "element" of the continuum, on the contrary, is opposed to any such separation; it gains its content only from relations to the totality of the system, to which it belongs, and apart from it loses all meaning. Thus a line can be defined as an infinite manifold of points; but this definition is only possible, because the "point" itself has been previously conceived as the expression of a pure relation of position, and thus the relation of spatial "coexistence" with other similar elements is contained in it. In the same sense, it can be said that the law of experience only "results" from the particular cases because it is already tacitly assumed in them. The individual empirical judgment contains within it, as an undeveloped demand, the thought of the thorough-going determinateness of natural processes as a final result of the completed system of experience. Every assertion regarding a mere coexistence of empirical determinations points to the thought that these determinations are somehow grounded in each other, although the form of this dependence is not directly known, but is only to be gained progressively. Just as the relational character of position and distance inheres in the individual point, so the character of a universal law inheres in the individual experience. The individual cannot be experienced save in connection with other spatial and temporal, near or remote elements; and this kind of connection presupposes a system of spatial and temporal positions, as well as a unitary whole of causal coördinations. The fact a is only accessible to us in a functional form as f (α), ϕ (β), ψ (γ), in which f, ϕ, ψ, represent the different forms of spatial-temporal and causal connection. The logical act of "integration," which enters into in every truly inductive judgment, thus contains no paradox and no inner difficulty; the advance from the individual to the whole, involved here, is possible because the reference to the whole is from the first not excluded but retained, and only needs to be brought separately into conceptual prominence.

Induction and the theory of invariants. The tendency to something unchanging, to something permanent in the coming and going of sensuous phenomena, is thus characteristic of inductive thought no less than of mathematical thought. The two do not differ in their goal, but in their means for reaching this goal. It is possible to trace in the development of geometrical methods how the manifold tendencies of modern geometrical thought can be unified, in so far as they are brought under the general standpoint of the "theory of invariants" and characterized by it. Each special form of geometry is then coördinated with a definite group of transformations as its appropriate theory of invariants, and these can be strictly defined and set over against each other. The conceptions of constancy and change are seen to be mutually conditioned by each other; the permanent connections affirmed by geometry can only be formulated with reference to possible changes. (See above p. 88 ff.) This fundamental logical relation now appears in a new light. Each assertion concerning an empirical connection of elements is meant to be valid independently of any absolute point of space or time. Maxwell incidentally gives the general "causal law" a turn that expresses this demand. The proposition, that like causes always bring forth like effects, he explains, has no sharply defined meaning, until we establish what is to be understood by like causes and like effects. As each occurrence only takes place once, and is thus fully individualized by the time of its occurrence and distinguished from all others, the likeness here in question cannot be meant in the sense of absolute identity, but only relatively to a definite point of view, such as needs to be expressly isolated and formulated. The real kernel of the causal principle lies in the assertion that, when the causes are merely distinguished from each other with reference to absolute space and absolute time, the same is true of the effects: "the difference between two occurrences does not depend on the pure differences of the times or places, in and at which they take place, but only on differences in the nature, configuration or move-ment of the bodies involved."[5] Here it becomes clear, that the

---

[5] Maxwell, *Matter and Motion*, Art. XIX.

content, on which the physical judgment is directed, is at first subjected in thought to a certain change, and the judgment separates out those moments unaffected by this change, that can be constantly affirmed. Just as we characterize as the geometrical properties of a certain form all those properties, that belong to it independently of its absolute position in space and of the absolute magnitude of its parts, so here an analogous mode of consideration is applied with regard to time. A functional relation f (a, b), that is only directly established for one point of time t_o, or for a plurality of discrete times t_1, t_2, t_3 ..., is freed from this limitation and removed from dependency on any individual moment of observation. So far as the remaining conditions are unchanged, any arbitrary moment of time may be taken as equivalent to that first given to us, so that the present moment involves a decision for the past and the future. Thus all experience is directed on gaining certain "invariant" relations, and first in these reaches its real conclusion. The conception of the empirical natural object originates and is grounded in this procedure; for it belongs to the concept of this object, that it remains "identical with itself" in the flow of time. We must indeed conceive each natural object as subject in principle to certain *physical* changes, called forth by external forces; but the reaction to these forces could not be represented in the form of law, if we were not able to recognize the object as logically permanent and provided with the same properties. In the midst of the temporal chaos of sensations, we produce fixed connections and coördinations by abstracting from time; and it is these fixed connections, which constitute the fundamental frame-work of empirical factuality.

Induction and analogy. Thus it is always a function of judgment, that assures us of the permanence of empirical being. This fact finds its expression not only in the inductions of mathematical physics, but also is clearly shown within descriptive natural science. Deeper analysis reveals here also how far the apparently purely receptive classification of particulars is dominated by ideal presuppositions, referring to the structure of the whole. Claude Bernard especially has thrown light on this reciprocal relation of idea and observation for physiology and the whole field of "experimental medicine." The order of the "real" and factual cannot be discovered without an ideal standpoint of comparison, without a conceptual anticipation of a *possible* order. Although it is to experience that we owe the definitive establishment of this order, beyond doubt, yet though there again must work out the scheme of experience beforehand. The induction of the descriptive sciences is thus always a "provisional deduction." "We can, if we will, call the tentative thought of the experimental investigator inductive, while we characterize the apodictic assertions of the mathematician as deduction; but this is only a difference affecting the certainty or uncertainty of the starting-point of our inferences, and not the manner in which these inferences themselves proceed." The principle of inference remains the same in the two cases, although its course may run in two directions: "there is for the mind only one way of concluding, as there is for the body only one way of going."[6] This unity of principle is especially clear in those limiting fields, where mathematical thought comes in contact, as it were, with experimental investigation. We saw how the progress of geometrical thought tends more and more to allow the subordination of the particular intuitive figures in the proof. The real object of geometrical interest is seen to be only the relational connection between the elements as such, and not the individual properties of these elements. Manifolds, which are absolutely dissimilar for intuition, can be brought to unity in so far as they offer examples and expressions of the same rules of connection. The conceptual construction of exact physics proves to be dominated by a corresponding logical procedure. Hitherto the analogical inference has been considered an essential part of the method of physics, and especially of the inductive procedure; no less a thinker than Kepler commends it as his truest guide and preceptor, from which no secret of nature is concealed.[7] But the scientific value of analogy remains incomprehensible, as long as one merely bases it on a sensuous similarity between the individual cases. Indeed the precise task of theoretical physics, by which it is distinguished from the naïve view, is to separate cases, that appear in direct perception as similar and analogous, by penetrating analytically further into the conditions

---

[6] Claude Bernard, *Introduction à l'étude de la médicine expérimentale*, Paris, 1865, esp. p. 83 ff.

[7] Kepler, *Paralipomena in Vitellionem*, Ch. IV, 4 (Op. II, 187). On the concept of analogy, *cf.* Mach, *Erkenntnis und Irrtum*, p. 218 ff.

of their origin.[8]   The genuine and truly fruitful analogy is not based
on a sensuous agreement of properties, but on a conceptual agreement
in relational structure.   When we consider electricity and light as
phenomena of the same sort, in the electro-magnetic theory of light,
this assertion is not based on an agreement capable of being grasped
by perception, but on the *form of the equations*, which we establish in
both as a quantitative expression of the phenomena, as well as on
the relations between the numerical constants characteristic of the
two fields.   (See above p. 163).   The comparison thus does not rest
on a mere indefinite similarity, but upon a true *identity* in the mathematical system of conditions; as in pure mathematics, this identity
is isolated as a logical "invariant" and considered for itself.   Thus
"analogy," which at first seems concerned with the sensuous particular, goes over more and more into mathematical "harmony;" the
doctrine of Kepler offers the classical example of this.   "Analogy"
passes into a view of the unitary quantitative structural laws which,
according to the assumptions of exact physics, rule the whole of
being, and thus unify what is most diverse.

II

Induction - and analysis, "compositive" and "resolutive" methods.
The first result claimed by the inductive "concept," in the strict

[8] Here is an example from Duhem, *La Théorie Physique*, p. 32 f: "La Physique expérimentale nous fournit les lois toutes ensemble et, pour ainsi dire, sur
un même plan. . . . Bien souvent, ce sont des causes tout accidentelles,
des analogies toutes superficielles qui ont conduit des observateurs à rappro-
cher, dans leurs recherches, une loi d'une autre loi.   Newton a fixé dans un
même ouvrage les lois de la dispersion de lumière qui traverse un prisme et les
lois des teintes dont se pare une bulle de savon simplement parce que des
couleurs éclatantes signalent aux yeux ces deux sortes de phénomènes.   La
théorie, au contraire, en développant les ramifications nombreuses du raisonnement déductif qui relie les principes aux lois expérimentales, établit parmi
celles-ci un ordre et une classification; il en est qu'elle réunit, étroitement
serrées, dans un même groupe; il en est qu'elle sépare les unes des autres et
qu'elle place en deux groupes extrêmement éloignés. . . . Ainsi, près des
lois qui régissent le spectre fourni par un prisme, elle range les lois auxquelles
obéissent les couleurs de l'arc-en-ciel; mais les lois selon lesquelles se succèdent
les teintes des anneaux de Newton vont, en une autre région, rejoindre les
lois des franges découvertes par Young et par Fresnel. . . . Les lois de
tous ces phénomènes que leurs éclatantes couleurs confondaient les uns avec
les autres aux yeux du simple observateur, sont, par les soins du théoricien,
classées et ordonnées."

sense of this word, is that it transforms the manifold of observations, which at first appears as a mere unrelated coexistence of particular elements, into some fixed serial form. We can realize the meaning of this task by referring to certain elementary problems of arithmetic, which form an exact example and analogue of the logical relation here involved. If any sequence of numbers is given, which are connected according to a definite rule at first unknown, then in order to discover this rule, the given sequence must be resolved into a complex of series, that obey relatively simpler laws. If we have before us, for example, the sequence of fourth powers $1, 16, 81, 256, 625, \ldots$, then we can establish the relation connecting the particular serial members, by finding first the differences between them, and further the differences between these differences *etc.*, until we finally reach a simple arithmetical series with a *constant* difference between its individual members. Thus we are led back to a completely known serial type, and the way is indicated, by which from this fundamental form we can again reach the given series through continually more complicated steps. This given series is now plainly revealed to us in the conditions of its structure, and in all the particular phases of its construction; by reducing it to the stages of its construction, the series becomes, as it were, transparent, and shows the same character of necessity in its advance from member to member, as was shown by the primitive series. The same "resolutive" method, that is here used in the field of numbers, is characteristic of the true scientific inductive inference. The given fact, as offered by immediate observation, appears at first to thought as if impenetrable. It can be simply established, but not determined according to its simple beginnings, *i.e.*, not deduced from identical rules of progress. But genuine, theoretically guided induction is never satisfied merely with establishing facts as given. It replaces the factual coexistence of sensuous data by another kind of connection, which indeed seems poorer in elements, when considered purely materially, but which can be more clearly surveyed according to the principle of its construction. Every experiment we institute, and on which we base our inductive inferences, works in this direction. The real object of scientific investigation is never the raw material of sensuous perception; in place of this, science substitutes a system of conditions constructed and defined by itself.

Experiment as the means of analysis. Strictly speaking, the experiment never concerns the real case, as it lies before us here and now in all the wealth of its particular determinations, but the experiment rather concerns an ideal case, which we substitute for it. The real beginnings of scientific induction furnish the classical example of this. Galileo did not discover the law of falling bodies by collecting arbitrary observations of sensuously real bodies, but by defining hypothetically the concept of uniform acceleration and taking it as a conceptual measure of the facts.[9] This concept provides for the given time-values a series of space-values, such as proceed according to a fixed rule, that can be grasped once for all. Henceforth we must attempt to advance to the actual process of reality by a progressive consideration of the complex determinations, that were originally excluded: as, for example, the variation of acceleration according to the distance from the centre of the earth, retardation by the resistance of the air, *etc.* As in the arithmetical example, we advance from the simple series, in which the differences between the members are constant, to series of the second, third and fourth order; so we analyse the real into various orders of relations, connected according to law and progressively conditioning each other. The sensuous appearance of simplicity in the phenomenon yields to a strictly conceptual system of superior and subordinate relations. However, as opposed to the mathematical concept, there is the characteristic difference that, while in mathematics the construction reaches a fixed end, in experience it is in principle incapable of completion. But no matter how many "strata" of relations we may superimpose on each other, and however close we may come to all particular circumstances of the real process, nevertheless there is always the possibility that some coöperative factor in the total result has not been calculated, and will only be discovered with the further progress of experimental analysis. Each result established has thus only the relative value of a preliminary determination; and as such only holds what is gained in order to use it as a starting-point for new determinations. The uncertainty, that seems to remain, does not affect the relations within the particular series, but only appears where the whole theoretical construction is compared with the actual observations. A contradiction here would not be resolved by our giving up the principles of

---

[9] *Cf.* more particularly *Erkenntnisproblem* I, 294; also Hönigswald, *Beiträge zur Erkenntnistheorie u. Methodenlehre,* Leipzig, 1906.

the earlier investigations, but by our adding to these investigations new factors, such as correct the first result yet enable us to retain it in a new meaning. In so far as we abstract from subjective errors of observation, the *truth* of the individual determinations remains in general unaffected. What is always being questioned anew is only the *sufficiency* of these determinations for explaining the complicated factual relations of reality. But the inductive concept proves, precisely by the fact that it leaves this question open, that its direction does not lead away from reality, but more closely to it. The "universal relations," which are at first brought out, do not indeed contain the special properties in themselves, but neither do they deny them. They rather leave room for them from the beginning, and anticipate their future possible determination. Galileo's law of falling bodies needs no correction of its content, in order to represent the phenomenon of falling in the case of a definite resistance of air, but merely requires a conceptual extension already admitted and foreseen in principle.

The relation of "universal" and "particular" relations. In this sense, the "concept" of natural science in no way abstracts from the particular, but brings out the particular all the more sharply. Each universal relation, that it establishes, has a tendency to connect itself with other relations, and by this combination to become more and more useful in the mastery of the individual. Each of the fundamental series, from which a more complex law is built, signifies in itself indeed only a certain sphere of conditions. However, it cannot be said, that these conditions are only partially or inexactly fulfilled by the concrete process, which results from their totality; rather they must all be fulfilled perfectly and without limitation, if the given case is to be possible. Thus the subsequent addition of a new standpoint, which places the phenomenon under consideration in relation to a new circle of facts, changes nothing in the meaning and value of the earlier determinations. Only one thing is demanded, *viz.*, that the relations progressively established in this way shall be compatible with each other. And this compatibility is assured in principle by the fact, that the determination of the particular case takes place on the basis of the determination of the general case, and tacitly assumes the validity of the latter. If we conceive a particular process as a synthesis of various laws, the question as to how the particular can "participate" in the universal ceases to be a *metaphys-*

ical problem; for the universal is no longer something substantially present, which somehow enters as a factual part into the individual, but it is a logical moment, implicitly assumed in a more inclusive system of relations. We determine a particular natural occurrence A, by inserting it into various functional connections f (A, B, C ...), ϕ (A, B', C' ...) ψ (A, B'', C'' ...) *etc.*, and conceive it as subordinated to the rules of all these connections. The "participation" of the individual in the universal thus appears no more of a riddle than the logical fact that, in general, various conditions can be united intellectually into a unitary result, in which each is fully retained. The question now is not how the individual in its "substance" issues from the universal and is separated from it, but how it is possible for knowledge to bring the rules of universal connections into relation, and to determine them reciprocally by each other, so that conceptual insight into the particular relations of physical reality results. (*Cf.* above p. 226 ff.)

"Isolation" and "superposition." The real problem of induction lies here, as is shown in many places in the theory of knowledge of modern natural science. The two tendencies of scientific thought, which Galileo contrasted as the "resolutive" and the "compositive" methods, are sometimes distinguished in modern discussion as the principles of "isolation" and of "superposition."[10] The first goal of experimental enquiry is to gain a *pure* phenomenon, that is, to free the phenomenon under consideration from all accidental circumstances. While reality presents us a mixture of heterogeneous circumstances seemingly inseparably interwoven and confused, thought demands the separate consideration of each particular moment and the exact determination of the part it plays in the structure of the whole. But this goal can be achieved only through a technical separation of what is factually connected, only by establishing special conditions, which enable us to trace the individual factors for themselves in their action. It is only when this separation is strictly carried through, that the constructive unification is seen in its clarity and sharpness. The complete picture of the total process arises when we connect the partial system and place them as it were upon each other. This picture of the total process appears not merely as a unitary intuition, but as a differentiated conceptual whole, in which the type of de-

[10] See Volkmann, *Erkenntnistheoretische Grundzüge der Naturwissenschaft*, Leipzig, 1896.

pendency between the individual elements is exactly defined. If the problem of physical induction is conceived in this way, it appears anew that the mathematical point of view is not so much the opposite as the necessary *correlate* of inductive conceptual construction. For that very synthesis of relations, which is required and which forms an essential part of experimental enquiry, has its ultimate abstract foundation and the guarantee of its general validity in the system of mathematics. As we have seen, the object of mathematics in the full universality of its concept is not the compounding of magnitudes, but the connection and reciprocal determination of relations. (See above p. 95 ff.) Thus the two scientific tendencies here have a point in common. Though experiment is necessary to analyse an originally undifferentiated perceptual whole into its particular constitutive elements, on the other hand, to mathematical theory belongs the determination of the form by which these elements are combined into a unity of law. The system of "possible" relational syntheses already developed in mathematics affords the fundamental schema for the connections, which thought seeks in the material of the real. As to which of the possible relational connections are actually realized in experience, experiment, in its result, gives its answer. But this answer can only be given when the question has previously been clearly stated; and this process of stating the question goes back to conceptions, which analyse immediate intuition according to conceptual standpoints. If the real is represented as the result of the interpenetration of elementary series of dependencies, then in principle it has gained the form of a mathematically determinable structure.[11]

[11] As I subsequently discover, this relation has received new confirmation and an unusually clear presentation from a modern physicist. Bouasse, *Physique générale*, in the collective volume *De la Méthode dans les sciences*, Paris 1909: "La physique ne sépare pas l'étude des *formes* de l'étude des *faits;* la déduction prévoit les faits que l'expérience confirme." "*Qu'est-ce donc qu'expliquer? C'est tout uniquement faire rentrer un fait dans une forme.* Le fait est expliqué lorsqu'il apparaît identique à un des phénomènes qu'engendre un des ces sorites indéfinis que nous appelons théories ou formes. . . . La physique n'est donc pas mathématique, parce qu'on y trouve des algorithmes algébriques; toute expérience devant, en définitive, entrer dans une forme, toute forme se développant naturellement sous les symboles mathématiques, toute physique est mathématique (op. cit. p. 76 f., 91, 100).

Thus when the principle of "isolation" and "superposition" is explained by the fact that all reality represents only the sum of the manifestations of particular laws of nature, and is to be conceived as having issued from these laws, the real epistemological meaning of the thought is concealed.[12] We cannot be concerned here with the origin of things, but only with the origin and character of our insight into things. The "real," as it is grasped in the sensuous impression, is not in and for itself a "sum" of various elements, but first appears to us as an absolutely simple and unanalysed whole. This original "simplicity" of intuition is only transformed into a manifold by the logically analytic work of the concept. The concept is here just as much a source of plurality as it elsewhere appears to be a source of unity. Since we conceive a particular process as successively introduced into different systems, whose general structure can be mathematically deduced, we thereby increase its determinateness, in so far as we define its position in the general plan of our thought with increasing exactness. The advance of experiment goes hand in hand with the advancing universality of the fundamental laws, by which we explain and construct empirical reality.

Laws and rules. The methodic difference has occasionally been indicated, which subsists between the mere "rules" of nature and the truly universal "laws" of nature. The inductions of Kepler on planetary motion express only generalized "rules" of process, while the fundamental law, on which they rest, was first formulated in Newton's theory of gravitation.[13] In Newton's theory we find the ellipse not merely as a real form of the orbit of Mars, but we survey at a glance the whole of "possible" orbits. The Newtonian concept of a centripetal force, that diminishes according to the square of the distance, leads to a perfect disjunction of the empirical cases in general. The transition between these cases is henceforth exactly predetermined; the magnitude of the initial velocity of a moving body decides (independently of the direction of this velocity) whether the form of its path is to be an ellipse or a hyperbola or a parabola. Thus the "law" of gravitation contains in itself the field of facts, which it rules, and ascribes strict division to its field; while the merely empirical rule of planetary motion allows the particular cases to stand in loose conjunction without sharp delimitation. In the

[12] See Volkmann, *op. cit.*, p. 89.
[13] See Volkmann, *op. cit.*, p. 59.

actual progress of science, however, the two views thus logically distinguishable are never strictly separated, but blend imperceptibly into each other. The "rule" already contains a tendency to raise itself to the form of law; while on the other hand, the conceptual perfection, which the law achieves, remains a mere provisional assumption, since it always contains a *hypothetical* element. Here we face the apparent circle, that everywhere confronts us in the relation of law and fact. If we conceive the motion of the planets to be determined by gravitational forces, which work in the inverse square of the distance, then it is evident that the form of a conic section is *necessary* for their path; however, it cannot be established that this determination "really" holds with regard to the kind and magnitude of attraction except through this methodic necessity just referred to, *i.e.*, by the power of this assumption to unify observations and to give them definite meaning. (*Cf.* above p. 146 ff.) If experience forced us to forego this assumption hereafter by adducing new material, then the pure concept as such could not prevent it; but in this case also, the *form* of the empirical concept would by no means be submerged along with its particular content. We require the same intellectual limit for the new field, that is revealed, and seek to establish this limit in a new law, to which the earlier is subordinated. The changed material conditions a changed manner of connection, while the general function of this connection, *viz.*, the deduction of the particulars from a supreme serial principle, remains the same.

This very function, and not its temporary and varying concrete expression in particular theorems, is assumed in the concept of experience itself; and thus belongs to the real "conditions of its possibility." When any series of observations $a_1 \, a_2 \, a_3 \ldots a_{11}$ is given, a double problem is forthwith offered for consideration. On the one hand, we can seek to enrich the material of this series by interpolation and extrapolation, by inserting between the given members hypothetical intermediate members, or by tracing the series beyond its original limits. On the other hand, our concern is to unite the manifold of members into an ultimate identity, by providing a rule, by which the transition from a_1 to a_2, from a_2 to a_3 *etc.* is determined and subjected to a fixed principle. If we may call the first procedure preëminently that of "induction" and the second that of "deduction," then it is evident that the two mutually coöperate and condition each other. The completion of the series by the introduction of new individual members always follows the direction of a unitary law of deduction, which hovers as a problem before thought. The selection and sifting of the material is under the direction of an active standard of judgment. We seek to trace a law of known conceptual structure through a given sequence of observations, and we measure the truth of this law by its success in characterizing the places left vacant by immediate perception, and in anticipating their filling-in by future investigation. In this sense, Kepler successively connected the facts regarding the positions of Mars, left by Tycho de Brahe, by the most diverse geometrical curves; and with these curves as known ideal norms, he compared the facts, until he finally reached the ellipse as that line, which enabled him to deduce the greatest number of observations from the relatively simplest geometrical principle of progression. Due to the nature of the problem, this work never reached an absolute limit; for no matter how many points of the path of the planet might be given, it is nevertheless always possible to connect them by any number of lines of different and more complicated form. Only the methodic requirement remains permanent, that in the last analysis the processes of nature be reduced to definite simple rules, no matter how necessary complex assumptions may prove for the representation of a limited concrete field of facts. This reduction to simple rules is comparable to the manner in which we gradually reduce arithmetical series of any higher order to the fundamental type of a series with a *constant* difference between its members.

The concept of the "fundamental" relation and the relation of mathematical necessity. This reduction of the manifold and ceaselessly changing material of perception to ultimate *constant relations* must be granted without limitation by even the most radical "empiricism." For the assumption of this fundamental relation is all that remains for empiricism of the concept of the "object," and thus of the concept of nature. "A body," Mach says, "looks differently in each illumination, gives a different optical image in each position, gives a different tactual image with each temperature, *etc.* However, all these sensuous elements so hang together, that with the same position, illumination and temperature, the same images recur. It is thus a permanency of the *connection* of sensuous elements, that is involved here. If we could measure all the sensuous elements, then we would say that the body subsists in the realization of certain equations, that hold between the sensuous elements. Even where we cannot measure, the expression may be retained symbolically. These equations or relations are thus the really permanent." The logical development of natural science tends more and more to a recognition that the original, naïve representations of matter are superfluous; at most we grant them the value of pictorial representations, and recognize the quantitative relations, that prevail between phenomena, as what is truly substantial in them. "To the extent that the *conditions* of a phenomenon are known, the impression of materiality is reduced. We recognize the relations between condition and conditioned, the equations which rule a larger or smaller field, as constituting what is really permanent and substantial, and as constituting that, the discovery of which makes possible a stable picture of the world."[14] So far modern empiricism is in complete agreement with the critical interpretation of the meaning and progressive tendency of natural science. The *Critique of Pure Reason* clearly taught that all that we know of matter are mere relations; but among these relations there are some that are independent and permanent, by which a definite object is given us.[15] Opposition sets in, however, when this concept of permanence, to which the concept of the object is reduced, is to be determined more closely in its logical meaning and logical origin. Is the permanence a *property* of the sensuous impressions, immediately inhering in them, or is it the result of intellectual work, by which we gradually transform the given according to definite logical requirements? In the light of our earlier developments, the answer cannot be doubtful. Permanence is never found ready made in the sensuous experience as such, for the sensuous experience gives only a conglomeration of the most diverse impressions, limited to a single moment of time and never recurring in exactly the same way. Permanence only appears to the extent that we are able to transform the sensuous manifold into a mathematical manifold, *i.e.*, in so far as we let it issue from certain fundamental elements according to rules held as unchangeable. The kind of certainty, that belongs to these rules, is clearly distinguished from the certainty of a particular sensation. Even from the standpoint of a mere "phenomenology" of the facts of consciousness, it is something entirely different whether the various contents of consciousness merely follow each other factually, or whether the

---

[14] Mach, *Die Prinzipien der Wärmelehre*, Leipzig 1896, p. 422 ff.
[15] *Kritik der reinen Vernunft*, 2nd ed., p. 341.

succeeding ones are known "from" their predecessors according to some pervading logical principle. Leibniz illustrated this difference incidentally by pointing to the example of the theory of numbers, which sharply marks the general relation involved here. If we take the series of squares as given, for example, then by making numerous tests, we can establish purely empirically the fact that the differences between the individual members can be represented by the progressive series of the odd numbers 1, 3, 5, 7 . . . On the basis of this fact, we may expect that when we proceed from the last given member of the series of squares, and add the corresponding odd number to this member, that a square will result; but nothing justifies us in taking this psychological expectation as the same as a logical necessity. No matter how many members have been tested and found to correspond to the rule, it always remains possible that at a certain point the previously constant type of progress may be broken off. No accumulation of observations regarding particular numbers, no matter how great, could ever enable us to reach a new form of certainty, that would give us more assurance in this regard. This form of certainty is to be gained when we proceed from the "universal" member of the series, *i.e.*, from its identical law of construction, rather than from the enumeration of its particular members. The formula $(n + 1)^2 - n^2 = 2n + 1$ shows at one stroke, and without the necessity of several tests, the constant and necessary relation, that subsists between the progression of the squares and that of the odd numbers. When once it is grasped, this formula holds of any arbitrary n, since in its deduction and proof no account is taken of the particularity of any definite number, and thus the particular value can vary arbitrarily without affecting the meaning of the proof itself. The totality of squares and that of the odd numbers are now taken up into one system, in which the one is known *through* the other, while hitherto, no matter how far we traced their reciprocal correspondence, the two were merely in conjunction.

This same fundamental difference can also be shown in the case of every true physical law. If we consider a law, such as Galileo's law of falling bodies or the law of Mariott, we discover that the correlated values of space and time, of pressure and volume, are not simply registered in conjunction, but are taken as conditioned by each other. So far as the mere facts are concerned, it is true that a list of numerical values might accomplish purely materially all that the mathematical functional rule could ever give us, provided that for each particular pressure-value the corresponding volume-value was noted in the list, and for each time-value the corresponding space-value. And yet such an accumulation of particular numerical data would lack precisely the characteristic element, in which rests the meaning of law. For the decisive element would be gone; the *type of determination* by which one magnitude is conceived to issue from another, would be left in darkness, even if the *result* were correctly stated. In the quantitative equation, this type of determination is clearly evident; for the quantitative equation shows by what purely algebraic operations from universal, established rules, the value of the dependent variable is to be gained from the value of the argument. And to this mathematical connection, physical theory adds a corresponding, objective and causal connection. Here the values of the function, along with those of the independent variables, belong to a common system of causes and effects, of conditions and conditioned; and they are thus connected with each other in such a way, that the assumption of one draws the others necessarily after it. Here also we do not place the particular values of the one series simply in conjunction with those of the other; but we seek, at least hypothetically, to grasp the two series in their law of construction and thus in the totality of their possible determinations, and to compare them with each other. Methodically guided induction tends to this goal; it represents what experience knows only as a complex coexistence of members, as being a resultant of simpler series of dependency, and these for their part progress according to the strict relation of the mathematical "ground" to the mathematical "consequence."

The two fundamental types of knowledge. This application of the concept of ground and consequence is obviously free from all metaphysical implications. Here again, in opposition to the theory of "description," it is necessary to maintain the purely *logical* character of the relation of ground and consequence, without making this character ontological. (*Cf.* above p. 137 f.) To "describe" a natural process in quantitative equations, in the sense of mathematical phenomenalism, means also to "explain" it in the only scientifically admissable sense; for the equation is the pattern of a purely conceptual insight. If we represent mathematically a given totality of observations by the "superposition" of several series, we do not indeed thereby increase our knowledge of the absolute and transcendent causes of the process; but we have raised ourselves to a new *type of knowledge*. It is true we do not grasp the force in things, by which a definite effect issues from a definite cause; but we comprehend the progress of each step of the theory to the next as strictly and as exactly as we grasp the transformation of any quantitative relation into another, that is logically equivalent to it. The concept of "description" is only justified and admissible when we take it in an active sense. To describe a group of phenomena, then, means not merely to record receptively the sensuous impressions received from it, but it means to transform them intellectually. From among the theoretically known and developed forms of mathematical connection (for instance, from among the forms of pure geometry), a selection and combination must be made such that the elements given here and now appear as constructively deduced elements in the system, which arises. The logical moment given here cannot be denied even in the theories of empiricism, or under whatever names it may be concealed. The "adjustment of ideas to reality" presupposes the very concept of this reality, and thus a system of intellectual postulates. It is in the principle of the unambiguity (*Eindeutigkeit*) of natural processes above all, that all these postulates are ultimately combined. "I am convinced," says Mach himself, "that in nature only so much happens as can happen, and that this can happen only in *one* way." All physical process is thus completely determined by the momentarily effective circumstances, and can thus take place in only one way.[16] However, if we analyse the grounds of this conviction, we are implicitly led back to all those fundamental conceptions, which the sensationalistic explanation explicitly denies. The conception of the unambiguity and "stability" of being obviously is not in the content of perception as given in our first immediate experience, but it indicates the goal, which the intellectual labor of science seeks to have this content approximate. This goal can only be reached, if we are able to establish certain permanent relations in the flux of sensations, which differ and have their truth limited to a single moment of time; the rules of these permanent relations we can call to mind independently of change of the momentary material. To the extent that this takes place, the scientific concept of *nature* develops and is confirmed. The biologi-

---

[16] Mach, *Prinzipien der Wärmelehre*, p. 392 f.; *cf. Analyse der Empfindungen* 2nd. ed., p. 222 ff.

cal theory of knowledge, on the other hand, seeks to preserve the constancy of being through the theory that all knowledge is a progressive adaptation to being; yet it is not able to ground the assertion of this constancy in an appropriate instrument of knowledge. A permanence of our environment is spoken of, out of which a corresponding permanence of thought is to develop. But it is overlooked that ultimately nothing else is meant by this permanence of the environment than the persistence of fixed functional relations, which are formulable between the elements of experience. Along with the content of these elements, there is recognized a form of their connection, which at any rate can in no way be reduced to the material oppositions of sensation, *i.e.*, to the bright and the dark, the sweet and the bitter, *etc*. But all conflict is herewith in principle removed. From the beginning, it has been admitted and emphasized that the conception of a constant law is indispensable in the definition of a natural object. It only remains to perceive that this conception is a perfectly *independent* element of knowledge, and resists all reduction to the assumed "simple" sense impressions. Progressive analysis leads to increasingly exact confirmation of this fundamental difference. The logical character of the pure concepts of relation is revealed to the extent that they arrange themselves in a fixed system and are exhibited in the whole wealth of their ramifications and mutual dependencies.

III

The problem of laws of nature. It has been shown that the two fundamental moments of induction: the gaining of particular "facts" and the connecting of these facts into laws, go back to the same motive of thought. In both cases, the problem is to raise from the flux of experience elements, that can be used as *constants* of theoretical construction. Even the establishment of a particular, temporally limited occurrence reveals this fundamental feature: *viz.*, that in the changing process, certain permanent connections can be grasped and retained. (p. 243 ff.) The scientific explanation of any involved group of phenomena by means of the "isolation" and "superposition" of simpler relations carries the task imposed here a step further. We discover in the ultimate empirical "laws of nature" what may be called *constants of a higher order*, such as rise above the mere existence of the individual fact, that is established in a definite magnitude. Nevertheless, the general procedure, which is here everywhere effective, only reaches apparent completion in this result. The "fundamental laws" of natural science seem at first to represent the final "form" of all empirical processes, but regarded from another point of view, they serve only as the material for further consideration. In the further process of knowledge, these "constants of the second level" are resolved into variables. They are only valid relatively to a certain sphere of experience, and must be ready, when this sphere is extended, to change their import. Thus we stand before a ceaseless progress, in which the fixed form of being and process that we believed we had gained, seems to escape us. All scientific thought is dominated by the demand for unchanging elements, while on the other hand, the empirically given constantly renders this demand fruitless. We grasp permanent being only to lose it again. From this standpoint, what we call science appears not as an approximation to any "abiding and permanent" reality, but only as a continually renewed illusion, as a phantasmagoria, in which each new picture displaces all the earlier ones, only itself to disappear and be annihilated by another.

Laws and constants. This very comparison, however, points to a necessary limit in radical scepticism. The images in the presentational life of the individual, to which the particular phases of science are here compared, always have a certain inner form of connection with each other, no matter how variegated and diverse in their succession; and without this inner connection, they could not be grasped as contents of the same consciousness. They all stand at least in an ordered temporal connection, in a definite relation of earlier and later; and this one feature suffices to give them a fundamental common character through all diversity of individual form. No matter how much the particular elements may differ from each other in their material content, they must nevertheless agree in those determinations, on which the *serial form* rests, in which they all participate. Even in the loosest, most slack succession of members, the preceding member is not absolutely destroyed by the entry of the succeeding member; but certain fundamental determinations persist, on which rest the homogeneity and uniformity of the series. In the successive phases of science, this postulate is most purely and perfectly realized. Each change in the system of scientific concepts places in a clear light the permanent structural elements to be ascribed to this system, as it is only under the assumption of these elements that it can be described. If we take as given the whole of experience, as it is represented in any definite stage of knowledge, this whole is never a mere aggregate of perceptual data, but is divided and unified according to definite theoretical points of view. It has already been shown on all sides that, without such points of view, no single assertion concerning facts, in particular no single concrete measurement, would be possible. (*Cf.* above p. 140 ff.) Thus if we view the totality of empirical knowledge at any point of time, we can represent it in the form of a function, which reproduces the characteristic relation according to which we conceive the individual members arranged in mutual dependency. Generally speaking, we have some such form as $F(A, B, C, D \ldots)$, in which it is to be remembered that what appears in this expression as an *element* may prove, otherwise considered, to be a very complex system, so that the member A would be replaced by $f(a_1, a_2, \ldots a_n)$, the member B by $\phi(b_1 b_2 \ldots b_n)$ *etc.* Thus there arises a complex whole of overlapping (*ineinandergreifender*) syntheses, which stand in a certain mutual relation of superordination and subordination. Two fields of phenomena A and B are first united each according to a particular law $\psi_1(\alpha_1, \alpha_2, \alpha_3)$, $\psi_2(\beta_1, \beta_2, \beta_3 \ldots)$; these laws are again connected among themselves by a new relation $\Phi(\psi_1, \psi_2)$, until we finally reach the most general relation, which ascribes to each individual factor its definite place with regard to the others. The fundamental form F is analysed for thought into a structure of mutually dependent determinations, which would be symbolically represented, for example, by an expression $F[\Phi_1(\Psi_1 \Psi_2), \Phi_2(\Psi_3 \Psi_4)\Phi_3 \ldots]$. If it is found that some wholly assured observation does not agree with the determinations, that are to be expected and calculated on the basis of this most general theoretical formula, then this formula needs correction. But this correction cannot involve the removal of any element indiscriminately from the formula, for the correction must proceed according to a definite principle of methodic advance. The transformation occurs, as it were, "from within outwards;" at first, while retaining the more inclusive relations F, Φ_1, Φ_2, *etc.*, the special relations Ψ_1, $\Psi_2 \ldots$ would be transformed; and in this way, we would attempt to establish again the unbroken agreement of theory and observation. The insertion of intermediate terms, and the institution of new experiments occurs as part of the intellectual tendency to verify and to "save" the more inclusive laws, by deducing the diverging result from the laws themselves through addition of a new determining factor.

The general form of experience. The preservation of a general "form" of experience is evident here; it is found even when the necessary revision goes beyond the "facts" and their purely empirical "rules" of connection to the principles and universal laws themselves. Also such principles as, for example, those on which Newton founds his mechanics, do not need to be taken as absolutely unchanging dogmas; they can rather be regarded as the temporarily simplest intellectual "hypotheses," by which we establish the unity of experience. We do not relinquish the content of these hypotheses, as long as any less sweeping variation, concerning a *deduced* element, can reëstablish the harmony between theory and experience. But if this way has been closed, criticism is directed back to the presuppositions themselves and to the demand for their reshaping. Here it is the "functional form" itself, that changes into another; but this transition never means that the fundamental form absolutely disappears, and another absolutely new form arises in its place. The new form must contain the *answer* to *questions*, proposed within the older form; this one feature establishes a logical connection between them, and points to a common forum of judgment, to which both are subjected. The transformation must leave a certain body of principles unaffected; for it is undertaken merely for the sake of preserving this body of principles, and these reveal its real goal. Since we never compare the system of hypotheses in itself with the naked facts in themselves, but always can only oppose one hypothetical system of principles to another more inclusive, more radical system, we need for this progressive comparison an ultimate constant standard of measurement of supreme principles of experience in general. Thought demands the identity of this logical standard of measurement amid all the change of what is measured. In this sense, the critical theory of experience would constitute the *universal invariant theory of experience*, and thus fulfill a requirement clearly urged by inductive procedure itself. The procedure of the "transcendental philosophy" can be directly compared at this point with that of geometry. Just as the geometrician selects for investigation those relations of a definite figure, which remain unchanged by certain transformations, so here the attempt is made to discover those universal elements of form, that persist through all change in the particular material content of experience. The "categories" of space and time, of magnitude and the functional dependency of magnitudes, *etc.*, are established as such elements of form, which cannot be lacking in any empirical judgment or system of judgments. Also the method followed here shows the same "rational" structure as was found in mathematics. Here as there, it was not necessary, in order to prove the independence of a conceptual relation from certain changes, that we should actually carry out these changes, but it was sufficient merely to grasp the *direction* of the change once for all, in making the decision.[17] We prove that the *meaning* of certain functions of experience is not affected in principle by a change in their material content. For example, the validity of a space-time dependency of the elements in natural processes expressed by universal causal law, remains unaffected by any change in the *particular* causal principles. The goal of critical analysis would be reached, if we succeeded in isolating in this way the ultimate common element of all possible forms of scientific experience; *i.e.*, if we succeeded in conceptually defining those moments, which persist in the advance from theory to theory because they are the conditions of any theory. At no given stage of knowledge can this goal be perfectly achieved; nevertheless it remains as a demand, and prescribes a fixed direction to the continuous unfolding and evolution of the systems of experience.

The concept of the a priori and the "invariants of experience." From this point of view, the strictly limited meaning of the *"a priori"* is clearly evident. Only those ultimate logical invariants can be called *a priori*, which lie at the basis of any determination of a connection according to natural law. A cognition is called *a priori* not in any sense as if it were *prior* to experience, but because and in so far as it is contained as a necessary premise in every valid judgment concerning facts. If we analyse such a judgment, we find, along with the immediate contents of sensuous data and elements differing from case to case, something permanent; we find, as it were, a system of "arguments," of which the assertion involved represents an appropriate functional value. In fact, this fundamental relation has never been seriously denied by even the most convinced "empiricism." When, for instance, the evolutionary

17 *Cf.* above p. 239 ff.

theory of experience lays weight on the fact that the sensation of time and the idea of time evolve "in adjustment to the temporal and spatial environment," this uncontested and incontestable proposition contains the very factors here in question, in the concept of "environment," which it presupposes. It is herein assumed, that there "is" a fixed, objective order of time, and that events do not succeed each other in arbitrary, capricious fashion, but proceed "out of each other" according to a definite rule. The truth of this assumption must remain fixed, if the conception of evolution is to have any justification, indeed any meaning. And it is to the truth of this judgmental connection, and not to the existence of any sort of presentations in us, that the concept of the *a priori* is alone applicable in its purely logical meaning. We are not concerned with the existence of psychic contents, but only with the validity of certain relations and with their superordination and subordination. According to the critical theory of knowledge, space and not color is "*a priori*," because only space forms an invariant in every physical construction. But the more sharply the opposition appears between truth and reality, the more clearly it is shown to contain an unsolved problem. Necessary as their separation is, a mediation between them on the other hand, must be assumed, if knowledge is to be brought into a unitary system. It must now be asked whether there is a path within knowledge itself, leading from the pure logical and mathematical systems to the problem of reality. And if such a path can be shown, what new meaning does the problem gain thereby, and in what direction is its solution to be pointed out to thought?

CHAPTER VI

THE CONCEPT OF REALITY

The separation of "subjective" and "objective" reality. The characteristic procedure of metaphysics does not consist in transcending the field of knowledge in general,—for beyond this field there would not be even material for a possible question,—but in separating correlative standpoints within the field of knowledge itself, and thus transforming what is logically correlative into an opposition of things. (*Cf.* above, p. 237 f.) At no point is this feature so significant as in the old question as to the relation of thought and being, of the subject and object of knowledge. This one opposition conceals all the others within it and can progressively develop into them. If once "things" and the "mind" become conceptually separated, they fall into two separate spatial spheres, into an inner and an outer world, between which there is no intelligible causal connection. And the conflict constantly grows sharper. If the objects only exist as a plurality, then for the subject the postulate of unity is essential; if the element of change and motion belongs to the essence of reality, then it is identity and unchangeableness, that is demanded of the true concept. No dialectical solution can ever fully transcend these separations, which are already present in the original formulation of problems; the history of metaphysics wavers between opposing tendencies, without being able to deduce the one from the other, or to reduce them to each other.

And yet at least the system of empirical knowledge has an original unity, which persists in spite of all these oppositions. The constant progress of science is not diverted from its goal by the varying fortunes of metaphysics. It must be possible to gain clarity regarding the direction of this advance, without presupposing the dualism of the metaphysical concepts. In so far as this dualism is applicable to experience, it must make itself intelligible entirely on the basis of experience and its characteristic principles. Thus the question is no longer what absolute separation underlies the opposition of the "inner" and the "outer," the "presentation" and the "object;" the question merely is from what standpoints and by what

271

necessity does *knowledge itself* reach these divisions. The separation and reconciliation of these concepts has been a problem throughout the whole history of philosophy; are these concepts merely intellectual phantoms, or do they have a fixed meaning and function in the structure of knowledge?

The development of the concepts of objectivity and subjectivity. If we consult immediate experience unmixed with reflection, the opposition of the "subjective" and the "objective" is shown to be wholly foreign to it. For such experience, there is only one plane of "existence," and this includes all contents uniformly and without distinction. What is grasped by consciousness here and now "is," and is precisely in the form offered by direct experience. Between the experiences, in particular, that refer to the individual's own body, and those that refer to "outer" things, there is still no fixed line of division. The temporal limits of the particular experiences are vague; the past, in so far as it is taken up into memory, is just as given and as "real" as the present. The various contents are arranged, as it were, in one plane; as yet there is no definite point of view, that would establish any preëminence of one over the other. If we characterize this level at all by the opposition of the subjective and the objective (and it can only be done in an extravagant and unreal sense), then we must ascribe to it thorough-going objectivity; for the contents still have that passivity, that indubitable givenness, which we customarily connect with the conception of the "thing." But the very first beginning of logical reflection destroys this impression of perfect unity and completeness. The division, whose growth starts here, is already present, though concealed, in the first attempts at a scientific view of the world. The fundamental tendency of this view is not simply to receive sensuous data, but to distinguish their *values.* The fleeting, unique observation is more and more forced to the back-ground; only the "typical" experiences are to be retained, such as recur in a permanent manner, and under conditions that can be universally formulated and established. When science undertakes to shape the given and to deduce it from definite principles, it must set aside the original relation of *coördination* of all the data of experience, and substitute a relation of superordination and subordination. Every critical doubt, however, that is directed against the universal validity of any perception, bears within it in germ the division of being into a "subjective" and an "objective" sphere. The analysis of the concept of experience has already led to that opposition, which is destined to resolve the metaphysical division of subject and object by taking up into itself its essential conceptual meaning.

Changing and constant elements of experience. The goal of all empirical knowledge lies in gaining ultimate invariants as the necessary and constitutive factors in each empirical judgment. From this standpoint, however, the various empirical assertions appear of very different value. Along with the loose, associative connection of perceptions united only under particular circumstances (as, for example, under definite physiological conditions), there are found fixed connections, which are valid for any whole field of objects, and belong to this field independently of the differences given in the particular place and definite time of the observation. We find connections, which hold their ground through all further experimental testing and through apparently contrary instances, and remain steadfast in the flux of experience while others dissolve and vanish. It is the former, that we call "objective" in a pregnant sense, while we designate the latter by the term "subjective." We finally call objective those elements of experience, which persist through all change in the here and now, and on which rests the unchangeable character of experience; while we ascribe to the sphere of subjectivity all that belongs to this change itself, and that only expresses a determination of the particular, unique here and now. The result of thus deriving the distinction between the subjective and the objective, is that it has merely *relative* significance. For there are no absolutely changeable elements of experience at any stage of knowledge we have reached, any more than there are absolutely constant elements. A content can only be known as changeable with reference to another, with which it is compared, and which at first claims permanent existence for itself. At the same time, the possibility always remains that this second content will be corrected by a third, and thus may no longer hold as a true and perfect expression of objectivity, but as a mere partial expression of being. Thus we are not concerned here with a fixed line of division, separating two eternally sundered fields of reality, but with a moving limit, which constantly shifts in the progress of knowledge. The present phase of experience appears just as "objective" in contrast to the past, as it appears "subjective" in contrast to the later phases. Only this mutual act of correction remains standing, only this function of comparison persists, while the material content of the two fields is in constant flux. The spatial expression of the division, *i.e.*, the analysis of being into an inner and an outer world, is thus insufficient and misleading, because it obscures this fundamental relation; it establishes instead of a living, reciprocal relation realized along with advancing knowledge, a fixed and absolutely closed division of things. The opposition involved here is not of spatial, but rather of dynamic nature; it signifies the differing power of empirical judgments to withstand continuous testing by theory and observation, without thereby being altered in content. In this perpetually renewed process, groups, that were originally taken as "fixed," are always being separated out; and now that they cannot satisfy the test, they lose this character of fixity, which constitutes the fundamental property of all objectivity. But it now becomes clear that our concern with this transition into the subjective is not with a change in the substance of things, but merely with a change in the critical evaluation of cognitions. "Things" are not thereby degraded into mere "presentations," but a judgment, that previously seemed to hold unconditionally, is now limited to a certain sphere of conditions.

The subjectivity of the sensuous qualities. This relation can be made plain, if we consider the best known example of this transformation of objectivity into subjectivity, *viz.*, the discovery of the "subjectivity of the sensuous qualities." Even for Democritus, who made this discovery, it means fundamentally nothing but the fact that the colors and tones, the smells and tastes have a peculiar character for knowledge, by which they are excluded from the scientific construction of reality. They pass from the γνησίη γνώμη into the σκοτίη γνώμη; they are separated from the pure mathematical ideas of space, of form and motion, to which alone henceforth physical "truth" is ascribed. However, this division does not mean that they are denied all part in being in general; rather, a narrower field is marked off for what previously passed as absolute witness of reality; nevertheless within this narrower field, the sense qualities retain their full validity. The seen color, the heard tone, is and remains something "real;" only this reality does not subsist in isolation and for itself, but results from the interaction of the physical stimulus and the appropriate organ of sensation. Thus when the qualities are explained as subjective, they fall outside the world of "pure forms," that mathematical physics constructs, but not outside of nature as such; for precisely this relation of physical and physiological conditions, on which the sense qualities rest, is itself a part of "nature," indeed, the concept of nature is first realized in the reciprocal causal dependency of particular elements. The same holds true when we go beyond the sphere of secondary qualities to the illusions and deceptions of the senses. When the straight stick appears in water as if broken, this is no unreal appearance but a phenomenon "well-grounded" in the laws of the refraction of light; and as such expresses with perfect adequacy a complex connection of elements of experience. Error only begins when we transfer a determination, which holds of a particular member, to the total complex; we thus apply a judgment, that has been found valid under certain limitations, to experience as a whole, apart from any limiting condition. That the stick is broken, is a valid empirical judgment, in so far as the phenomenon it refers to can be confirmed and deduced as necessary;—only we must add to this judgment, as it were, a logical index signifying the special conditions of its validity, from which it cannot be abstracted.

The series of degrees of objectivity. If we survey the whole of these considerations, the series of degrees of objectivity is clearly revealed. As long as we remain at the stage of the metaphysical distinction of the inner and outer, we have an opposition that permits absolutely no mediation. Here there is only a simple "either-or." Just as a thing cannot be simultaneously at two different places in space, so the "inner" cannot in any aspect be at the same time an outer, and conversely. In the critical formulation of the question, on the contrary, this limitation is set aside. The opposition is no longer between two members, but between many members; as we have seen, the same content of experience can be called subjective and objective, according as it is conceived relatively to different logical points of reference. Sensuous perception, as opposed to the hallucination and the dream, signifies the real type of the objective; while measured by the schema of exact physics, sense perception can become a phenomenon that no longer expresses an independent property of "things," but only a subjective condition of the observer. In truth, we are always concerned here with a relation, holding between a relatively narrower and a broader sphere of experience, between relatively dependent and independent judgments. Instead of a mere duality of determinations a series of values is given, which progresses according to a definite rule. Each member points to a successor, and requires this successor for its own completion. Even in the popular and prescientific view, the first, important phases of this development can be recognized. When we characterize a sensuous impression, that is given us here and now in a definite nuance, as "red" or "green," even this primitive act of judgment is directed from variables to constants, as is essential to all knowledge. Even here the content of the sensation is separated from its momentary experiencing (*Erlebnis*) and is opposed as independent; the content appears, over against the particular temporal act, as a permanent moment, that can be retained as an identical determination. But this intellectual permanence, which is latent even in the individual impression, lending it a real existence, is nevertheless far behind that in the thing-concept of ordinary experience. Here it is not sufficient simply to combine the sensuous perceptions, but along with this mere unification must go an act of *logical completion*. The object of experience is conceived as a continuous being, whose persistence in every point of the continuous sequence of moments of time is postulated as *necessary*. Direct perception, on the contrary, always offers us only isolated fragments, only entirely discrete values, which in no combination constitute a continuous whole. The truly "seen" and "heard" furnishes only disconnected, temporally separated masses of perception; while the concept of the "object" requires the perfect filling of the time series, and thus, strictly speaking, requires the assumption of an infinite totality of elements. Thus at this second level, the general procedure is clearly revealed for transforming and enriching the given, on the basis of the logical demand for its thorough-going connection. It is on the continuation of this procedure, that science bases its definition of nature and the natural object. The logical tendencies found in the concept of experience of the ordinary view of the world are now consciously taken up and carried further with methodic purpose. The "things," that arise henceforth, prove,—the more distinctly their real meaning is comprehended,—to be metaphorical expressions of permanent connections of phenomena according to law, and thus expressions of the constancy and continuity of experience itself. This fixity and continuity is never fully realized in any sensuously perceptible object; so in order to reach it, thought is led to a hypothetical substructure of empirical being, which however has no other function than to represent the permanent order of this being itself. (*Cf.* above p. 164 ff.) Thus there is an unbroken development from the first stages of objectification to its completed scientific form. The process would be completed as soon as we succeeded in advancing to the ultimate constants of experience in general, which, as we have seen, constitute both the presupposition and the goal of investigation. The system of these unchanging elements constitutes the type of objectivity in general,—in so far as this term is purely limited to a meaning wholly comprehensible to knowledge.

The logical gradations of the contents of experience. True, it remains an insoluble problem how the "thing-in-itself" passes into the mere "presentation," how absolute existence changes into absolute knowledge; but with this question we have nothing to do in the critically clarified conception of the opposition of the subjective and the objective. Here we do not measure presentations with respect to absolute objects; but different partial expressions of the same total experience serve as standards of measurement reciprocally for each other. Each partial experience is accordingly examined as to what it means for the total system; and this meaning determines its degree of objectivity. In the last analysis, we are not concerned with what a definite experience "is," but with what it "is worth;" *i.e.*, with what function it has as a particular building-stone in the structure of the whole. Dream experiences also are not distinguished from waking experiences by any specific character (*Dingcharakter*) attached as a permanently recognizable property. They also have a sort of "being," in so far as they are grounded in definite physiological conditions, in "objective" bodily states; but this being does not extend beyond the sphere and time-span within which these conditions are fulfilled. Insight into the subjective character of the dream means nothing else than the reëstablishment of a logical gradation among the contents of consciousness, which for a time threatened to disappear. Thus in general, the opposition of the subjective and objective serves in its development for the increasingly strict organization of experience. We seek to gain permanent contents in place of changing contents; but at the same time we are conscious, that every attempt in this direction only partially fulfills the fundamental demand, and hence requires completion in a new construction. We thus gain a sequence of superordinated and subordinated moments, representing, as it were, various complementary phases of the solution of the same problem. None of these phases,—not even those most remote from the goal,—can be entirely dispensed with; but, on the other hand, no one of them represents an absolutely unconditional solution. Thus it is true we can never compare the *experience* of things with the *things themselves*, as they are assumed to be in themselves separate from all the conditions of experience; but we can very well replace a relatively narrower aspect of experience by a broader, so that the given data are thereby ordered under a new, more general point of view. The earlier results are not thereby rendered valueless, but are rather confirmed within a definite sphere of validity. Each later member of the series is necessarily connected with the earlier ones, in so far as it answers a question latent in them. We face here a perpetually self-renewing process with only relative stopping-points; and it is these stopping-points, which define the concept of "objectivity" at any time.

The problem of transcendence. The direction of this progress of experience is also directly opposed to what would be expected on the ordinary metaphysical presuppositions. From the standpoint of these presuppositions, it is the subject, it is the presentations in us, which alone are given to us in the beginning, and from them we gain access to the world of objects only with difficulty. The history of philosophy shows, however, how all attempts of this sort fail. If we have once enclosed ourselves in the circle of "self-consciousness," no labor on the part of thought (which itself belongs wholly to this circle) can lead us out again. On the other hand, the criticism of knowledge reverses the problem; for it, the problem is not how we go from the "subjective" to the "objective," but how we go from the "objective" to the "subjective." It recognizes no other and no higher objectivity than that, which is given in experience itself and according to its conditions. Thus it does not ask whether the *whole* of experience is objectively true and valid,—for this would presuppose a standard that could never be given in knowledge,— but it only asks whether a special, particular content is a permanent or transitory part of this whole. We are not concerned with estimating the absolute value of the system in its totality, but with establishing a difference in value among its particular factors. The question as to the objectivity of experience in general rests ultimately on a logical illusion, of which the history of metaphysics offers many examples. Such a question is on the same plane in principle with the question, for instance, as to the absolute place of the world. Just as in this latter we falsely carry over to the universe as a whole a relation, which is only valid for the particular parts of the universe in their reciprocal relations, so in the former question a conceptual opposition, proper for distinguishing the particular phases of empirical knowledge, is applied to the conceived *totality* of these phases and their sequence. The full measure of "objectivity" belongs to each particular experience, as long as it is not displaced and corrected by another. To the degree that this continual testing and self-correction is carried on, there is an increasing exclusion of material from the final scientific conception of reality, although the material retains its right within a limited sphere. Elements, that at first seemed necessary and constitutive in the concept of empirical being,— such as the specific content of the particular sensations,—lose this dominant position, and henceforth have no central significance but only a peripheral one. Thus the designation of an element as "subjective" does not belong to it originally, but presupposes a complicated use of intellectual and empirical controls, which is only gained at a relatively high level. This designation first arises in the reciprocal criticism of experiences, where the changing existence is separated from the permanent. The "subjective" is not the self-evident, given starting-point out of which the world of objects is constructed by a speculative synthesis; but it is the result of an *analysis* and presupposes the permanence of experience and hence the validity of fixed relations between contents in general.

The meaning of judgment. The progress of this analysis is evident when we turn back to the relation of the "universal" and "particular," which appeared in the definition of the inductive judgment. We saw that each particular judgment originally claims unconditional validity for itself. As a judgment, it does not intend to describe only momentary sensations in their individual peculiarity, but to establish a matter of fact held to be valid in itself, independent of all particular temporal circumstances. The judgment as such, by virtue of its logical function, looks beyond the circle of what is given at any moment, while it affirms a universally valid connection between the subject and predicate. (See above p. 246 ff.) Only special motives are able to lead thought to deviate from this first demand, and to limit its assertions expressly to a narrower circle. This limitation only takes place in so far as there is a *conflict* between different empirical assertions. Assertions, that would be incompatible in content if taken absolutely, are now set in harmony with each other by being related to different subjects;—thus at least one of them confines itself to expressing the "nature" of things, not absolutely, but only from a special point of view and under certain limiting conditions. As the particular geometrical form, according to a well-known proposition of the Kantian theory of space, is only gained by limitation of the "one" all-inclusive space, so the particular judgment of experience issues from a limitation of the one system of experiential judgments in general, and presupposes this system. The particular judgment of experience arises when a plurality of spheres of experience, each one conceived to be entirely determined according to law, intersect and mutually determine each other. There is no road to the law from the absolutely isolated "impression," where every thought of logical relation is extinguished; on the other hand, it is entirely intelligible how, owing to the general demand for a thorough ordering of experiences according to law, we are led at first to exclude particular contents, such as cannot apparently be brought into the general plan, in order to deduce them subsequently from a particular complex of conditions.

The "transcending" of sensuous experience. It is thus a logical differentiation of the contents of experience and their arrangement in an ordered system of dependencies, that constitutes the real kernel of the concept of reality. This connection is confirmed anew when we consider the logical character of scientific investigation more closely, for this is indeed the genuine witness of empirical reality. The scientific experiment never makes a simple report regarding the present, momentary facts of perception, but it only gains its value by bringing the particular data under a definite standpoint of judgment, and thus giving them a meaning not found in the simple sensuous experience as such. What we observe, for instance, is a definite deflection of the magnetic needle under certain conditions; what we assert, on the contrary, as the result of the experiment is always an objective connection of physical propositions, which far transcend the limited field of facts accessible to us at a particular moment. As Duhem has admirably explained, the physicist, in order to reach a real result in his investigations, must always transform the actual case before his eyes into an expression of the ideal case, which theory assumes and requires. Therewith the particular instrument before him changes from a group of sensuous properties into a whole of ideal intellectual determinations. It is no longer a definite tool, a *thing* of copper or steel, of aluminum or glass, to which he refers in his assertions; but in its place are concepts, such as that of the magnetic field, the magnetic axis, the intensity of the current, *etc.*, which, for their part, are again only the symbol and husk of universal mathematico-physical relations and connections.[1] The characteristic merit of experiment rests on the fact that in it one stroke establishes a thousand connections. The limited circle of facts, that is sensuously accessible, expands before our intellectual vision into a universal connection of phenomena according to natural law. The immediate indications of the moment are transcended on every hand; in their place appears the conception of a universal order, of such a sort that it has equal validity in the smallest as in the greatest, and can be reconstructed again from any particular point. It is only by means of this enrichment of its immediate import, that the content of perception becomes the content of physics, and thus becomes an "objectively real" content.

Thus we have to do here with a type of "transcendence;" for the particular given impression does not remain merely what it is, but becomes a symbol of a thorough-going systematic organization, within which it stands and to a certain extent participates. This change in the empirical impression does not, however, alter its metaphysical "substance," but merely its logical form. What at first appeared in isolation, now combines and shows a reciprocal reference; what previously passed as simple, now reveals an inner wealth and multiplicity, in so far as we can reach from it other and still other data of experience, by a continuous progress according to definite rules. In connecting the particular contents with each other, —as it were, ever with new threads,—we give them that fixity, which is the distinguishing property of empirical objectivity. What imprints the mark of true objectivity is not the sensuous vividness of the impression, but this wealth of inner relations. What elevates the "things" of physics above the things of sense and lends them their peculiar kind of "reality," is the wealth of consequences proceeding out of them. Such things of physics only indicate different ways of advancing from one experience to another, so that the totality of being can be finally surveyed as the totality of the system of experience.

[1] Duhem, *La Théorie Physique*, p. 251 f., *cf.* above p. 190 ff.

The concept of "representation." The concept and term, *representation*, has, in spite of all the attacks against it, persistently maintained a central position in the history of the theory of knowledge; and this concept here receives a new meaning. In metaphysical doctrines, the "presentation" (*Vorstellung*) refers to the object, which stands behind it. Thus the "sign" here is of an entirely different nature than the signified, and belongs to another realm of being. Precisely in this lies the real riddle of knowledge. If the absolute object were already otherwise known to us, then at least it might be understood how we could indirectly read off its particular properties from the sort of presentations, that arise from it. Once we have assured ourselves of the existence of two different series, we might attempt by an analogical inference to carry over the relations we find in one series to the other; on the other hand, it is incomprehensible how we should succeed in inferring the existence of one series, from data belonging exclusively to the other. As soon as we have even a general certainty of transcendent things beyond all knowledge, we can seek, in the immediate content of experience for signs of this reality, which are given at least in concept; how this concept itself arises and what makes it necessary, is, on the other hand, not explained by the theory of signs. This difficulty continually reappears in the development of the concept of representation. In ancient atomism, the "images" of things, which inform us of their being, are conceived as material parts, which separate themselves from things and undergo many physical changes on the way to our sense-organs. It is, although in reduced measure, the real substance of the body, which enters us in sense-perception and blends with our own being. But this materialistic account cannot reach the logical goal for which it was undertaken; for here again the unity of experience is only apparently preserved. Even if things gave off a part of themselves, as it were, in order to become known, it would still remain as obscure as before how this part could possibly be taken, not as what it was in and for itself, but as an expression of an including whole. This reference to the whole would always require a peculiar function, that is not deduced but presupposed. Thus the Aristotelian and scholastic theory of perception seems to come closer to the real psychological fact, when, instead of explaining this function, it postulates it from the beginning. The whole content of the "immaterial species," by which we grasp the being of things, is accordingly reduced to the act of representation. We do not know any determinations of the species itself, yet only *through* it do we know the relations of outer things; *"cognoscimus non ipsam speciem impressam, sed per speciem."* The "similarity," which is to be assumed between the sign and the signified, is thus not to be conceived as if both belonged to the same logical category. The species agree in no single actual property with the object to which they refer, for they are characterized only by this operation of reference, and not by any property in which they could be similar to other things. The conception, that they are *"formales similitudines ac veluti picturae objectorum,"* is expressly contested and rejected,—at least in the most mature and consistent exposition of the theory by Suarez. "The assimilation of consciousness with the object does not mean for Suarez that elements are thereby brought into consciousness, and that these elements stand in objective relation to other functions of consciousness, and appear to these as the object; but rather his meaning is that the whole consciousness becomes an instrument of knowledge, and in so far an image (better, an expression, *species expressa*). Consciousness performs an act, assumes a peculiar property, which of itself directs consciousness on the real object. The living activity of consciousness, a perceiving, not a perceived, which knows the extra-conscious object by means of the species, is said to be similar to the object."[2] A weighty new distinction appears here, although concealed in scholastic terminology. The fact, that an element "refers" to another and indirectly represents it, is now no longer explained by a particular property of this element itself, but is reduced to a characteristic function of knowledge, and in particular to the judgment. It is true this insight could not be maintained with all strictness; for, on the contrary, the functional relation of *expression*, to which the analysis leads, always threatens to change into a substantial relation of the *participation* of things in certain objective properties. Thus the species again become the "traces" of things although they no longer possess complete existential import, but only a faded "essence." The conflict of these two conceptions finally deprives the concept of "representation" of its clear and definite meaning. In order to let the operation of expression be seen, the content, which serves as a sign, must be stripped more and more

[2] H. Schwarz, *Die Umwälzung der Wahrnemungshypothesen durch die mechanische Methode.* Leipzig 1895, I, p. 25; *cf.* p. 12 ff.

of its thing-like character; at the same time, however, the objectifying meaning ascribed to the content seems to lose its best support. Thus the theory of representation always threatens to lapse into skepticism; for what assurance have we that the symbol of being, which we believe we have in our presentation, genuinely reproduces the content of being, and does not misrepresent its essential features?

Transformation of the concept of representation and progress to the "whole of experience." The new meaning, which the criticism of knowledge gives to the concept of representation, removes this danger. It is now recognized that each particular phase of experience has a "representative" character, in so far as it refers to another and finally leads by progress according to rule to the totality of experience. But this reference beyond concerns only the transition from one particular serial member to the totality, to which it belongs, and to the universal rule governing this totality. The enlargement does not extend into a field that is absolutely beyond, but on the contrary, aims to grasp as a definite whole the same field, of which the particular experience is a part. It places the individual in the system. But if we ask further, as to whence the particular empirical content has this capacity of representing the whole, we are involved by the question in a reversal of the problem. The connection of the facts and their reciprocal relation is what is primary and original, while their isolation represents merely the result of a technical abstraction. Hence if we understand "representation" as the expression of an ideal rule, which connects the present, given particular with the whole, and combines the two in an intellectual synthesis, then we have in "representation" no mere subsequent determination, but a constitutive condition of all experience. Without this apparent representation, there would also be no presentation, no immediately present content; for this latter only exists for knowledge in so far as it is brought into a system of relations, that give it spatial and temporal as well as conceptual determinateness. Just as the necessity for positing a being outside of all relation to knowledge can not be deduced from the mere concept of knowledge, so must the concept of knowledge necessarily have within it that postulate of connection to which the critical analysis of the problem of reality leads. The content of experience becomes "objective" for us when we understand how each element is woven into the whole. If we attempt to characterize this whole itself as an illusion, it is a mere play of words; for the difference between reality and appearance presupposed here is itself only possible within the system of experience and under its conditions. (C above p. 273 ff.) The question as to the "similarity" of the empirical sign with what it signifies offers no further difficulty. The particular element, which serves as a sign, is indeed not materially similar to the totality that is signified,—for the relations constituting the totality cannot be fully expressed and "copied" by any particular formation,—but a thoroughgoing logical community subsists between them, in so far as both belong in principle to the same system of explanation. The actual similarity is changed into a conceptual correlation; the two levels of being become different but necessarily complementary points of view for considering the system of experience.

Association as a principle of explanation. Indeed the sensationalistic theory of knowledge might attempt to bring this fact into its own explanation, without contesting it, by reducing it to the psychological concept of "association." The concept of association seems to offer all the principles for the formulation and solution of the problem of reality, for it proceeds from the content of the particular impressions to the fixed connections between them. The defect of the explanation is shown, however, when we analyse more closely the *form of connection* assumed here and which, according to the concepts of associationistic psychology, appears alone admissible. The "connection" between the particular members of the series here means nothing but their frequent empirical coexistence. And this coexistence of particular presentations produces not so much a connection between them, as rather an appearance of one. No conceptual principle, such as can be expressed and established in strict logical identity, unifies the elements of association. The paths from one element to another are in and for themselves unlimited. Which of these paths is followed in real psychological thought depends merely on the preceding psychical "dispositions," and thus upon a circumstance, which is to be regarded as variable from moment to moment, and from individual to individual. Here that constancy and exactitude is lost, which is the distinguishing feature of the conception of reality. But it is only through the critical evaluation of concepts that the formation of the "object" comes to light. If we proceed from the particular content of experience at a particular moment of time, there are given in it not only certain elements but also certain *lines of direction*, according to which thought can gradually expand the particular phase into the whole system. The advance is not left to individual caprice, but is demanded according to law. As science grasps the whole of these demands more strictly and exactly, it progressively gains the concept of the real. It has already been shown on all hands that this development must everywhere transcend the sphere of mere association. Association, understood in the most favorable sense, is merely able to formulate the question; the answer lies only in the universal *serial principles*, which predetermine and order the possible logical transitions from member to member, in accordance with certain points of view. The specific meaning of these points of view must remain constant, if the advance is not to lose itself in the indefinite. The necessary guiding concepts of association cannot arise from association itself, but belong to another field and logical origin.[3]

In general it appears, that the further we advance into the particular conditions of the problem of reality, the more clearly it unites with the problem of truth. If it is once understood how knowledge attains a constancy of certain predicates and establishes judgmental connections, then the "transcendence" of the object as opposed to the mere presentation no longer offers any difficulty. And the means used by knowledge are shown to be the same in both fields of problems. Just as the real achievement of the concept is not in "copying" a given manifold abstractly and schematically, but in constituting a law of relation and thus producing a new and unique connection of the manifold, just so it is shown to be the form of connection of experiences that transforms changeable "impressions" into constant objects. In fact, the most general expression of "thought" is the same as the most general expression of "being." The opposition, that metaphysics could not reconcile, is resolved by going back to the logical function from the application of which both problems arose, and in which they must finally find their explanation.

II

The concept of objectivity and the problem of space. In the history of scientific and speculative thought, the problem of reality has always been inseparably connected with the *problem of space*. So close is

---

[3] *Cf.* above esp. p. 14 and p. 261 ff.

this connection, and so exclusively has it governed logical interest, that all questions connected with the concept of the real are considered solved, as soon as the question of the reality of the "external world" has been finally decided. Even the *Critique of Pure Reason* could only approach its real theme by starting with a transformation of the theory of space. Thereby, however, the fate of its historical effect was in large part decided; in the minds of contemporaries and successors, what was intended to be a criticism of the concept of experience was misapprehended as a metaphysics of the concept of space. In truth, here again the order of problems must be reversed. We cannot proceed from a fixed view of the "subjective" or "objective" character of space and from it determine the concept of empirical reality in general; but it is the supreme and universal principles of empirical knowledge, that must finally decide the question as to the "nature" of space.

The theory of projection and its defects. The empirical-physiological view, which was founded by Johannes Müller, conflicts with this, in so far as it starts with an axiom of unconcealed metaphysical character. It is presupposed, that what we perceive are not things themselves in their real shape and real mutual position and distance, but are immediately only certain determinations of our own body. The object of visual sensation is not the external objects, but parts of the retina, which we can grasp in their real spatial magnitude and extension. The problem of the physiology of vision is to describe the transition, by which we pass from this consciousness of the retinal images to knowledge of the spatial order of objects. It must be shown how we come to place the sensations given "in us" in the outer world, and how we come to comprehend them as a self-subsisting spatial world. However, if we take the problem in this form, it soon proves to be insoluble. All attempts to reduce the peculiar process of "projection," which is assumed, to "unconscious inferences," move in a circle; they always assume a general knowledge of the "external," yet this is what is to be deduced. In fact, there is no phase of experience, in which sensations are given as inner states and separated from all "objective" reference. Sensation in this sense is no empirical reality, but only the result of an abstraction, resting upon very complex logical conditions. We pass from the seen objects to the assumption of certain nerve excitations and the sensations corresponding to them; we do not proceed in the reverse direction from the sensations known in themselves to the objects, which may correspond to them.[4] Thus the general form of spatiality, the coexistence and externality of the particular elements is no mediated result, but is a fundamental relation posited with the elements themselves.[5] It cannot be asked how this form arises in and for itself; it can merely be enquired how the form is more closely determined and specialized in empirical knowledge. What needs explanation is not the fact that we go from the inner to the outer,—for the absolutely "inner" is a mere fiction,—but how we are led to regard certain contents of the original external world as finally "in us," *i.e.*, how we are led to determine them not only as in general spatial, but as in necessary correlation with our bodily organs, with certain parts of our retina or our brain. (*Cf.* above p. 274 ff.) It is not localization in general, but this particular localization, that is to be explained; and any such explanation must obviously take the general relation of spatiality as a basis.

Concept and perception distinguished (*Helmholtz*). To determine the concept of "reality" means also to find a *motive of differentiation*, enabling us to separate the originally homogeneous totality of experiences into groups of different value and import. For example, if we conceive the different perceptual images, which we receive from one and the same "object" according to our distance from it and according to changing illumination, as comprehended in a series of perceptual images, then from the standpoint of immediate psychological experience, no property can be indicated at first by which any of these varying images should have preëminence over any other. Only the *totality* of these data of perception constitutes what we call empirical knowledge of the object; and in this totality no single element is absolutely superfluous. No one of the successive perspective aspects can claim to be the only valid, absolute expression of the "object itself;" rather all the cognitive value of any particular perception belongs to it only in connection with other contents, with which it combines into an empirical whole. And yet this thoroughgoing connection does not mean the complete equivalence of the particular factors. We only gain perception of a definite

[4] *Cf.* more particularly for the psychological refutation of the "projection hypotheses," Stumpf, *Über den psychologischen Ursprung der Raumvorstellung*, p. 184 ff., as well as James, *Principles of Psychology*, II, 31 ff.

[5] *Cf.* later Ch. 8.

spatial form, when we break through this equivalence. As Helmholtz explains incidentally, when we ask what is to be understood by a body extended in three dimensions, we are led psychologically to nothing else than a series of particular visual images, which mutually pass into one another. Closer analysis, however, shows that the mere succession of these images, no matter how many we assume, would never of itself give us the presentation of a corporeal object, if the thought of a rule were not added, by which a certain order and position is ascribed to each in the total complex. In this sense, the presentation of the stereometric form plays "the *rôle* of a concept compounded from a great series of sense-perceptions, which, however, could not necessarily be construed in verbally expressible definitions, such as the geometrician uses, but only through the living presentation of the law, according to which the perspective images follow each other." This ordering by a concept means, however, that the various elements do not lie alongside of each other like the parts of an aggregate, but that we estimate each of them according to its *systematic* significance. We also distinguish the "typical" experiences, such as we assume to occur uniformly, from "accidental" impressions only found in individual circumstances. And it is the "typical" experiences, that we exclusively apply in the construction of the "objective" spatial world, while endeavoring to exclude all contents in conflict with them.

Helmholtz' account has illumined this process in detail. Here the general rule is first established "that we always perceive as present in the field of vision such objects as ought to be present to produce, under ordinary normal conditions of the use of our eyes, the same impression on the nervous apparatus." A stimulation, occurring under unusual conditions, is at first given the meaning that would belong to it if it were conceived to arise in the ordinary way. "To use an example, let us assume that the eye-ball is mechanically stimulated at the outer corner of the eye. We then think that we see a light on the nasal side of the field of vision. In the ordinary use of our eyes, where they are stimulated by a light coming from without, the external light must enter the eye from the nasal side, if a stimulation of the retina in the region of the outer corner of the eye is to occur. It is thus in accordance with the previously established rule, that we place a luminous object in such a case in this aforementioned position in the field of vision,—in spite of the fact, that the mechanical stimulus comes neither from in front in the field of vision nor from the nasal side of the eye, but on the contrary, from the external surface of the eye and more from behind."[6] The particular observations are thus adjusted to a certain set of conditions, that we regard as constant. In these conditions we have a fixed system of reference, to which each particular experience is tacitly referred. And it is only through this peculiar interpretation of the material of sensation, that the whole of objective visual and tactual space arises. This whole of space is never a mere dead copy of particular sense perceptions, but it is a construction, following from certain universal rules. To the extent that the unchanging elements of experience are separated from the changing elements in accordance with these rules, there results a division into an objective and a subjective sphere. And here there is no doubt, that the knowledge of subjectivity does not represent the original starting-point, but a logically mediated and later insight. Helmholtz expressly emphasizes that the knowledge of *objects* precedes the knowledge of *sensations*, and that it is far superior in clarity and distinctness. Under ordinary psychological conditions of experience, the sensation is so exclusively directed on the object and enters into it so completely, that it disappears behind it, as it were. The apprehension of sensation *as* sensation is always the work of subsequent conscious reflection. We must first always learn to direct our attention on our particular sensations, "and we ordinarily learn this only for the sensations, that serve as means of knowledge of the outer world." "While we attain an extraordinary degree of niceness and sureness in objective observation, we not only do not gain this for subjective observation, but we rather gain to a high degree the capacity to overlook these subjective sensations, and to proceed in the estimation of objects independently of them, even where they could easily enough be noticed by reason of their strength."[7]

The division into circles of objectivity. What is described here as a merely negative achievement, as an act of overlooking and forgetting, is really that highly positive function of the concept already seen on every side. It is the retention of the *identical* relations in the varying content of presentation, that characterizes every stage, even the earliest stages, of objectively valid knowledge. What is absolutely

[6] Helmholtz, *Handbuch der Physiolog. Optik*, §26, p. 428.
[7] Helmholtz *op. cit.*, p. 432.

changeable falls away from the momentary content, and only that remains, which can be fixed in permanent conceptions. By the central tendency of thought, a certain circle of experiences, which satisfy certain logical conditions of constancy, is raised out of the flux of experience and distinguished as the "fixed kernel" of being. In this first construction and characterization of the "real," the relatively fleeting contents, on the other hand, in which no thorough determinateness of experience is expressed, can be left unnoticed. Deeper reflection, however, teaches that these elements also do not fall absolutely outside the circle of experience, but that they also claim a place in it in so far as their variation is itself not arbitrary, but subjected to definite rules. Thus the variable itself, by a new interest of knowledge, is made the object of consideration. *This* knowledge of the "subjective" means an *objectification of a higher level,* in that it discovers an element of determinability in a material, that was at first left absolutely undetermined. The given is now divided into wider and narrower circles of objectivity, which are distinctly separated and arranged according to definite points of view. Each particular experience is henceforth determined not only by the material content of the impressions, but through the characteristic function by which some experiences serve as a fixed point of reference for measuring and interpreting others. In this way we produce definite, conceptually distinguished centers, around which the phenomena are ordered and divided. The particular phenomena do not flow by uniformly, but are limited and separated from each other; the initial surface-picture gains foreground and background, as it were. The real origin of the concept of object proves to be, not the "projection" of the inner and outer, but the division of phenomena into partial fields, distinguished by their systematic meaning. Each particular receives an index, which signifies its position in the whole, and in this index its objective value is expressed. For the naïve view, it is the "thing," that is given from the beginning, and that is expressed and partially copied in each of our perceptions. Thus the naïve view also presupposes a whole, with which we compare each particular experience and thereby measure its value. The demand made by the naïve view still holds from the standpoint of critical consideration. The defect of the naïve view is not that it makes this demand, but that it confuses the demand and its fulfillment;—it takes the problem, that knowledge has to work out, as already solved.

The whole, which we seek and on which the concept is directed, must not be conceived as an absolute being outside of all possible experience; it is nothing else than the ordered totality of these possible experiences themselves.

"Projection" and "selection." The modern psychology of space perception has replaced the projection theory by another view, which expresses the facts of knowledge more purely and sharply, since it represents them independently of all metaphysical assumptions. According to it, the presentation of "objective" space is not a work of "projection" but of "selection;" it rests on an intellectual selection made among our sense perceptions, especially in the field of visual and tactual impressions. In the homogeneous mass of these impressions, we only retain those contents, that correspond to "normal" physiological conditions; while we more and more suppress others, that arise under extraordinary conditions and lack the same repeatability as the first. Because our apperception separates a certain group of experiences in this way from the flux of the others, this group tends to be emphasized. These pass as reality absolutely, while all the other contents only have value in so far as they refer as "signs" to this reality. It is thus no absolute difference in being, but a difference of emphasis, that separates the objective from the subjective. The construction of spatial reality has in it a process of logical elimination, and without this its result could not be understood. The mass of spatial "perceptions" is gradually organized according to a definite plan, and in this organization gains fixed form and structure.[8] From the standpoint of logic, it is of especial interest to trace the function of the concept in this gradual process of construction. Helmholtz touches on this question when he affirms, that even the presentation of a connection of contents in temporal sequence according to law would not be possible without a conceptual rule. "We can obviously learn by experience what other sensations of vision, or some other sense, an object before us would give us, if we should move our eyes or our bodies and view the object from different sides, touch it, *etc.* The totality of all these possible sensations comprehended in a total presentation is our presentation of the body; this we call perception when it is supported by present sensations, and memory-image when it is not. In a certain sense,

---

[8] *Cf.* more particularly on the theory of "selection," James, *Principles of Psychology* II, 237 ff.

although contrary to ordinary usage, such a presentation of an individual object is already a *concept*, because it comprehends the whole possible aggregate of particular sensations, that this object can arouse in us when viewed from different sides, touched or otherwise investigated."[9] Here Helmholtz is led back to a view of the concept that is foreign to traditional logic and that at first appears paradoxical even to him. But in truth the concept appears here in no mere extravagant and derivative sense, but in its true and original meaning. It was the "serial concept," in distinction from the "generic concept," that was decisively revealed in the foundations of the exact sciences, and that, as is now seen, has further applications, proving itself to be an instrument of objective knowledge.

III

The function of judgment; permanence and repetition. The psychological analysis of the idea of space thus confirms the concept of objectivity derived from the logical analysis of knowledge. The mysterious transition between two different, essentially separated spheres of being now disappears; in its place appears the simple problem of the connection of the particular, partial experiences into an ordered totality. The particular content, in order to be called truly objective, must expand beyond its temporal narrowness and become the expression of the total experience. Henceforth the content stands not only for itself but for the laws of this experience, which it brings to representation. The particular moment is the starting-point of an intellectual construction, that determines and includes in its narrower and broader consequences the whole of experiential reality. The procedure of this logical "integration" is discoverable in outline in the simplest judgment about "facts." Everywhere, even where only a concrete, particular property is ascribed to an individual thing, a connection once established logically persists. This persistence is posited with the very form of judgment, and is found even where the content, on which the judgment is directed, is changeable. In the simplest schematic characterization of this relation, it is enough to point to the fact, that in any assertion *a is b* an element of permanence is included, since a dependency is therewith established, and this dependency is intended to be valid

[9] *Handbuch der physiolog. Optik*, 2nd ed., p. 947 f.

not only for a particular instant, but is regarded as identically transferable to the whole sequence of moments of time. The "property" b belongs to the "thing" a not only at a definite point of time t_0, in which it is apprehended by the act of perception, but it is maintained for the whole series $t_1 t_2 t_3 \ldots$ (*Cf.* above p. 243 ff., 268.) Thus at first one and the same determination is posited as absolutely capable of repetition and as fixed in judgment. To this original act, however, another may be added, in which the change of the particular elements is conceived as logically determined. As judgment has ascribed to the subject a, at the time t_1, the predicate b, so it can ascribe to it at the time t_1 the predicate b', at the time t_2 the predicate b'', in so far as this change of properties does not take place without rule, but is conditioned and demanded according to law by appropriate changes in another series. Only thus does the general schema for the concept of the empirical "object" result; for the scientific concept of a definite object comprises in its ideal completion not only the totality of its properties given here and now, but also the totality of necessary *consequences*, that could develop from it under definite circumstances. We connect a series of different, temporally separate, circumstances by means of a unitary complex of causal rules; and this connection gives the individual, in Platonic language, the seal of being. The content of the particular time-differential gains objective meaning, in so far as the total experience can be reconstructed from it according to a definite method.

The problem of the "transsubjective." In contrast to this continuous process, by which the originally fragmentary and disconnected experiences are gradually shaped into a system of empirical knowledge, the metaphysical view at some point necessarily reaches a chasm, which thought cannot bridge even though it can leap across it. This defect appears most clearly and sensibly exactly where the most strenuous effort is made to break through the limits of the mere "world of presentation," in order to reach the world of real "things." When "transcendental realism" undertakes to show the inferences, that lead from the field of the subjective (at first the only accessible field), to the realm of the "transsubjective," a barrier between thought and being is already raised by this way of stating the question, that no logical efforts can remove. It is taken as an assumption needing no further scrutiny, that all consciousness is first directed on the subjective states of our ego and that nothing except these states is immediately given.   There is a sphere of "immanence," which never goes beyond these first, original data; there is a type of self-consciousness, which limits itself merely to taking in passively the content of particular, present impressions, without adding any new element or logically judging the content according to a definite conceptual point of view.   Only this is asserted, that this first stage, while sufficient for the consciousness of the ego, in no way suffices to ground the consciousness of the object.   The object of natural science especially cannot be treated by these primitive means.   The objects of science, "mass" and "energy," "force" and "acceleration" are strictly and unmistakably distinguished from all contents of immediate perception.   He who grants science the right to speak of objects and of the causal relations of objects, has thereby already left the circle of immanent being and gone over into the realm of "transcendence."

The correlation of the consciousness of the ego and the consciousness of the object.   One can fully accept these conclusions so far;—but it is a strange error to believe that they affect not only psychological, presentational idealism but also critical idealism in its ground and root.   Critical idealism is distinguished from the "realism" here advocated, not by denying the intellectual postulate at the basis of these deductions of the concept of objective being, but conversely by the fact that it grasps this intellectual postulate more sharply and demands it for every phase of knowledge, even the most primitive.   Without logical principles, which go beyond the content of given impressions, there is as little a consciousness of the ego as there is a consciousness of the object.   From the standpoint of critical idealism, what is to be contested is not so much the concept of "transcendence" as rather the concept of "immanence," that is here presupposed.   The thought of the ego is in no way more original and logically immediate than the thought of the object, since both arise together and can only develop in constant reciprocal relation. No content can be known and experienced as "subjective," without being contrasted with another content, which appears as objective. (*Cf.* above p. 273.)   Thus the conditions and presuppositions of "objective" experience cannot be added as a supplement, after the subjective world of presentations has been completed, but they are already implied in its construction.   The "subjective" signifies only an abstract moment of a conceptual distinction, and as such has no independent existence, because its whole meaning is rooted in its logical correlate and opposite.

The separation of thought and experience. Although this relation seems very clear, when once the metaphysical difference of subject and object is changed into a methodological distinction, nevertheless it is worth while to pause over it. For the kernel of all the misconceptions among the various epistemological tendencies is found here. The deeper reason, that the outer and inner world are opposed as two heterogeneous realities, lies in an analogous opposition between experience and thought. The certainty of pure experience is taken to be entirely different from that of thought. And as both are different in their origin, they are accordingly related each to a separate sphere of objects, within which they possess unique and exclusive validity. It is pure experience, free from any admixture of the concept, which assures us of the states of our ego; while all knowledge of the outer object is really guaranteed by the necessity of thought. Inner perception, by which the ego comprehends itself, has a peculiar and, of its sort, unsurpassable "evidence" (*Evidenz*); but this "evidence" is purchased at the price of conceiving the content as absolutely individual, and as merely given in its unique properties here and now. When from our presentations, "all intellectual necessity, all logical order is removed, and they exist only in a blending of the similar and the like:" then and only then can they be taken as self-evident. Thus the beginning of every theory of knowledge must involve, that we divest ourselves of all connection with the realms of intellect and nature, of all commerce with the common values of civilization, so that we merely retain our individual consciousness "in its whole emptiness and nakedness." Only in this way do we reach a type of certainty, in which thought in no way participates,—only to find, that we cannot remain with such a certainty, that it must be broadened by logical assumptions and postulates, by means of which we posit an *object* of knowledge.[10] This apparently presuppositionless beginning, however, contains a premise, that is unjustified both from the standpoint of logic and from that of psychology. The division here between perception and thought negates the concept of consciousness no less than the objective concept of experience. All consciousness demands some

[10] *Cf.* Volkelt, *Die Quellen der menschlichen Gewissheit.* München 1906; esp. p. 15 ff.; *cf. Erfahrung und Denken,* Leipzig 1878, Ch. I.

sort of connection; and every form of connection presupposes a relation of the individual to an inclusive whole, presupposes the insertion
of the individual content into some systematic totality. However
primitive and undeveloped this system may be conceived to be, it
can never wholly disappear without destroying the individual content
itself. An absolutely lawless and unordered something of perceptions is a thought, that cannot even be realized as a methodological
fiction; for the mere possibility of consciousness includes at least the
conceptual anticipation of a possible order, even though the details
may not be made out. Therefore if we characterize every element,
that goes beyond the mere givenness of the individual sensation, as
"transsubjective," then the paradoxical proposition holds, that not
only the certainty of the object, but also the certainty of the subject
conceals a "transsubjective" element. For even the mere "judgment
of perception" gains its meaning only through reference to the *system*
of empirical judgments, and must therefore acknowledge the intellectual conditions of this system. (*Cf.* above p. 245 f.)

   The concept of the object in critical idealism. If we determine the
object not as an absolute substance beyond all knowledge, but as the
object shaped in progressing experience, we find that there is no
"epistemological gap" to be laboriously spanned by some authoritative decree of thought, by a "transsubjective command."[11] For
this object may be called "transcendent" from the standpoint of a
psychological individual; from the standpoint of logic and its supreme
principles, nevertheless it is to be characterized as purely "immanent." It remains strictly within the sphere, which these principles
determine and limit, especially the universal principles of mathematical and scientific knowledge. This simple thought alone constitutes
the kernel of critical "idealism." When Volkelt repeatedly urges in
his criticism, that the object is not given in mere sensation, but that
it is first gained on the basis of *intellectual necessity*,[12] he defends the
most characteristic thesis of this idealism. The ideality, which is
here alone asserted, has nothing in common with the subjective
"presentation;" it concerns merely the objective validity of certain
axioms and norms of scientific knowledge. The *truth* of the object—
this alone is affirmed—depends on the truth of these axioms, and has

---

[11] *Cf.* Volkelt, *Quellen der menschlichen Gewissheit*, p. 46 f., etc.; *Erfahrung
und Denken*, p. 186 ff.

[12] *Quellen der menschlichen Gewissheit*, p. 32 ff., etc.

no other and no firmer basis. It is true that there results, strictly speaking, no absolute, but only a relative being. But this relativity obviously does not mean physical dependency on particular thinking subjects, but logical dependency on the content of certain universal principles of all knowledge. The proposition, that being is a "product" of thought, thus contains no reference to any physical or metaphysical causal relation, but signifies merely a purely functional relation, a relation of superordination and subordination in the validity of certain judgments. If we analyse the definition of the "object," if we bring to clear consciousness what is assumed in this concept, we are led to certain logical necessities, which appear as the inevitable constitutive "factors" of this concept. Experience and its object are conceived as dependent variables, which are successively reduced to a sequence of logical "arguments;" and it is this pure dependence in content of functions on their arguments, which is characterized in the language of idealism as the dependence of the "object" on "thought." (*Cf.* above esp. p. 267 ff.)

The objectivity within pure mathematics. This sort of dependence is so unmistakable, that it is also attested and emphasized by the opposing side. It is finally admitted, that the actual necessity by which we go beyond the circle of particular disconnected sensations and rise to the conception of continuous objects connected by strict causal rules, is ultimately a logical necessity. "It is in the name of *reason*, that the certainty of actual necessity dominates me and compels me to make transsubjective assumptions. Everything that we call judgment, reflection, thought, understanding, reason, science would seem uprooted if we acted contrary to this certainty." By validity of being, nothing is to be understood except "the transsubjective significance, which we give to the contents of judgment by virtue of intellectual necessity."[13] Thus it is according to the universal rules of reason, that we construct the concept of being and determine it in detail. The justification, as well as the limits of every sort of "transcendence," is thereby exactly characterized. This limitation of transcendence is most clearly seen when we compare the object of experience with the object of pure mathematics. The object of pure mathematics also can in no way be reduced to a complex of sensations; it is also characteristic for it to transcend the given in an intellectual construction, that has no direct correlate in

---

[13] Volkelt, *Die Quellen der menschlichen Gewissheit*, p. 33 and 37.

any particular content of presentation. And yet the objects of mathematical knowledge, the numbers and the pure forms of geometry, do not constitute a field of self-subsisting, absolute existences, but are only the expression of certain universal and necessary ideal connections. If this insight is once gained, it can be carried over to the objects of physics, which, as we have seen, are nothing but the result of a logical work, in which we progressively transform experience according to the demands of the mathematical concept. The "transcendence," which we ascribe to the physical objects, in distinction from the flowing, changing content of perception, is of the same sort and rests on the same principles as the distinction between the mathematical idea of the triangle or of the circle and the particular intuitive image, by which it is represented here and now in a real presentation. In both cases, the momentary sensuous image is raised to a new logical significance and permanence; but no entirely foreign being is grasped by us by virtue of this division in either case, but a new character of conceptual necessity is imprinted on certain contents. The same conditions, that were fundamental in the transition from the empirical data of touch and vision to the pure forms of geometry, are necessary and sufficient for the transformation of the content of mere perception into the world of empirical-physical masses and movements. Here, as there, a constant standard of measurement is introduced, to which the changeable is henceforth referred; and on this fundamental function rests the positing of any sort of objectivity.

Thus "realism" is justified when it urges, that what makes a judgment a judgment, and what makes knowledge knowledge, is not something given, but something that is added to the given. "We could never mean anything, if we were limited merely to the given; for all attempts to remain purely within the given in the case of meaning, of judgment, become tautologies and lead to meaningless propositions. Judgment and knowledge go beyond the given in their meaning; what is meant in them is transcendent to the given and thus to them in so far as they are considered only as given, as present psychical contents. Each thought . . . . is thus transcendent to itself, in so far as it can never mean itself."[14] These propositions are entirely satisfactory; but only a slight change in their

[14] W. Freytag, *Der Realismus und das Transzendenzproblem*, Halle 1902, p. 123.

formulation is needed in order to draw an entirely different consequence from them than the one here drawn. If all thought is really
"transcendent to itself," if it belongs to its original function not to
remain with the present sensations but to go beyond them,—then
the converse inference holds. The "transcendence," which can
be grounded and proved by thought, is no other than that which is
posited and guaranteed in the fundamental function of judgment.[15]
The "object" is just as much and just as little transcendent as
judgment is. Thereby, however, the correlation of knowledge and
the object is admitted in the critical sense; for, much as judgment
transcends the mere content of present, sensuous perception, we can
not affirm that it stands beyond the logical principles of knowledge.
It was dependence on these principles, and not on any concrete
psychic contents or acts, that methodic idealism alone represented
and demanded. "Immanence" in the sense of psychologism must
indeed be overcome in order to reach the concept of the physical
object; but this object itself, while it transcends the sphere of sensation, gains its existence in conceptual relations, from which its essence
and its definition are inseparable. To the psychological immanence
of impressions is opposed, not the metaphysical transcendence of
things, but rather the logical universality of the supreme principles
of knowledge. The fact must be granted unconditionally, that the
particular "presentation" reaches beyond itself, and that all that is
given means something that is not directly found in itself;[16] but it has
already been shown that there is no element in this "representation,"
which leads beyond experience as a total system. Each particular
member of experience possesses a symbolic character, in so far as the
law of the whole, which includes the totality of members, is posited
and intended in it. The particular appears as a differential, that is
not fully determined and intelligible without reference to its integral.

---

[15] Cf. Freytag, op. cit., p. 126: "In this general conviction of the objective
nature of truth, however, the transcendence of judgment is involved as a necessary presupposition. For if judgment were not transcendent, had it no meaning going beyond what is given in it, did all its meaning consist in what it is
as a psychic process, then the truth of the judgment would be directly established; whether I judge a is b or a is not b, each judgment would be correct in
itself, because in the first precisely the a was meant, which is actually judged
to be b, and thus would also have been given; in the second the a is meant, that
is actually judged as not being b, and therefore would be also thus given."

[16] Cf. e.g. Uphues, Kant und seine Vorgänger, Berlin 1906, p. 336.

Metaphysical "realism" misunderstands this shift of logical meaning when it conceives it as a sort of transubstantiation. "Whatever is to mean anything," it is here argued, "must mean something else than it is; for what it is, that it is, and therefore does not need to mean."[17] But this "other" need not at all be something actually heterogeneous; rather we are here concerned with a relation between different empirical contents, which belong to a common order. This relation brings it about, that we can pass from a given starting-point through the whole of experience in a progress according to rule, but not that we can pass beyond experience. The constant reaching out beyond any given, particular content is itself a fundamental function of knowledge, which is satisfied in the field of the objects of knowledge. Among the philosophical physicists, it is Fechner, who has clearly grasped the problem involved here. "The fact, that the world of phenomena is such that one phenomenon can exist only in and through another, can easily lead one (and has led some) to deny real existence to phenomena in general; and can lead them to assume, as a basis for the varying multiplicity of phenomena, independently existing, fixed things in themselves behind the phenomena; these things in themselves can never appear as they *are* among phenomena, but they rather produce the whole dependent appearance of phenomena either by external interaction among themselves or by their inner working. For it is said: if one phenomenon is always only to appeal to another with reference to the ground of its existence, then a ground for all existence is ultimately lacking; if A says, 'I can only exist in so far as B exists,' and B in return, 'I can only exist in so far as A exists,' then both of them are founded on nothing . . . . But instead of seeking for the ground of the existence of A and B (which neither can find in the other) in something behind them, that gives a ground and kernel to their appearance, it is to be sought in the totality, of which they are both members; the whole is the basis and kernel of the whole and of all that is in it. . . . . In the whole, the basis of the individual is not to be sought in something particular, in something other behind the individual, concerning whose ground we would have to ask anew; but we can investigate by what rules the individual is united to the whole, and what are the ultimate elements. . . . . What we can find objective in a material thing never consists in an obscure thing behind and independent of

[17] E. v. Hartmann, *Das Grundproblem der Erkenntnistheorie*, p. 49.

perceptions or phenomena, but consists of a unified connection according to law of the same, of which connection each phenomenon realizes a part, but which connection goes beyond the particular perceptions or the particular phenomena, which the thing affords."[18] Although these propositions draw the line between metaphysics and physics very distinctly, Fechner still betrays an inner obscurity in the definition of the object of physics.

The unity of the physical world. In order to avoid the conception of matter as a completely unknown, indeterminate something, which "lies at the basis" of the sensuously perceivable properties, Fechner defines matter by these very sensible properties; the matter of the physicist is "in entire agreement with the most common usage;" it is nothing else than what is perceptible by touch. Matter is thus equivalent to the "tangible." What may lie behind the data of touch and feeling need not concern the physicist; to him, tangibility alone is able to be pointed out, to be grasped by experience and to be traced further; and this suffices to give the concept the fixed foundations necessary for his purposes.[19] Here the attempt to exclude the metaphysical elements of the concept of matter has again led to setting aside the *logical* element, that is characteristic of it. The critical view stands between the two views. It defines the object of natural science by relation to the "whole of experience;" but it is also aware that this whole can never be represented and grounded as a mere sum of particular sense data. The whole gains its form and system only by the assumption of original relations, of which no one can be pointed out as "tangible" like a given sensuous content;— and it is one of the manifold expressions of these relations, that is presented in the concept of matter, as also in that of force or energy. (*Cf.* above p. 169 f.)

IV

The historical transformation of the "thing." The reduction of the concept of thing to a supreme ordering concept of experience disposes of a dangerous barrier to the progress of knowledge. For the first naïve view of reality, it is true, the concept of thing offers no riddle or difficulty. Thought does not need to reach things gradually and by

[18] Fechner, *Über die physikalische und philosophische Atomenlehre.* 2nd ed., Leipzig 1864, p. 111 ff.
[19] *Op. cit.*, p. 106 f.

complicated inferences; but possesses them immediately, and can grasp them just as our bodily organs of touch grasp the corporeal object. But this naïve confidence is soon shattered. The *impression* of the object and the *object itself* are separated from each other; instead of identity, the relation of representation appears. No matter how complete our knowledge may be in itself, it never offers us the objects themselves, but only *signs* of them and their reciprocal relations. An increasing number of determinations, that were formerly taken as belonging to being itself, are now changed into mere expressions of being. Following the well-known path of the history of metaphysics, the thing is to be conceived as free not only from all the properties belonging to our immediate sensations (for the thing is in itself neither luminous nor odorous, neither colored nor sounding), but also as having all spatial-temporal properties, such as plurality and number, change and causality, stripped from it. All that is known, all that is knowable, is set over in opposition against the absolute being of the object. The same ground, that assures us of the existence of things, shows them to be incomprehensible. Skepticism and mysticism are united in this one point. No matter how many new relations of "phenomena" scientific experience may teach us, the real objects seem not so much revealed as rather the more deeply concealed by these relations.

All this doubt and suspicion disappear, however, when we reflect that what appears here as the uncomprehended residuum of knowledge really enters into all knowledge as an indispensable factor and necessary condition. To know a content means to make it an object by raising it out of the mere status of givenness and granting it a certain logical constancy and necessity. Thus we do not know "objects" as if they were already independently determined and given *as objects,*—but we know *objectively*, by producing certain limitations and by fixating certain permanent elements and connections within the uniform flow of experience. The concept of the object in this sense constitutes no ultimate limit of knowledge, but is rather the fundamental instrument, by which all that has become its permanent possession is expressed and established. The object marks the logical possession of knowledge, and not a dark beyond forever removed from knowledge. The "thing" is thus no longer something unknown, lying before us as a bare material, but is an expression of the form and manner of conceiving. What metaphysics ascribes as a *property* to things in themselves now proves to be a necessary element in the process of objectification. While in metaphysics the permanence and continuous existence of objects is spoken of as distinguishing them from the changeableness and discontinuity of sense perceptions, here identity and continuity appear as *postulates*, which serve as general lines of direction for the progressive unification of laws. They signify not so much the known properties of things, but rather the logical instrument, by which we know. In this connection the peculiar changeableness is explained, that is manifest in the content of scientific concepts of objects. According as the *function* of objectivity, which is unitary in its purpose and nature, is realized in different empirical material, there arise different concepts of physical reality; yet these latter only represent different stages in the fulfillment of the same fundamental demand. Merely this demand is unchanging, not the means by which it is satisfied.

Helmholtz' theory of signs. Thus natural science, even where it retains the concept of the absolute object, can find no other expression for its import than the purely formal relations at the basis of experience. This fact appears in significant form in Helmholtz' theory of signs, which is a typical formulation of the general theory of knowledge held by natural science. Our sensations and presentations are *signs*, not *copies* of objects. From a copy we demand some sort of similarity with the object copied, but we can never be sure of this in the case of our presentations. The sign, on the contrary, does not require any actual similarity in the elements, but only a functional correspondence of the two structures. What is retained in it is not the special character of the signified thing, but the objective relations, in which it stands to others like it. The manifold of sensations is correlated with the manifold of real objects in such a way, that each connection, which can be established in one group, indicates a connection in the other. Thus by our presentations we do not know, indeed, the real absolutely in its isolated, self-existent properties, but we rather know the rules under which this real stands and in accordance with which it changes. What we discover clearly and as a fact without any hypothetical addition is the law in the phenomenon; and this law is a condition of our comprehending the phenomenon, and is the only property, which we can carry over directly to things themselves.[20] But in this view also no entirely new content is

---

[20] Helmholtz, *Handbuch der physiolog. Optik*, 2nd ed., p. 586 ff., 947 f.

posited, but only a double expression is produced for one and the same fact. The lawfulness of the real means ultimately nothing more and nothing else than the reality of the laws; and these exist in their unchanging validity for *all* experience, in abstraction from all particular limiting conditions. When we declare connections, that at first could appear as mere regularities of sensation, to be laws of things, we have merely produced a new designation for the universal meaning ascribed to them. The known facts are not changed in their nature by our choosing this form of expression, but are merely strengthened in their objective truth. Thinghood is always only such a formula of confirmation and, when separated from the whole of the empirical connections by which it is accredited, has no significance. The objects of physics are thus, in their connection according to law, not so much "signs of something objective" as rather objective signs, that satisfy certain conceptual conditions and demands.

The logical and the ontological conceptions of relativity. It follows of itself that we never know things as they are for themselves, but only in their mutual relations: that we can only establish their relations of permanence and change. But this proposition involves none of the skeptical consequences that are connected with it in realistic metaphysics. If we proceed from the *existence* of absolute elements, then it must indeed appear as a defect of thought, that this existence can never be mastered in its pure and separate form. According to this view, things *exist* for themselves; but they are only *known* to us in their interactions, and their interactions influence and obscure the nature of each. Helmholtz formulates his view thus: "Each property or quality of a thing is in reality nothing but its capacity to produce certain effects on other things. . . . . We call such an effect a property when we hold the reagent, on which it works, in mind as self-evident without naming it. Thus we speak of the solubility of a substance, that is its reaction to water; we speak of its weight, that is its attraction for the earth; and with the same justification we call it blue, since we thereby presuppose as self-evident, that we are merely concerned with defining its effect on a normal eye, If, however, what we call a property always involves relation between two things, then such an action naturally never depends on the nature of one of the agents alone, but always exists only in relation to and in dependence upon the nature of a second being, on which the effect is exerted."[21] These propositions, with their striking formulation of the general principle of relativity, have been appealed to as a ground for demanding the exclusion in principle of all ontological elements from natural science.[22] Yet in truth, they too contain an unmistakable ontological element. A closer interpretation of the principle of the relativity of knowledge does not find this principle to be a mere consequence of the universal interaction of things, but recognizes in it a preliminary condition for the concept of thing itself. This is the most general and most radical meaning of relativity. Its meaning is not that we can only grasp the relations between elements of being by conceiving these elements still as obscure, self-existent kernels, but rather that we can only reach the category of thing through the category of relation. We do not grasp the relations of absolute things from their interaction, but we concentrate our knowledge of empirical connections into judgments, to which we ascribe objective validity. Therefore the "relative" properties do not signify in a negative sense that residuum of things, that we are able to grasp, but they are the first and positive ground of the concept of reality. There is a circle in attempting to explain the relativity of knowledge from the thorough-going interaction of things, for this interaction is rather only one of those *ideas of relation*, that knowledge posits in the sensuous manifold in order to unify it.

The unity of the scientific view of the world. It is of especial interest to follow the manner in which this view gains methodic clarity and sharpness within modern physics. A characteristic example of it is lately offered by a prominent physicist in his exposition of the progress and general goals of the physical method. In a short sketch in his treatment of the unity of the physical world, Planck has defined the general points of view, that account for the continual transformation of physical theories. If the first stage of our physical definitions is marked by the fact that we try to reproduce the content of sensation directly in the concept, all further logical progress consists in removing this dependence more and more. Sensation as such contains an anthropomorphic element, in so far as it neces-

---

[21] Helmholtz, *Die neueren Fortschritte in der Theorie des Sehens.* (*Vorträge und Reden*, 4th ed., Braunschweig 1896, p. 321), *cf. Physiologische Optik*[2], p. 589.

[22] *Cf.* Stallo, *The Concepts and Theories of Modern Physics*, German ed. Leipzig 1901, p. 131, 186 ff.

sarily involves a relation to a definite sense-organ, thus to a specific physiological structure of the human organism. How this anthropomorphic element is constantly forced into the background, so that it entirely disappears in the ideal plan of physics: of this, the history of natural science furnishes a single continuous example.[23] It must now be asked, what compensation is offered by the scientific scheme of the world for these lost contents? What positive advantage is there at the basis of its significance and necessity? It is soon made clear that the required compensation cannot be based on anything material, but purely on a formal moment. While science renounces the wealth and varied multiplicity of immediate sensation, it regains its apparent loss in content through increased unity and completeness. The inner heterogeneity of the impressions has disappeared along with their individual particularity, so that fields, that are absolutely incomparable from the standpoint of sensation, can now be conceived as members of the same total plan and in reciprocal connection. The peculiar value of the scientific construction lies in this alone: that in the scientific construction there appears as connected by continuous conceptual intermediates, what in the first naïve view appears in mere conjunction and unrelated. The more purely this tendency is carried through, the more perfectly the investigation fulfills its task. "If we look more closely, we see that the old system of physics was not like a single picture, but rather a collection of pictures; there was a special picture for each class of natural phenomena. And these various pictures did not hang together; one of them could be removed without affecting the others. That will not be possible in the future physical picture of the world. No single feature will be able to be left aside as unessential; each is rather an indispensable element of the whole and as such has a definite meaning for observable nature; and, conversely, each observable physical phenomenon will and must find its corresponding place in the picture." We see that the characteristics of the true physical theory, as developed here, entirely coincide with

[23] See Planck, *Die Einheit des physikalischen Weltbildes*, Leipzig 1909.— The exposition of the development of the conceptual constructions of natural science in the fourth chapter (*cf.* esp. p. 164 ff.) was already completed when Planck's lecture appeared. It is so much the more welcome to me because our own result is confirmed in all essential points in the philosophical part of Planck's treatment and is thereby illumined from another point of view.

the criteria of empirical reality, as shown by epistemological analysis. Planck requires as the condition of any physical theory: "unity in respect of all features of the picture, unity in respect of all places and times, unity with regard to all investigators, all nations, all civilizations." The realization of all these demands is what constitutes the real meaning of the concept of the object. In opposition to the phenomenalistic view, which remains with the data of mere sensation, Planck can justly characterize his view as "realistic." This "realism," however, is not the opposite but the correlate of a rightly understood logical idealism. For the independence of the physical object of all particularities of sensation shows clearly its connection with the universal logical principles; and it is only with reference to these principles of the unity and continuity of knowledge, that the content of the concept of the object is established.

CHAPTER VII

SUBJECTIVITY AND OBJECTIVITY OF THE RELATIONAL CONCEPTS

The problem of the subjectivity and objectivity of relational concepts. The analysis of knowledge ends in certain fundamental relations, on which rests the content of all experience. Thought cannot go further than to these universal relations; for only within them are thought and its object possible. And yet with this answer it may seem as if we have moved in a circle. The end of the enquiry leads us back to the same point at which we began. The problem seems shifted, but not solved; for the opposition of the subjective and the objective still persists as sharply as ever. The pure relations also fall under the same question, that was previously directed upon sensations and presentations. Are pure relations an element of *being*, or are they mere constructions of *thought?* Is the nature of things revealed in them, or are they only the universal forms of expression of our consciousness, and thus valid only for consciousness and the sphere of its contents? Or does there exist a mysterious pre-established harmony between mind and reality, by virtue of which both must necessarily agree ultimately in fundamental character?

The universal functions of rational and empirical knowledge. But to state the problem in this way shows that it has been overcome in principle as the result of the preceding enquiry. At any rate, "common" ground for resolving the opposition of thought and being exists. Yet this common ground is not to be sought in an absolute ground of all things in general, but merely in the universal functions of rational and empirical knowledge. These functions themselves form a fixed system of conditions; and only relative to this system do assertions concerning the object, as well as concerning the ego or subject, gain an intelligible meaning. There is no objectivity outside of the frame of number and magnitude, permanence and change, causality and interaction: all these determinations are only the ultimate invariants of experience itself, and thus of all reality, that can be established in it and by it. The same point of view directly includes consciousness itself. Without a *temporal* sequence and order of contents, without the possibility of combining them into certain *unities* and of separating them again into certain *pluralities*, finally without the possibility of distinguishing relatively constant conditions from relatively changing ones, the conception of the ego has no meaning or application. Analysis teaches us very definitely *that* all these relational forms enter into the concept of "being" as into that of "thought;" but it never shows us *how* they are combined, nor *whence* they have their origin. In fact, every question as to this origin, every reduction of the fundamental forms to an action of things or to an activity of mind would involve an obvious *petitio principii;* for the "whence" is itself only a certain form of logical relation. If causality is once understood as a *relation*, all question as to the causality of *relations in general* disappears. With regard to relations in general, we can only ask what they *are* in their logical meaning, and not as to the manner in which and the beginnings from which they have arisen. After these relations are established in their meaning, we can seek out the origin of particular objects and processes with their help and with the guidance of experience. It is hopeless, on the contrary, to seek to trace them to more ultimate beginnings, to psychical or physical "forces," as is done with a changing empirical existence.

The reciprocal relation of the "form" and "matter" of knowledge. Thus the possibility disappears of separating the "matter" of knowledge from its "form" by referring them each to a different origin in absolute being; as when, for example, we seek the origin of one factor in "things," and of the other in the unity of consciousness.[1] For all the determinateness, that we can ascribe to the "matter" of knowledge, belongs to it only relatively to some *possible order* and thus to a formal *serial concept*. The particular, qualitatively specific sensation gains its character only by distinction from other contents of consciousness; it exists only as a serial member and only as such can be truly conceived. Abstraction from this condition would result not merely in a greater or less "indefiniteness" of its content,

---

[1] *Cf.* Riehl, *Der philosophische Kriticismus* (esp. II, 1, p. 285 ff.) and also the treatment by Hönigswald, *Beiträge zur Erkenntnistheorie u. Methodenlehre,* Lpz. 1906.—On the following, *cf.* my criticism of this writing, *Kant Studien* XIV, p. 91–98; on the problem of constant magnitudes, *cf.* above p. 230 ff.

but would absolutely annihilate it.[2] This indissoluble logical correlation withstands every attempt to explain the relation here by two separate causal factors, which are assumed to exist and to act in themselves. Matter *is* only with reference to form, while form, on the other hand, *is valid* only in relation to matter. If we abstract from their correlation, neither matter nor form has any "existence," concerning whose ground and origin we could enquire. The material particularity of the empirical content can thus never be adduced as proof of the dependence of all objective knowledge on an absolutely "transcendent" ground; for this determinateness, which as such undeniably exists, is nothing but a characteristic of knowledge itself, by which the concept of knowledge is first completed. If we give it its most purely scientific expression, it means ultimately nothing but the proposition, that universal rules of connection and universal equations of natural phenomena do not suffice for the construction of experience and the constitution of its object; but the knowledge of particular *constants* is also needed, such as can only be gained by experimental observation. Still it is not evident how these constants attest to more than the *empirical reality* of the objects of experience, or how they reveal anything regarding its absolute foundations. For the particularization of a law presupposes this law and is only intelligible with reference to it; hence the particular, fixed value always remains in the sphere of that concept of being, that is defined and limited by the universal principles of mathematics. This limitation, however, constitutes the true "ideality" of a value. Such ideality cannot be established by correlation with the presentations and acts of thought of psychological individuals, but only by correlation with the universal principles and conditions of scientific truth. (See above p. 296 f.)

The existence of the "eternal truths." But while the question as to the metaphysical origin of these conditions proves to be a misconception, if the problem as to whether they are to be deduced from the mind or from things or from an interaction between the two comes to nothing, the old opposition of the "subjective" and the "objective" has not yet been removed in every sense. On the contrary, it seems to press anew for solution when we enquire as to what are the instruments of knowledge, what are the forms of judgment and

[2] *Cf.* now especially: G. F. Lipps, *Mythenbildung und Erkenntnis*, Lpz. 1907, p. 154 ff.

of relating thought, in which we can present to ourselves *temporally*, in actual empirical experience, the pure timeless validity of the ideal principles. We may be tempted to reject this question entirely also, for the sake of the strictness and purity of the logical foundation. Leibniz declares (in close conjunction with Plato), that the "eternal truths" are valid in complete independence of every fact of reality, whatever its properties may be. These eternal truths merely represent hypothetical systems of consequences; they connect the validity of certain conclusions with the validity of certain premises, without considering whether examples of these abstract connections can be found in the world of empirical things, indeed, without asking whether there are individuals in whose thought the transition from the premises to the consequences is ever actually made. The truths of pure arithmetic would remain what they are, even if there was nothing that could be counted nor anyone who knew how to count.[3] In the real classics of idealism, the negation of any merely psychological foundation is expressed with the greatest clarity. These classical writers all incline to the same thought, that has found paradoxical expression in Bolzano's conception of a realm of "propositions and truths in themselves." The "subsistence" of truths is logically independent of the fact of their being thought. For example, the meaning of the propositions of pure geometry, as they issue from each other in a strict, necessary sequence, thus forming an ideal whole, can be perfectly deduced and expressed without going back to the psychological acts, in which we bring the content of these propositions to intuitive or conceptual presentation. No matter whether these acts are different in different individuals or whether they are uniform and have constant properties,—in no case is it *these* properties, that we mean, when we speak of the objects of geometry, of lines, surfaces and angles. The being, that we ascribe to these objects is no sort of temporal reality, such as belongs to concrete physical or psychical contents, but is merely their reciprocal being determined; it affirms an objective dependency in the realm of what is thought of, but no actual causal connection in the realm of thinking.

The concept of truth of modern mathematics. It is especially the modern *extension of mathematics*, that has fully revealed this fact

[3] See Leibniz, *Juris et.aequi elementa* (Mollat, *Mitteilungen aus Leibnizens ungedruckten Schriften.* Lpz. 1893, p. 21 f; *cf.* my ed. of *Leibniz' Hauptschriften zur Grundlegung der Philosophie.* Lpz. 1904 ff., II, p. 504 f.)

and has thereby prepared the ground anew for the logical theory that rests upon it. These structures, with which the general theory of the manifold is concerned, are the true and complete objects of mathematics, and their conceptions first represents mathematics in its full extension. These structures can be fully developed, however, without going into the complex and difficult psychological problems as to in what intellectual processes we present to ourselves the meaning of the infinite totalities here under consideration. Further, since all the properties of these groups are fixed by their original concept and belong to them in a necessary, unalterable way, there is no room left for any *arbitrary* activity of thought; on the contrary, thought is resolved entirely into its object and is determined and guided by it. "For," as a modern representative of mathematical logic says, "phrase it as you will, there is a world that is peopled with ideas, ensembles, propositions, relations, and implications, in endless variety and multiplicity, in structure ranging from the very simple to the endlessly intricate and complicate. That world is not the product but the object, not the creature but the quarry of thought, the entities composing it—propositions, for example,—being no more identical with thinking them than wine is identical with the drinking of it. The constitution of that extra-personal world, its intimate ontological make-up, is logic in its essential character and substance as an independent and extra-personal form of being. . . . . Just as the astronomer, the physicist, the geologist, or other student of objective science looks abroad in the world of sense, so, not metaphorically speaking but literally, the mind of the mathematician goes forth into the universe of logic in quest of the things that are there; exploring the heights and depths for facts—ideas, classes, relationships, implications."[4] These propositions define the problem very sharply, on the positive as well as on the negative side. The *necessity* of universal mathematical connections must remain undisputed; and this necessity constitutes a peculiar entity, an objective content, which is set over against the psychological activity of thought as an absolutely binding norm. But does this content of truth stand on the same plane as sensuous reality, of which we can have merely empirical knowledge? Are the "facts" of the mathematician nothing else, and have they no other meaning than those, for

[4] C. J. Keyser, *Mathematics—a lecture delivered at Columbia University*, New York, 1907, p. 25 f.

example, which the comparative anatomist and zoologist establish in the description and comparison of various bodily structures? The very logic of mathematics and of mathematical physics definitely forbids any such identification of the *exact* and the *descriptive* methods. Necessities cannot be simply described, cannot as such be absolutely "found;" for what is merely found holds only for the moment for which it is established, and thus signifies an empirically unique fact. The question as to the intellectual operations, in which we comprehend these necessities, thus presses anew for solution. These operations can indeed never blend indistinguishably into what is known by them; the laws of the known are not the same as those of knowing. Nevertheless, the two types of law are related to each other, in so far as they represent two different aspects of a general problem. Thus there exists a deeper and more intimate mutual relation between the *object* and the *operation* of thought than between —the wine and the drinking of the wine. The wine and the drinking are not exactly correlated;—but every pure act of cognition is directed on an objective truth, which it opposes to itself, while on the other hand, the truth can only be brought to consciousness by these acts of cognition and through their mediation.

II

The relational concepts and the activity of the ego. Hence it is necessary, on the basis of the concept of "objectivity" gained from the analysis of scientific principles, to define the concept of "subjectivity" in a new sense. The general characterization of the object, resulting from such analysis, has implicit in it the answer to the question as to what are the intellectual means and methods by which we reach this knowledge. One aspect above all stands out decisively. As long as the object was simply the "thing" in the customary sense of naïve dogmatism, the particular "impression" or mere sum of such impressions might grasp it and copy it. But this sort of appropriation is denied, when the validity of certain logical relations is established as the necessary condition and real kernel of the concept of the object. For pure relations can never be represented in mere sensuous impressions; the similarity or dissimilarity, the identity or difference of the seen or touched, is never itself something that is seen or touched.[5] Everywhere we must go

[5] *Cf.* more particularly Ch. VIII.

back from the passive sensation to the activity of judgment, in
which alone the concept of logical connection and of logical truth is
adequately expressed. We may conceive the idea of the thing—
as a complex of sensuous properties—to have arisen through these
properties having been perceived for themselves, and having combined of themselves by virtue of an automatic mechanism of "association." But necessary connection, in order to be defined psychologically, requires reference to an independent activity of consciousness.
The progress of the judgment according to law is the correlate of the
unification of relations according to law in the object of knowledge.

Constancy and change in knowledge. The content of truth, and
with it the content of "being," seems once more to be in flux; for
according to this whole view, we can only realize what a certain
truth "is" by intellectually regenerating it, *i.e.*, by allowing it to
develop before us out of its particular conditions. But this "genetic"
view of knowledge is not opposed to the demand for permanence.
For the activity of thought, to which we here recur, is not arbitrary,
but a strictly regulated and constrained activity. The functional
activity of thought finds its support in the ideal *structure* of what is
thought, a structure that belongs to it once for all independently of
any particular, temporally limited act of thought. The two elements of structure and function in their interpenetration determine
the complete concept of knowledge. All of our intellectual operations are directed on the idea of a "fixed and permanent" realm of
objectively necessary relations. It is evident, therefore, that all
knowledge contains a *static* and a *dynamic* motive, and only in their
unification is its concept realized. Knowledge realizes itself only
in a succession of logical acts, in a series that must be run through
successively, so that we may become aware of the rule of its progress.
But if this series is to be grasped as a unity, as an expression of an
identical reality, which is defined the more exactly the further we
advance, then we must conceive the series as converging toward an
ideal limit. This limit "is" and exists in definite determinateness,
although for us it is not attainable save by means of the particular
members of the series and their change according to law. A different
view results, according as we choose our standpoint at the conceived
limit or within the series and its progress; yet each of the two aspects
requires the other for its completion. The change is directed toward
a *constancy*, while the constancy, on the other hand, can only come
to consciousness in the *change*. There is no act of knowledge, which

is not directed on some fixed content of relations as its real object; while, on the other hand, this content can only be verified and comprehended in acts of knowledge.

The independence of logical truths of the thinking subject. At this point, the general tendencies dominating present epistemological discussion divide most distinctly. On the one hand, attempt is made to maintain the pure objectivity of the logical and the mathematical, by giving up all reference in principle to thought and to the "thinking mind." If we analyse the ideal structure of mathematics and if we clearly represent the whole of its definitions, axioms and theorems, it is urged, we do not find the concept of the thinking subject to which this whole system is given, among the "logical constants" that ultimately remain. Therefore, this concept of the subject does not belong to the field of pure logic and mathematics; it is rather to be accounted one of those "entirely meaningless" conceptions, that have only gotten into science by the help of philosophy.[6] Thus all closer relation is lacking between the ideal truths of mathematics and logic and the activity of thought; it is rather emphasized that the mind, in grasping these truths, receives them only receptively as given material. The mind is as completely passive in cognizing a definite system of inferences as sense is—according to the ordinary view—in the perception of sensuous objects.[7] "In short, all knowledge must be recognition, on pain of being mere delusion; arithmetic must be discovered in just the same sense in which Columbus discovered the West Indies, and we no more create numbers than he created the Indians. The number *2* is not purely mental, but is an entity which may be thought *of*. Whatever can be thought of has being, and its being is a precondition, not a result, of its being thought of."[8] The "objectivity" of pure concepts and truths is accordingly put on a plane with that of physical things. Nevertheless, the difference between them is sharply revealed again, when we recall that in the sphere of logic and mathematics we cannot reach absolute "objects," but always only relative objects. Not *number*, but rather the numbers constitute a true "entity." Here

---

[6] *Cf.* Russell, *The Principles of Mathematics*, I, p. 4: Philosophy asks of Mathematics: What does it mean? Mathematics in the past was unable to answer, and Philosophy answered by introducing the totally irrelevant notion of mind. But now Mathematics is able to answer, so far at least as to reduce the whole of its propositions to certain fundamental notions of logic.

[7] Russell, *op. cit.*, §37, p. 33.

[8] Russell, *op. cit.*, §427, p. 451.

the individual gains its meaning and content only from the whole;—but this whole cannot be presented like a quiescent object of perception, but, in order to be truly surveyed, must be grasped and determined in the law of its construction. In order to comprehend the number series *as a series* and thus to penetrate into its systematic nature, there is needed not merely a *single* apperceptive act (such as is considered sufficient for the perception of a particular thing of sense), but always a manifold of such acts, which reciprocally condition each other. Thus a *movement of thought* is always demanded, yet a movement which is no mere change of presentation, but in which what is first gained is retained and made the starting-point of new developments. Thus from the activity itself flows the recognition of a fixed body of truths. Within the act of production, there arises for thought a permanent logical *product,* in so far as thought is aware that this act does not proceed arbitrarily but according to constant rules, which it cannot avoid if it is to gain certainty and definiteness.

The problem of pragmatism. The "spontaneity" of thought is thus not the opposite but the necessary correlate of "objectivity," which can only be reached by means of it. If this fundamental relation between spontaneity and objectivity is not fully grasped, if a single element of it is stressed one-sidedly, a reaction will ensue such as to endanger the constancy of logic itself. From this general motive, we can perhaps most easily understand the attack, that has been made by "pragmatism" on "pure logic." In so far indeed as pragmatism consists of nothing but the identification of the concepts "truth" and "utility," we can confidently abandon it to the general fate of philosophical catch-words. What has been advanced in defence of this view is clad almost entirely in a rhetorical and polemical style, and melts away when we attempt to translate it into the sober language of logic. The very concept of utility defies all attempt at exact definition: now it is the particular individual with his special wishes and inclinations, now it is some common, generic structure of man, with reference to which utility is defined and measured. The first case leaves unsolved precisely the decisive problem, the possibility of exact scientific knowledge; for a science of nature is no more constructed from individual feelings and tendencies than from individual sensations, and is rather directed on the exclusion of all purely "anthropomorphic" elements from its system of the world. (*Cf.* above p. 306 ff.) In the second case, a constant psycho-physical subject is assumed having a permanent organization, which develops under conditions themselves possessing objective laws; thus the whole concept of being, which should be deduced, is in fact presupposed. There is "utility" only in a world, in which everything does not issue arbitrarily from everything, but where certain consequences are connected with certain presuppositions. The standpoint of utility is only intelligible and applicable inside of *being* and a definite order of process. (*Cf.* above p. 264 f.)

Truth and the "practical." However, such considerations do not affect the finer and more subtle interpretation that pragmatism has received especially from Dewey and his school. Here the problem is freed at least from those obscurities and ambiguities that attached to it in popular philosophical discussion. Here it clearly appears, that pragmatism is concerned with the relation to be assumed between the objectively valid propositions of science and the activity of thought. For, as closer examination shows, thought has here become the pure and complete expression of "doing." Our inferences and conclusions, our investigations and experiments are "practical" not because they are necessarily directed on gaining an *outer* purpose, but merely in the sense that it is the *unity of all thought*, which stands constantly before us as an ultimate goal and directs our cognition. The truth of any particular proposition can only be measured by what it contributes to the solution of this fundamental problem of knowledge, by what it contributes to the progressive unification of the manifold. We can never set a judgment directly over against the particular outer objects and compare it with these, as things given in themselves; we can only ask as to the function it fulfils in the structure and interpretation of the totality of experience. We call a proposition "true," not because it agrees with a fixed reality beyond all thought and all possibility of thought, but because it is verified in the process of thought and leads to new and fruitful consequences. Its real justification is the effect, which it produces in the tendency toward progressive unification. Each hypothesis of knowledge has its justification merely with reference to this fundamental task; it is valid to the degree that it succeeds in intellectually organizing and harmoniously shaping the originally isolated sensuous data.[9]

---

[9] *Cf. Studies in Logical Theory*, edited by Dewey (The Decennial Publications of the University of Chicago, First Series, Vol. III, Chicago 1903).

The critical concept of truth. The *critical* view of knowledge and of its relation to the object is not affected, however, by these developments; for in them a thought is merely spun out, which the critical view has recognized and taken as a basis from the beginning. For the critical view, too, concepts do not gain their truth by being copies of realities presented in themselves, but by expressing ideal orders by which the connection of experiences is established and guaranteed. The "realities," which physics affirms, have no meaning beyond that of being ordering concepts. They are not grounded by pointing out a particular sensuous being, that "corresponds" to them, but by being recognized as the instruments of strict connection and thus of thorough-going relative determinateness of the "given." (*Cf.* above p. 164 ff.) The recognition of this fact, however, involves none of the consequences, that pragmatism is accustomed to attach to it. No matter how far we may grant and urge the "instrumental" meaning of scientific hypotheses, we are obviously concerned here with a purely *theoretical* goal, that is to be sought with purely theoretical means. The will, that is here to be satisfied, is nothing but the will to logic. The direction of the progress of knowledge is not determined by any sort of individual needs, such as vary from one subject to another, but by the universal intellectual postulate of unity and continuity. In fact, this consequence has occasionally come to light very clearly,—in spite of all the ambiguity in the concept of the "practical." James himself emphasizes the fact that our knowledge is subject to a double compulsion; just as we are bound to the properties of our sensuous impressions in our knowledge of facts, so our thought is determined by an "ideal compulsion" in the field of pure logic and mathematics. Thus, for example, the hundredth decimal of the number π is ideally predetermined, even if no one has actually calculated it. "Our ideas must agree with realities, be such realities concrete or abstract, be they facts or be they principles, under penalty of endless inconsistency and frustration."[10] It is clear that the assumption of such "coercions of the ideal order" does not differ from the assumption of an objective, logical criterion of truth; both are only different expressions of the same fact. What is accomplished is no refutation of "pure logic," but at any rate a further development of the thought on which it rests. It is not a new solution that is offered, but a new problem,

[10] James, *Pragmatism*, New York 1907, p. 209 ff.

such as could be neglected at first in the general attempt to ground knowledge. The universal truths of logic and mathematics are not only deprived of any empiristic grounding, but they also seem to lack any *relation* to the world of empirical objects. Their apriority is based on their "freedom from existence," and only holds to the extent that this condition is satisfied. The moment thought turns to the empirical *existence* of the objects, it seems to separate itself from the real foundation of its certainty. True insight into the necessity of a connection can only be reached, where we renounce all attempts to assert anything regarding the reality of the elements, that enter into the relation.[11] Nevertheless we cannot hold to this unconditional separation,—however unavoidable it may appear at first from a methodological standpoint; for the mere possibility of a mathematical *science of nature* contradicts it. For in mathematical science of nature the two types of knowledge, that are here opposed, are once more immediately related to each other. It is empirical being itself, that we seek to grasp in the form of rational mathematical order. That this demand is never ultimately and definitely satisfied, results from the nature of the task itself. For the material, that is here subjected to mathematical treatment, is never found finished as a complete store of "facts," but is only shaped in the process, and in the process constantly assumes new forms. It is no constant, but a variable datum and is to be comprehended in its variability, in the possible transformation that it can undergo through new observations and investigations. But this variability is of the essence of the empirical and involves no element of "subjective" arbitrariness. The change is itself determined and necessary, since we do not go arbitrarily from one stage to another but according to a definite law. It is customary to appeal especially to the purely relative validity of our astronomical system of the world, in order to prove the relativity of the concept of empirical truth. It is argued, that since the absolute motions of the heavenly bodies are not given to us in experience and can never be given to us, and since therefore we can never compare our astronomical constructions with these motions themselves and thus test them, there is no meaning in granting to any one system (for instance, the Copernican) the merit of "truth" above all others. All systems are equally true and equally real,

---

[11] See above p. 319 f.; *cf.* Meinong, *Über die Stellung der Gegenstandstheorie im System der Wissenschaften*, Lpz. 1907, section 5 ff.

since they are all equally remote from the absolute reality of things
and signify only subjective conceptions of phenomena, which can
and must differ according to the choice of the intellectual and spatial
standpoint. The defect of this inference, however, is plain; for the
abolition of an absolute standard in no way involves the abolition
of *differences in value* between the various theories. The latter
remain in all strictness in so far as the general presupposition is
retained that the changing phases of the concept of experience do
not lie absolutely outside of each other, but are connected by logical
relations. The system and convergence of the series takes the place
of an external standard of reality. Both system and convergence
can be mediated and established, analogously to arithmetic, entirely
by comparison of the serial members and by the general rule, which
they follow in their progress. This rule, on the one hand, is given by
the fact that the *form* of experience persists; thus the particular
configurations in space, which we take as a basis for our construction
of the system of the world, vary,—while space and time, number and
magnitude are retained as instruments of any construction. (*Cf.*
above p. 266 ff.) Further there are also certain *material* features of
the system, that are unaffected by the transition from one stage to
the following: the variation does not absolutely cancel the earlier
stage, but allows it to remain in a new interpretation. All the
observations of Tycho de Brahe enter into the system of Kepler,
although they are connected and conceived in a new way. The
claim of any such connection, however, is not measured with respect
to things themselves, but by certain supreme *principles* of the knowledge of nature, that are retained as logical norms. The spatial
order is called "objective," that corresponds to these principles, that
e.g., has been constructed by us in accordance with the presupposition and the requirements of the law of inertia. (*Cf.* above p. 183 ff.)
Going back to such supreme guiding principles insures an inner
homogeneity of empirical knowledge, by virtue of which all its
various phases are combined in the expression of *one* object. The
"object" is thus exactly as true and as necessary as the logical unity
of empirical knowledge;—but also no truer or more necessary. Yet
little as this unity ever appears as complete, much as it remains rather
a "projected unity," its concept is none the less definitely determined.
The demand itself is what is fixed and permanent, while every form of
its realization points beyond itself. The one reality can only be

indicated and defined as the ideal limit of the many changing theories; yet the assumption of this limit is not arbitrary, but inevitable, since only by it is the continuity of experience established. No single astronomical system, the Copernican as little as the Ptolemaic, can be taken as the expression of the "true" cosmic order, but only the whole of these systems as they unfold continuously according to a definite connection. Thus the instrumental character of scientific concepts and judgments is not here contested. These concepts are valid, not in that they copy a fixed, given being, but in so far as they contain a plan for possible constructions of unity, which must be progressively verified in practice, in application to the empirical material. But the instrument, that leads to the unity and thus to the truth of thought, must be in itself fixed and secure. If it did not possess a certain stability, no sure and permanent use of it would be possible; it would break at the first attempt and be resolved into nothing. We need, not the objectivity of absolute things, but rather the objective determinateness of the *method of experience*.

The reconciliation of permanence and change. The real *content* of the object of thought, to which knowledge penetrates, corresponds therefore to the active *form* of thought in general. In the realm of rational knowledge, as in that of empirical, the same problem is set. In the process of knowledge, the conception is established of a body of ideal relations, that remain self-identical and unaffected by the accidental, temporally varying conditions of psychological apprehension. The affirmation of such a constancy is essential to every act of thought; only in the way in which this affirmation is proved is founded the difference of the different levels of knowledge. As long as we remain in the field of purely logical and mathematical propositions, we have a fixed whole of truths, that remain unchanged. Each proposition is always what it once is; it can be supplemented by others being added to it, but it cannot be transformed in its own import. Pure empirical truth, however, seems in principle to be deprived of this determinateness; it is different today from what it was yesterday, and merely signifies a temporary hold, that we gain amid the flux of presentations only to relinquish again. And still both motives, in spite of their opposition, are ultimately united in a unitary type of knowledge. It is only in abstraction that we can separate the absolutely permanent element from the changing and oppose them to each other. The genuine, *concrete* problem of knowledge is to make the permanent fruitful for the changing. The body of eternal truths becomes a means for gaining a foothold in the realm of change. The changing is considered as if it were permanent, when we attempt to understand it as the result of universal theoretical laws. The difference between the two factors can never be made to disappear entirely; still in the constant assimilation between them consists the whole movement of knowledge. The changeableness of the empirical material proves in no way to be a hindrance, but rather a positive furthering of knowledge. The opposition between mathematical theory and the totality of observations known at any time would be irreducible, if there were fixed, unchanging data involved on both sides. The possibility of removing the conflict only opens up, when we become aware of the conditional character of our empirical cognitions and thus of the plasticity of the empirical material with which knowledge works. The harmony of the given and the demanded is established by our investigating the given in the light of the theoretical demands, and thus broadening and deepening its concept. The constancy of the ideal forms has no longer a purely static, but also and especially a dynamic meaning; it is not so much constancy in *being*, as rather constancy in logical *use*. The ideal connections spoken of by logic and mathematics are the permanent lines of direction, by which experience is orientated in its scientific shaping. This function of these connections is their permanent and indestructible value, and is verified as identical through all change in the accidental material of experience.

The double form of the concept. Identity and diversity, constancy and change appear from this side also as correlative logical moments. To establish an absolute, real opposition between them would be to cancel not only the concept of being, but that of thought. For thought is not limited to separating out the analytically common element from a plurality of elements, as has been shown but reveals its true meaning in its necessary progress from one element to another. Difference and change are not points of view in principle "alien to thought,"[12] but belong in their fundamental meaning to the characteristic function of the intellect and represent this function in its full scope. If this correlative double form of the concept is misunderstood, then

---

[12] For the concept of 'alienation from thought' ("*Denkfremdheit*"), see Jonas Cohn, *Voraussetzungen u. Ziele des Erkennens*, Lpz. 1908, esp. 107 ff.

a gap, that cannot be bridged, must arise between knowledge and phenomenal reality. We reach once more the fundamental view of Eleatic metaphysics, which has had in fact an interesting and significant revival in modern criticism of knowledge. In order to *understand* reality by our mathematico-physical concepts, it is concluded that we must *destroy* it in its real nature, *i.e.*, in its multiplicity and changeableness. Thought does not tolerate any inner heterogeneity and change in the elements, from which it constructs its form of being. The manifold physical qualities of things are resolved into the one concept of the other, which is itself nothing else than the hypostatization of empty *space* devoid of qualities. The living intuition of the temporal flow of events is transfixed by thought into the permanence of ultimate constant magnitudes. To explain nature is thus to cancel it as nature, as a manifold and changing whole. The eternally homogeneous, motionless "sphere of Parmenides" constitutes the ultimate goal to which all natural science unconsciously approaches. It is only owing to the fact that reality withstands the efforts of thought and sets up certain limits, that it cannot transcend, that reality maintains itself against the logical leveling of its content; it is only by such opposition from reality, that being itself does not disappear in the perfection of knowledge.[13] Paradoxical as this consequence may seem, it is nevertheless exactly and consistently deduced from the assumed explanation of the intellect and its peculiar function. But this explanation requires a limitation. The identity, toward which thought progressively tends, is not the identity of ultimate substantial things, but the identity of functional orders and correlations. These functional correlations, however, do not exclude the element of diversity and change, but are determined only in it and by it. It is not the manifold as such that is cancelled, but only one of another dimension; the mathematical manifold is substituted in scientific explanation for the sensuous manifold. What thought demands is not the extinction of plurality and change in general, but its mastery by the mathematical continuity of serial laws and serial forms. In establishing this continuity, thought requires the standpoint of diversity no less than it requires the standpoint of identity. The former also is not absolutely forced upon it from without, but is grounded in the character and task of scientific "reason." When analysis resolves the given, sensuous

[13] See E. Meyerson, *Identité et Réalité*, esp. p. 229 ff.

qualities into a wealth of elementary *motions*, when the reality of the "impression" becomes the reality of the "vibration," then it is seen that the path of investigation does not merely consist in going over from plurality to unity, from motion to rest, but that it also leads in the reverse direction, and that the setting aside of the apparent constancy and simplicity of perceptual things is no less necessary. Only by thus setting aside the simplicity of the perceptual things can we reach the new meaning of identity and permanence, which lies at the basis of scientific laws. The complete concept of thought thus reëstablishes the harmony of being. The inexhaustibleness of the problem of science is no sign of its fundamental insolubility, but contains the condition and stimulus for its progressively complete solution.

CHAPTER VIII

ON THE PSYCHOLOGY OF RELATIONS

I

Logical relations and the problem of self-consciousness. The problem of knowledge, instead of leading us to a metaphysical dualism of the subjective and the objective worlds, has led us to a totality of relations, that contains the presupposition of the intellectual opposition of the "subject" and the "object." Confronted with this totality, the customary separation is seen to be impracticable. This totality is objective, in so far as all the constancy of empirical knowledge rests upon it as well as the whole possibility of objective judgment, while, on the other hand, it can only be comprehended *in judgment* and thus in the activity of thought. From this, it is evident that the characterization of this totality is subject to a double method and can be attempted in a twofold way. What these relations *are* in their purely logical meaning, can only be learned from the meaning they gain in the total system of science. Each particular proposition is bound and connected with another within this system; and the position, that the proposition thus gains within the whole of possible knowledge, indicates the measure of its certainty. The question, as to how this whole is *realized* in the knowing individuals, must be subordinated, as long as we are concerned with understanding the pure system of foundations and deducing it in its truth. The development of science itself urges us to leave this question in the background. Science advances from one objectively valid proposition to another for which it claims the same form of validity, without being diverted from this path by psychological considerations and psychological doubt. And yet precisely this independent progress ultimately creates a new problem for psychology also. It is evident, that in so far as psychological analysis proceeds from mere sensuous experience and seeks to remain in this type of experience, it can in no way do justice to the problems, that are continually raised from the side of science. The object, that science presents to us as distinct and certain, requires new psychological means for its description. Thus the general demand for a *psychology of relations* leads to a transformation of psychological methods in general. This transformation in the principles of psychology constitutes a weighty problem of epistemology. It is clear that here, as in other fields, it is the type of conceptual construction, that undergoes a characteristic change.

Plato's psychology of relations. For a long time modern psychology seemed completely to have lost sight of the peculiarity of the pure relational concepts. Only in relatively recent times and by remarkable, round-about ways, has it begun to approach them again. From a historical point of view, there is a strange anomaly in this; for what the modern psychologist may easily consider the end of his science, really constitutes its historical beginning. The conception of scientific psychology goes back historically to Plato. With Plato, the *concept of the soul* emerges for the first time from the general *concept of nature*, and receives peculiar and independent features. The soul is no longer the mere breath of life, which has in it the principle of its preservation and self-movement, but the soul passes over from this general meaning to that of *self-consciousness*. This transition, however, is only possible owing to the fact that Plato has already made sure of his necessary intermediate terms, in pure logic as well as in pure geometry and arithmetic. There is no path from mere perception as such to the new concept of the "self," that is to be established. For perception seems a mere part of the processes of nature; as represented by Empedocles and all the older philosophy of nature, perception seems nothing else than the adjustment, which takes place between our body and the material things of its environment. In order to know the corporeal things in perception, the soul must be of the same nature and constitution as they. In the development, which Plato has given the proposition of Protagoras in the *Theatetus*, this view is still plainly re-echoed. "Subject" and "object" are related as two forms of motion directed on each other; these forms of motion we can never separate in their purity and independence, but can only grasp in their reciprocal determination by each other. We always grasp only the result, without being able to analyse it into its real components. To this view, Plato conforms for a short space—as long as he is concerned with the analysis of sensation,—but leaves it behind when he turns to the analysis of the pure concepts. The image and analogy of *physical* action and reaction now has to be abandoned. Unity and diversity, equality and inequality are not corporeal objects, pressing in upon us with corporeal forces. The manner, in which they are presented to the ego, is fundamentally new and characteristic. The eye may distinguish the bright and the dark, and touch may distinguish the light and the heavy, the warm and the cold,—but the whole of knowledge is never exhausted in the totality of such sensuous differences. It is knowledge when we say of color or tone, that each of them *is*, that one of them is *different* from the other, or that both of them united are *two*. But although *being* and *non-being*, *similarity* and *dissimilarity*, *unity* and *plurality*, *identity* and *opposition* are objectively necessary elements of every assertion, they cannot be represented by any content of perception. For just this is their function, that they go beyond the particularity of sensuous contents in order to establish a connection between them; and while both the sensuous contents participate in this connection in the same sense, the latter can be pointed out directly in neither of the two particular elements as such. The relation between the heterogeneous fields of sense-perception could not be attained, if there were no structures which remained outside of their special characters and thus outside of their qualitative oppositions. These universal elements neither require nor are connected with any special organ; rather the soul gains them purely from itself in free construction. The concept of the unity of consciousness here first gains a firm foothold and sure foundation. If we rest in the content of the particular sensation, nothing is offered us but a chaos of particular experiences. The perceptions are packed together in us like the heroes in the wooden horse,[1] but nothing is found that refers them to each other and combines them into an identical *self*. The true concept of the self is connected with the concepts of the one and the many, the like and the unlike, being and non-being, and finds its true realization only in these. When we comprehend the perceptions under these concepts, we combine them *into one idea*,—whether or not we designate this unity as "soul." The "soul" is thus, as it were, conceived and postulated as the unitary expression of the content and as the systematic arrangement of pure relational concepts. The fundamental problem of psychology is defined with reference to the fundamental problems of pure logic and mathematics; and it is this connection, which definitely frees the Platonic concept of the soul from the Orphic speculation and that

---

[1] *Cf.* esp. *Theatetus*, 184 C. ff.

of the philosophy of nature with which it at first seems closely connected.

Aristotle's Doctrine of the Κοινόν. The Platonic view is undoubtedly influential in Aristotle's doctrine of the Κοινόν; but already its center of gravity has shifted. In his division of sense-perceptions, Aristotle starts from the fact that there is a particular content belonging to each sense, peculiar to it alone and distinguishing it from all the other senses. Thus color belongs to vision, tone as such, an ἴδιον belongs to hearing, while touch includes indeed a plurality of qualities, but is related to each of them in the same way as any sense is related to its specific content. But this sort of relation does not suffice, when we are determining the psychological correlate of concepts like motion and rest, magnitude and number. These concepts represent something truly "common," that reaches beyond all particular differences. As Aristotle further concludes, a universality of the receptive organ must correspond to the universality of the object. When, for example, we correlate the white with the sweet or oppose them to each other, it is necessarily *sense itself* that performs this act of comparison. For with what other faculty could we grasp contents of purely sensuous nature? But here sense no longer functions in any special capacity, as mere vision or mere taste, but as a "common sense" in an inclusive meaning. To this common sense, all the individual data of perception are referred, to be collected in it and related to each other.[2] Thus what was conceived by Plato as a spontaneous and free function of "consciousness itself," here appears as a power at once abstract and sensuous, wherein is combined all in which the different types and fields of perception agree. This psychological view of Aristotle's corresponds to his fundamental logical view; it rests on the view of the "concept" as nothing but a sum of sensuous properties, which are found uniformly in a plurality of objects.

"Thoughts of relation" in modern psychology. At first, modern psychology makes only isolated attempts to reach a new interpretation of the problem. Leibniz harks back directly to Plato when he urges, that the contents, which the traditional doctrine ascribes to the "common sense" (in particular, extension, form and motion), are *ideas of the pure understanding:* and although they are formed on the occasion of sense-impressions, can never be completely grounded

[2] *Cf.* especially Aristotle, περὶ ψυχῆς II, 6, 418a; III, 2, 426b.

in them. In modern German psychology, it is Tetens especially, who takes up this suggestion and develops it into a theory of the pure "thoughts of relation." But, on the whole, Locke's schema is dominant throughout; and according to it a concept can only be taken as truly understood and deduced, when we are able to show the simple sense-contents out of which it is compounded. The ideas of "reflection" also, although they seem at first to occupy a special position, are finally measured by this standard. They possess truly *positive* content only in so far as they can be expressed immediately in individual, perceptually given presentations. This reliance on a perceptual criterion appears most clearly in the concept of the *infinite*, which is criticized for no other reason than because it is clear that what it means is never *realized* in actual presentations, but subsists merely with reference to a possible, unlimited intellectual progress. Even if, from the standpoint of logic and mathematics, the general rule of this progress constitutes the existence and truth of the infinite, nevertheless, from the psychological view, this rule carries the stamp of the merely negative. For within this psychological view no adequate expression has been discovered for the validity and peculiarity of relations. Nevertheless the thought of these relations constantly returns, no matter how much it may be repressed. Like a ghost of uncertain nature and origin, it mixes among the clear, certain impressions of perception and memory. No matter how much one may ridicule, with Berkeley, the infinitesimal magnitudes of mathematics as the "ghosts of departed quantities," these ghosts cannot be laid to rest. Analysis here strikes an ultimate remainder, which it can neither comprehend nor set aside. The concepts, which prove effective and fruitful in actual scientific use, are never contained in those elements that the psychological view recognizes as the unique bearers of "objectivity." The significance of such scientific concepts rests on the fact, that they persistently transcend the type of reality that is here taken as a pattern.

The concept of substance. The deeper ground of this conflict, however, rests in the fact that psychological criticism has by no means freed itself from the presuppositions it struggles against. The concept, which Locke attacks most sharply and thoroughly, is the concept of *substance*. All the weapons of ridicule are summoned against the assumption of an independent, separate "something" without properties, which is assumed to be the "bearer" of sensuous qualities. He shows repeatedly that the real validity of knowledge is overthrown by the assumption of substance, since what is best known to us and most certain from the standpoint of experience is explained by means of something entirely without content and unknown. A mysterious "I know not what" is made the conceptual ground of all truly knowable qualities and properties. In this polemic against the concept of substance, Locke believes he has struck the real kernel of all metaphysics and of all scholastic explanation of reality. And after Hume has transferred its result from outer to inner experience, the work of criticism seems ended. The substance of the *ego* seems now explained away like the substance of the *thing;* mere associative connections of presentations take the place of both. In spite of all this, the view of physical and psychical reality, that is constructed on these foundations, has in it the general *category* of substantiality in its decisive meaning. Only the applications of this category have been changed, while it itself retains its old rank and position unnoticed. The substantiality of the "soul" is only apparently done away with; for it lives on in the substantiality of the sensuous "impression." After, as before, the conviction prevails, that only that is truly "real" and the ground of the real, which stands for itself alone and is intelligible purely of itself as an isolated existence. Here what is unchanging and essential in known reality lies before us, while all connections, which are established subsequently between the particular contents, form a mere addition of the mind. They thus only express an arbitrary tendency of the imagination, but not an objective connection of things themselves. This result constitutes, as it were, the negative proof of the stability that the substantialistic view still possesses in spite of everything. When the attempt is made to conceive the pure concepts of connection (in particular the concepts of cause and effect) not as impressions and copies of objects,—their logical import disappears. What is not "impression" is by that very fact a mere fiction. Nor does this fiction gain any inner value through the fact that it appears to be based in the "nature" of the mind and established with a kind of universality and regularity under definite conditions.

The doctrine of the "form-quality" in modern psychology. Modern psychology tried for a long time to avoid the skeptical consequences of Hume's doctrine, without submitting the premises on which it rests to a thorough-going change. In its own concepts, there developed a new form of "reality," which was at first naïvely and confidently accepted. All the peculiarities of psychological analysis were directly taken as properties of the psychical object. Thus, there resulted that self-deception discovered and described by James as the "psychologist's fallacy." The means used to represent a definite psychical fact and to make it communicable in a simple way were taken as real moments of this fact itself. Imperceptibly, the standpoint of analysis and reflective observation took the place of the standpoint of real experience.[3] A typical example of this whole view is the doctrine of the "simple" elements, out of which each state of consciousness is to be compounded. The ultimate parts which we can conceptually discriminate become the absolute atoms out of which the *being* of the psychical is constituted. But this being remains ambiguous in spite of everything. Properties and characteristics constantly appear in it, that cannot be explained and deduced from the mere summation of the particular parts. The further introspection advances and the further it follows its own way without prejudice, the more do new problems come to light. At first they are formulated only from a limited viewpoint and a special interest. The questions, which are combined in psychology under the concept of the *form-quality*, furnish the first stimulus to a renewed revision of the general concepts. Here it is shown in specially striking examples that not every spatial or temporal whole, that experience offers us, can be represented as a simple aggregate of its particular parts. When our consciousness follows and apprehends a simple melody, at first all the content present to it seems to consist in the perception of the particular tones. Closer observation shows, however, that such a description does not touch the real fact. We can, by transition into another kind of tone, allow all the individual tones originally in the melody to disappear, and we can replace them by others without ceasing to retain and to recognize the melody itself *as a unity*. Its specific character and properties thus cannot depend for us on the particularity of the elements, since the melody remains in spite of all change of this particularity. Two entirely different complexes of tone-sensations can give us the same melody while, on the other hand, two complexes of elements having the same content can lead to entirely different melodies, in so far as these elements are different from each other in their relative sequence. The same

[3] *Cf.* James, *Principles of Psychology*, I, 196 ff., 278 ff., *etc.*

thought can be carried over from the unity of the "tone-form" to the unity of the "spatial form." In space, we call certain figures "similar" to each other and conceive them as identical according to their mere geometrical concept even though they are built out of qualitatively wholly different spatial sensations. This consciousness of the identity of wholes that differ from each other in all their particular elements requires, if not a special explanation, at least a special psychological designation. The concept of form-quality contains this designation, though at first it only defines the problem without giving it a definite solution. What connects the manifold presentational contents into *one* psychological form, is not to be found in these contents themselves nor in their mere coexistence as an aggregate. But a new function is found here embodied in an independent structure of definite properties. The existence of such a structure and of the peculiar increase in content given with it is to be recognized as an empirical datum, no matter from what theoretical presuppositions one may interpret it.[4]

Such a theoretical interpretation is already found, when one attempts to explain the unity of the complex psychical structures, not from the mere *coexistence* of their parts, but from the reciprocal *action* they exert on each other. It is objected and is not to be disputed, that the melody appears as an independent content as opposed to the particular tones which enter into it. But in the explanation of this fact it is by no means necessary to introduce entirely new elements along with the ordinary elements of sensation and presentation and as added to them. To form a whole means psychologically nothing else than to *act* as a whole. Not only

---

[4] *Cf.* esp. Ehrenfels, *Über Gestaltqualitäten, Vierteljahrschr. f. wiss. Philos.*, XIV (1890), p. 249 ff., also Meinong, *Zur Psychologie der Komplexionen u. Relationen, Zeitschr. für Psychologie u. Physiologie der Sinnesorgane* II (1891), p. 245 ff.—The psychological explanation of the problem of the "form-qualities" would doubtless have been of more general significance, if it had gone into the corresponding logical problems more closely, that immediately present themselves here. As the psychological examples adduced always show, the concern here is with a general process of liberation and separate consideration of the relational content, which is particularly characteristic and of fundamental importance for wide fields of mathematics. (See especially Ch. III, p. 93 ff.) The possibility of retaining a relation as *invariant* in its meaning, while the members of the relation undergo the most various transformations, is only illumined and established from a new side in purely psychological considerations.

the parts as such, but also their whole complex always produces definite effects upon our feeling and presentations. And it is by virtue of these effects, which proceed from the complex and are thus dependent on the order of the elements within it, that we judge concerning the similarity or dissimilarity, the equality or inequality of wholes. This explanation seems in its general application to render unnecessary any assumption of special relational presentations and relational concepts. It is again the simple perception, which decides not only regarding sensuous properties such as color and tone, smell and taste, but which also informs us of unity and plurality, of permanence and change, and of the temporal sequence and the temporal persistence of contents. For all these determinations are distinguished from the simple sense impressions only by the fact that they are "effects" not of single stimuli, but of complexes of stimuli. Just as a certain ether vibration produces in us the impression of a definite color, so a certain composition and connection of stimuli affecting our consciousness produce in it an impression of similarity or difference, of change or permanence. Thus, e.g., when different tone sensations are aroused in us at definite temporal distances from each other, we can measure the length of the intervening pauses by each other and accordingly speak now of a more rapid, now of a slower succession. No special intellectual act is needed for the "comparison" of temporal intervals, but the simple assumption suffices that a certain effect issues from the complex of rapidly succeeding tones, that is different from the effect produced with a greater distance between the tones.[5] However, if we follow this attempted explanation to the end, we discover at once the epistemological circle that it involves. The totality of pure relational *thoughts* is reduced to an actual effect resulting from certain manifolds, while however the mere application of the *standpoint* of cause and effect involves a special relational thought. It is not the case that we can be led from the knowledge of certain causal connections to the understanding of relational concepts in general; rather, conversely, what these concepts mean must be already presupposed in order to be able to speak of causal connections of reality. When

---

[5] Schumann, *Zur Psychologie der Zeitanschauung. Ztschr. f. Psychol.* XVII (1899) p. 106 ff.—For a criticism of Schumann, *cf.* esp. Meinong's treatment, *Über Gegenstände höherer Ordnung u. deren Verhältnis zur inneren Wahrnehmung. Ztschr. f. Psychol.* XXI. (1899), p. 236 ff.

the psychological explanation proceeds from the factual elements and the action which they exert in the whole psychic process, it has already taken for granted everything the logical justification of which stands in question. It places a world of things, in which various objective relations prevail, at the starting point of its consideration,—while it had appeared at first as if this whole kind of reality could and should be deduced out of simple sensations as the unique data of pure experience, without the assumption of any other element. This reversal is indeed not strange; for it actually only reëstablishes the order of problems that was perverted in the first approach to the problem. What is truly known and given empirically in the field of consciousness, is not the particular elements, which then compound themselves into various observable effects, but it is rather always a manifold variously divided and ordered by relations of all sorts,—such a manifold as can be separated into particular elements merely by abstraction. The question here can never be how we go from the parts to the whole, but how we go from the whole to the parts. The elements never "subsist" outside of every form of connection, so that the attempt to deduce the possible ways of connection from them moves in a circle. Only the total result itself is "real" in the sense of experience and of psychological process, while its individual components have only the value of hypothetical assumptions. Their value and justification accordingly is to be measured by whether they are able to represent and reconstruct in their combination the totality of the phenomena.

Ebbinghaus's physiological account of relations. Thus within psychological investigation itself the speculative attempts to deny the peculiar character of pure relations and to replace them by mere complexes of sensations gradually die out. The ideal of conceptual "explanation," that was here taken as a standard, is in general retained; but it is recognized that our actual experience and our real empirical psychological knowledge prevent its realization. Just as we must assume certain classes of simple sensations as ultimate facts so, it is now explained, we must recognize also as fundamental and irreducible data of consciousness certain specific relations, such as unity and plurality, similarity and difference, spatial coexistence and temporal duration. "Naturally this does not mean to explain things," a representative of this view remarks, "but it means to prefer honest poverty to the appearance of wealth."[*] One way

<hr>

[*] Ebbinghaus, *Grundzüge der Psychologie*, 2nd ed., Lpz. 1905, p. 462.

seems still to remain, that would permit us to resolve the conceptual manifold of relations into the unity of a single causal origin. What is denied to the purely psychological type of consideration here seems to receive a physiological explanation. The general relational determinations, which are uniformly found with all sensations regardless of their qualitative difference, thus prove to be a *common condition* of sensation, for which a corresponding community is to be demanded in the appropriate physiological processes. This agreement in the physical foundations of each perception, no matter to what particular field it may belong, can easily be pointed out. At first the sense organs and their appropriate nervous centers appear to be very differently built and differently equipped apparatuses; nevertheless, they form a unity in so far as they are all built up out of the same material, out of nervous elements according to certain uniform principles. "If external stimuli affect them, then the processes called forth in them must naturally be different, in so far as the physico-chemical properties of the stimulus and the function of the receiving apparatus adapted to it are different. At the same time, those processes must be like or similar, in which the mode of connection of the stimuli into a whole in the outer world and the properties of the nervous matter within the sense organ and the general principles of its construction are the same. . . . . It is owing to the *special* character of the stimulations involved in seeing, hearing and tasting, that we sense their mental effects as something entirely disparate, as bright, loud, bitter;—it is due to the *agreeing* features of the same stimulations, that we are aware of these impressions as permanent or intermittent, as changing *etc.*, according to circumstances." The nervous processes, that lie at the basis of the "intuitions" of space and time, of unity and plurality, of constancy and change, "are thus found entirely in the same processes that are correlated with sensations; but they are not found in all features, but only in the common characteristics of these processes, which as yet cannot be indicated more closely."[7]

Criticism of the physiological explanation of relational concepts. At first glance, this explanation seems to operate wholly with the means of modern scientific investigation; nevertheless in principle it leads us back to the Aristotelian doctrine of the "common sense." It is true we have not a special organ at our disposal for apprehending

[7] Ebbinghaus, *op. cit.*, p. 442 f.

relations, as we have for the particular sense qualities; but yet there exists a sort of common organ by which we take up into our consciousness the real relations of outer objects. However, if this causal explanation is meant also to be a logical deduction of the validity of the relational concepts, then it would also involve that ὕστερον πρότερον in itself, which we have previously met. For in order to explain the *presentation* of equality or difference, of identity or similarity, such an account must obviously go back to similarity or difference in *things*, more specifically in the peripheral and central organs of perception. The concept of being, assumed at the outset, already contains in it all those *categorical* determinations that are subsequently drawn out of it in the course of the psycho-physiological deduction. The *truth* of these determinations has to be presupposed, although the manner in which they reach the consciousness of the individual subject may be open to an—indeed only hypothetical— explanation. Further, every purely physiological representation of the matter must leave in obscurity the point that is above all in question. An identity or community in the outer stimuli never suffices to explain the correlative expression of these relations in consciousness. The physically like must be recognized and judged *as* like, the actually different must be conceived *as* different for there to be that separation of the general content from the particular content of sensation, which is here assumed. Thus the pure conscious functions of unity and difference can never be dispensed with or be replaced by reference to the objective physiological causes. The demand is that they be represented by a type of explanation that remains within the field of the psychical phenomena themselves, without going back to their hypothetical grounds. Thus from all sides we are directed with increasing clarity to a second great field of psychological investigation,—a field that was at first neglected and subordinated. In contrast to the psychology of sensation, appears the psychology of thought. Such a psychology is dominated from the first by an entirely different formulation of the problem, and by a new order of value between the "absolute" and the "relative" elements of consciousness.

II

Meinong's theory of "founded contents." The general problems involved in the theory of the "form-qualities" receive sharper expression in the transformation which this doctrine has undergone in the theory of "founded contents" (*fundierten Inhalte*). Here it is clearly shown that the questions thereby introduced into psychology do not tend to a mere extension of its field, but to an inner transformation of its concept. Two forms of psychic "objects" are now definitely opposed to each other. Over the simple sensations and qualities of the different senses, there are built "objects of higher orders." These latter objects are borne by the elementary contents and require them as supports, but cannot be reduced to them. We cannot, indeed, speak of equality or difference, of unity or plurality, without thinking of equality or difference, unity or plurality of *something*. But this something may vary arbitrarily; it can appear as color or tone, as odor or taste, as concept or judgment,—of all of which, difference or unity can be predicated,— without the real meaning being thereby affected. Thus the dependence, which seems to belong to pure relations in their actual occurrence and, as it were, in their psychic existence, does not exclude a complete independence in their characteristic meaning. The universally valid relations do not exist as temporally or spatially bounded parts of psychical or physical reality; but they "subsist" absolutely by virtue of the necessity which we recognize in certain assertions. Whoever represents to himself four real objects does not represent fourness as a particular piece of reality along with them, although he claims a certain objective truth and validity for his judgment regarding the numerical relation. Thus in general, in contrast to the relations between existences, there appear *pure ideal relations:* and corresponding to this distinction, there is further a characteristic opposition in the value of the cognitions relating to these objects. Wherever the judgment refers to an object of actual reality and is meant to designate a particular determination of it, the judgment is necessarily limited to the here and now, thus to an assertion of merely empirical validity. To this case, in which we merely ascribe to an individual thing an individual property known by experience, there is opposed the other case, in which the type of dependency between the two elements *a* and *b* is determined and exactly prescribed by the "nature" of the members. Concerning ideal relations of this sort, judgments are possible that do not need to be tested by different successive cases in order to be grasped in their truth, but which are recognized once for all by insight into the *necessity* of the connection.

Along with the empirical judgments concerning objects of experience, there are thus *"a priori"* judgments concerning "founded objects." While the psychic "phenomenon," like color or tone, can simply be established in its occurrence and properties as a fact, there are judgments connected with the "metaphenomenal" objects, like equality and similarity, that are made with consciousness of timeless and necessary validity. In place of the mere establishment of a fact, there appears the systematic whole of a rational connection with elements that reciprocally demand and condition each other.[8]

"Objects of a higher order." However energetically this theory may strive to extend the field of psychological problems beyond the usual limits, it is still under the influence of the traditional theory of the concept at one point. It starts with the simple sensational contents as recognized data in order to proceed from them to the more complex structures. The "objects of a higher order" cannot be separated from whatever perceptual elements they may be founded in, without thereby losing all meaning. On the other hand, the converse of this proposition does not hold; while the *"superius"* is necessarily referred to the *"inferiora,"* the latter for their part are characterized by the fact that they exist for themselves alone and rest entirely in themselves. The relations, which are built up over the perceptual elements, appear as a subsequent result; their being or non-being adds nothing to the existence of the elements, and can neither ground nor imperil it. Sharper analysis, however, destroys this last appearance of the independence of the simple. In place of a succession, of a superordination and subordination of contents, analysis fixes a relation of strict correlativity. Just as the relation requires reference to the elements, so the elements no less require reference to a form of relation, in which alone they gain fixed and constant meaning. Each conceptual assertion regarding an *"inferius"* considers this *inferius* from the standpoint of some relation, which we correlate with the content in question. The "foundations" are always determinable and determined only as foundations of possible relations. What deceives us at first is the fact, that all the relational determinations into which an individual can enter are somehow contained in it but in no way actually realized from the beginning. In order to

[8] *Cf.* more particularly on the theory of "founded contents" and "objects of higher order" Meinong, *Z. f. Psychologie* XXI, 182 ff.; *cf.* also Höfler, *Zur gegenw. Naturphilosophie*, p. 75 ff.

transform the "potential" logical import into "actual" import, there is needed a series of involved intellectual operations and constantly renewed conceptual work. But the possibility of separating a content from this or that conceptual determination, and of considering it *before* this determination, as it were, must not lead us to divest it of all forms of determination in general. Consciousness *as consciousness* would be extinguished, not only if we conceived the sensuous phenomena, such as the colors and tones, the smells and tastes to be removed, but also if we conceived the "metaphenomenal" objects, such as plurality and number, identity and difference to be removed. The existence of consciousness is rooted merely in the mutual correlativity of the two elements, and neither is to be preferred to the other as "first" and original.

The conflict between empiricism and nativism. From this standpoint, new light falls on the old psychological controversy between "empiricism" and "nativism." This controversy we see is also rooted in an unclarified statement of the problem. It is asked whether the common determinations, which appear with the sensations are immediate properties of the sensations, whether their unity and plurality, their spatial arrangement, their longer or shorter temporal duration are properties just as immediate as the differences of the sensations themselves, and are apprehended at the same time as sensations,—or do they form a later product of *mental comparison*, which first imprints a definite form on the unordered perceptual material? In other words, is it a peculiar spiritual activity that leads to these determinations, or are they directly given in the first act of perception implicitly as elements? There are, however, two different standpoints, which unnoticed are identified with each other in these questions. The logical separation of the moments of knowledge is replaced by a temporal separation in the occurrence of certain psychic contents; and an attempt is made to solve both of these completely heterogeneous problems at the same time and by means of each other. When it is shown from the standpoint of nativism, that even the earliest state of consciousness, which can be assumed or conceived, exhibits some form of spatial-temporal or conceptual connection, it is believed that the logical value of the connections has therewith been reduced to that of mere sensation. It is concluded that there is an immediate consciousness of relations in the same way that there is an immediate consciousness of colors or tones. We apprehend in inner perception, in mere "feeling," the meaning of "and," "but," "if" and "therefore" just as truly as we gain information by perception concerning the content "blue" or "cold." The *actus purus* of the understanding thus proves to be unnecessary, for everything that it was to produce is in fact already contained in the first data of perception.[9] If we would be critically just to this conception, we must distinguish the general tendency ruling it from the special way in which it is worked out. What is meant to be emphasized above all is this: that the element of order is related to the element of content in no temporal relation of before and after, of earlier or later. Only analysis can discover this distinction in the originally unitary material of the "given." In this sense, it is correct that even the most elementary psychic state includes the general elements of form. The inference, however, that these elements therefore belong to the mere *passivity* of perception, is not thereby justified. On the contrary, the converse inference holds. The fact that there is no content of consciousness, which is not shaped and arranged in some manner according to certain relations proves that the process of perception is not to be separated from that of *judgment*. It is by elementary acts of judgment, that the particular content is grasped as a member of a certain order and is thereby first fixed in itself. Where this is denied, judgment itself is understood only in the external sense of a comparing activity, which subsequently adds a new predicate to a "subject" already fixed and given. Such an activity appears indeed as accidental and arbitrary, in contrast to the material with which it is connected. Whether or not such an activity takes place, this material remains what it is and retains the properties, which belong to it prior to all logical activities. In its real form, on the contrary, judgment does not signify such an arbitrary act, but is rather the form of objective determination in general, by which a particular content is distinguished as such and at the same time subordinated systematically to a manifold. From this form, we cannot abstract without thereby losing all the qualitative differences in content. Thus it may be that when we consider the pure *temporal* relation, relations will be "found" *at the same time* as the sensational contents; nevertheless it is true, that this very "finding" involves within itself the elementary forms of intellectual activity. If we consider these forms removed,

[9] James, *The Principles of Psychology* I, 244 f.; *cf.* esp. II, 148.

all possibility would thereby disappear of further application of the concept of consciousness. Whatever the content might be or mean in and for itself,—for us, for the unity of the self, it would not be present. For the self comprehends and constitutes itself only in some form of activity. It is always with certain types of the "apperception of unity," that the apprehension of definite relations between objects is necessarily connected psychologically.[10] Thus the inseparable correlation of sensations with the pure relations, when its consequences are consistently traced, implies the opposite of what was initially deduced from it; it shows not the passivity of the ego in apprehending these thoughts, but conversely the element of activity, that belongs to every process of perception in so far as it does not stand for itself alone, but belongs to the *whole* of consciousness and experience. We can, in fact, attempt to deduce the relations from sensation; but then we have already placed in sensation determinations, that go beyond the isolated impression. It is no longer the abstract "simple" sensation; but it rather signifies merely the initial, still unordered content of consciousness, to which, however, definite relations and connections are always essential, leading from it to other elements.

The psychology of the idea of space. This is all the clearer when we consider the special problem, that has always been the center of discussion between empiricism and nativism. The fate of the different theories is decided by the question as to the psychological origin and the psychological meaning of the idea of space. If it is possible to deduce space from absolutely non-spatial sensations differing only in their quality and intensity, then there is nothing in principle to prevent us from carrying through the same explanation for all the different types of relation in general. However, it is at once evident that the empiristic theory, when it undertakes to deduce the origin of spatial order from the mere material of perceptions and the simple forces of associative connection, is obliged to be untrue to its own methodic ideal. For there can be no doubt that such an origin, if indeed it is to be assumed, cannot be pointed out in our actual experience. Each experience, whatever its properties may be, reveals some primitive form of the "coexistence" of particular elements and thus the specific element in which every spatial con-

---

[10] *Cf.* above all the treatment by Th. Lipps, *Einheiten und Relationen, Eine Skizze zur Psychologie der Apperception*, Lpz. 1902.

struction, no matter how complex, is originally rooted. If we attempt to go behind this psychological fact, if we attempt to show how order itself arises and evolves out of the absolutely unordered, we surrender ourselves to an hypothesis, which goes beyond the limits of experience in two directions. Empirically we know just as little of a simple, absolutely *unlocalized* perception, as we know of a particular function of the soul, which transforms the previously formless into a form, on the basis of unconscious "inferences." However we may judge of the methodic value of such concepts, it would be dangerous and misleading to misunderstand them as an expression of concrete facts. Here the criticism of the "empiristic" theories of space in modern psychology (especially by Stumpf and James) is in the right, in so far as it urges that mere "association" as such can produce no new psychic content. No mere repetition and arrangement of contents could give them spatiality, if spatiality were not already somehow placed in them.[11] But here again the temporal

[11] *Cf., e.g.*, James, *op. cit.*, II, 270, 279, *etc.*—The effects of the general schema of associationistic psychology are also shown in its critics, however, in the fact that they can only affirm the originality of the spatial *order* by condensing it into a peculiar and original *content* of perception. Stumpf in particular explains, that the spatial order would not be intelligible without there being a positive, absolute content at its basis. This content first gives the spatial order its peculiar character, by which it is distinguished from other orders. "To distinguish the various orders from each other, we must recognize everywhere a special absolute content with reference to which the order is found. And thus space also is not a mere order, but is precisely what distinguishes the spatial order, adjacency (*Nebeneinander*), from other orders." (*Über den psychologischen Ursprung der Raumvorstellung*, p. 15, *cf.* p. 275.) Two different points of view, that are not strictly differentiated, are combined in this argument. It may be accepted that every relation is the relation of something, and therefore presupposes some "foundation," on which it is built,—although it is also to be remembered that the dependence is throughout reciprocal, so that the "foundation" requires the relation just as much as the latter requires the "foundation." But this does not involve that what constitutes the peculiar character of a definite principle of order must be able to be pointed out as a *property* of the ordered elements. For if we assumed this, we would finally have to ascribe to the content as many special "qualities" as there are ways of connecting the content with, and referring it to, others. A specific perceptual quality would be demanded not only for spatial order, but for temporal order and further for all sorts of quantitative or qualitative comparison. But in general it is not obvious how a mere difference in the content of the compared elements should serve to define and separate the various possible types of relation. If two orders are distinguished *as orders*, some means of conscious-

connection of the two factors in no way proves their logical equivalence. When the criticism of knowledge distinguishes the spatial and temporal form from the content of sensation and treats it as an independent problem, it does not require the conception of a *real separateness* of the two in some mythical stage of consciousness. What it affirms and defends is merely the simple thought, that the *judgments* based and constructed on these forms of relation have a characteristic validity of their own, which is denied to mere assertions regarding the existence of the sensation given here and now. The original unitary content is differentiated, when we recognize it as the starting-point of two different systems of judgments, which are separated according to their *dignity*. According to whether we emphasize the specific moment of a particular sensation (the blue and red, the rough and smooth, *etc.*), or merely consider the universal relations subsisting between these particular elements, different propositions arise, belonging to thoroughly different types of grounding. Psychology, indeed, within the limits of its task, cannot trace and survey this conceptual division in its totality, since psychology merely describes and analyses thought as a temporal process, but not the content of what is thought. The tendency of the process only becomes clear in its final result; only the fully developed *system of geometry*, constructed according to unitary rational principles,

---

ness must be given by which the *type of connection* can be grasped purely as such, and distinguished from all others. If we ascribe to consciousness the capacity of distinguishing the simple data of perception from each other, it is not obvious how we can deny it the same capacity in the case of the various original functions of order. The deeper ground of the difficulty here seems to lie not so much in psychology itself, as in the ordinary conception of logic. Logic, in its traditional form, is founded on the thought of *identity*, and seeks to reduce all types of connection and inference ultimately to identity. If, however, identity is taken as an expression of the relational form in general, then the *diversity* of relations (which in any case must be explained) can be founded merely in the content of the elements, that are related to each other. The modern form of logic, however, has destroyed the basis of this view; it has shown with increasing distinctness that it is impossible to reduce the diverse forms of judgment to the single type of identity. (For further particulars, *cf. e.g.*, Jonas Cohn, *Voraussetzungen und Ziele des Erkennens*, p. 85 ff.) Just as we are here forced to recognize an original plurality of diverse relational syntheses (R, R′, R″, *etc.*), that are not mutually reducible to each other, so psychological consideration must ultimately recognize differences belonging to the manner and way of "apperceptive connection" itself, without finding their expression in any particular quality of sensation.

contains the definite characterization of the elements of space.   But although psychology cannot establish this characterization, it need not contradict it at any point.   Its own treatment of the problem of relations leads it with inner necessity to a point where a new type and direction of consideration begins.   The separation of the element of relation from the element of content, to which psychology is forced, remains, so to speak, *proleptical* in it, and only receives full illumination in a broader field.

The psychology of thought.   Even the purely empirical-experimental view of mental phenomena significantly reveals such a tendency of the problem.   The effort more and more is to apply the method of experiment not only to the facts of sensuous perception, but with its help to discover the complex processes of conceptual *understanding* in their fundamental features.[12]   But in applying the method of experiment here, it becomes constantly clearer that it is not the intuitive factual presentations, not the direct perceptual images, that bear and support this process of the understanding. The understanding of the simplest proposition in its definite logical and grammatical structure requires, if it is to be apprehended *as a proposition*, elements, that are absolutely removed from intuitive representation.   The pictorial presentations of the concrete objects, of which the assertion holds, can vary greatly or even wholly disappear, without the apprehension of the unitary meaning of the proposition being endangered.   The conceptual connections, in which this meaning is rooted, must therefore be represented for consciousness in peculiar categorical acts which are to be recognized as independent, not further reducible factors of every intellectual apprehension. The manner, in which psychological investigation has reached this insight, is indeed noteworthy enough; and it further signifies the historical dependence of the methods and problems of psychology. "Thought" is not conceived and observed here in its independent activity; but the attempt is made to establish its peculiar character in the reception of a finished content from without.   Accordingly, the new factor thus gained appears rather as a paradoxical, incompletely understood *remainder* left by analysis, than as a positive and characteristic function.   The criticism of knowledge reverses this relation; for it, that problematic "remainder" is what is really first

---

[12] A neat and synoptical treatment of this tendency in psychological investigation is given by Messer, *Empfindung und Denken*, Lpz. 1908.

and "intelligible" and is the point of departure. It does not study thought where thought merely receptively receives and reproduces the meaning of an already finished judgmental connection, but where it creates and constructs a meaningful system of propositions. When psychology pursues this line of enquiry and considers thought equally in the concrete totality of its *productive* functions, the initial opposition of methods is more and more resolved into a pure correlation. Psychology in this sense gives the *approach* to problems, which must seek their solution in logic and in their application to science.

SUPPLEMENT

EINSTEIN'S THEORY OF RELATIVITY CONSIDERED FROM THE EPISTEMOLOGICAL STANDPOINT

AUTHOR'S PREFACE

The following essay does not claim to give a complete account of the philosophical problems raised by the theory of relativity. I am aware that the new problems presented to the general criticism of knowledge by this theory can only be solved by the gradual work of both physicists and philosophers; here I am merely concerned with beginning this work, with stimulating discussion, and, where possible, guiding it into definite methodic paths, in contrast to the uncertainty of judgment which still reigns. The purpose of this writing will be attained if it succeeds in preparing for a mutual understanding between the philosopher and the physicist on questions, concerning which they are still widely separated. That I was concerned, even in purely epistemological matters, to keep myself in close contact with scientific physics and that the writings of the leading physicists of the past and present have everywhere essentially helped to determine the intellectual orientation of the following investigation, will be gathered from the exposition. The bibliography, which follows, however makes no claim to actual completeness; in it only such works are adduced as have been repeatedly referred to and intensively considered in the course of the exposition.

Albert Einstein read the above essay in manuscript and gave it the benefit of his critical comments; I cannot let it go out without expressing my hearty thanks to him.

ERNST CASSIRER.

University of Hamburg.

CHAPTER I

Concepts of Measure and Concepts of Things

"The use, which we can make in philosophy, of mathematics," Kant wrote in the year 1763 in the Preface of his *Attempt to Introduce the Concept of Negative Magnitudes into Philosophy*, "consists either in the imitation of its methods or in the real application of its propositions to the objects of philosophy. It is not evident that the first has to date been of much use, however much advantage was originally promised from it. The second use, on the contrary, has been so much the more advantageous for the parts of philosophy concerned, which, by the fact that they applied the doctrines of mathematics for their purposes, have raised themselves to a height to which otherwise they could make no claim. These, however, are only doctrines belonging to the theory of nature. . . . . As far as metaphysics is concerned, this science, instead of utilizing a few of the concepts or doctrines of mathematics, has rather often armed itself against them and, where it might perhaps have borrowed a sure foundation for its considerations, we see it concerned with making out of the concepts of the mathematician nothing but fine imaginings, which beyond his field have little truth in them. One can easily decide where the advantage will fall in the conflict of two sciences, of which the one surpasses all others in certainty and clarity, the other of which, however, is only striving to attain certainty and clarity. Metaphysics seeks, *e.g.*, to discover the nature of space and the supreme ground from which its possibility can be understood. Now nothing can be more helpful for this than if one can borrow from somewhere sufficiently proved data to take as a basis for one's consideration. Geometry offers several data, which concern the most general properties of space, *e.g.*, that space does not consist of simple parts; but these are passed by and one sets his trust merely on the ambiguous consciousness of the concept, which is conceived in a wholly abstract fashion. . . . . The mathematical consideration of motion in connection with knowledge of space furnishes many data to guide the metaphysical speculations of the times in the track of truth. The celebrated Euler, among others, has given some opportunity for this, but it seems more comfortable to remain with obscure abstractions, which are hard to test, than to enter into connection with a science which possesses only intelligible and obvious insights."

The essay of Euler, to which Kant here refers the metaphysician, is the former's *Réflexions sur l'espace et le temps,* which appeared in the year 1748 among the productions of the Berlin Academy of Science. This essay sets up in fact not only a program for the construction of mechanics but a general program for the epistemology of the natural sciences. It seeks to define the concept of truth of mathematical physics and contrasts it with the concept of truth of the metaphysician. Materially, however, the considerations of Euler rest entirely on the foundations on which Newton had erected the classical system of mechanics. Newton's concepts of absolute space and absolute time are here to be revealed not only as the necessary fundamental concepts of mathematico-physical knowledge of nature, but as true physical realities. To deny these realities on philosophical, on general epistemological grounds, means, as Euler explains, to deprive the fundamental laws of dynamics—above all the law of inertia—of any real physical significance. In such an alternative, however, the outcome cannot be questioned: the philosopher must withdraw his suspicions concerning the "possibility" of an absolute space and an absolute time as soon as the reality of both can be shown to be an immediate consequence of the validity of the fundamental laws of motion. What these laws *demand,* also "is"—and it is, it exists in the highest sense and highest degree of objectivity which is attainable for our knowledge. For before the reality of nature as it is represented in motion and its empirical laws all logical doubt must be silent; it is the business of thought to accept the existence of motion and its fundamental rules instead of attempting to prescribe to nature itself from abstract considerations concerning what can or cannot be conceived.

This demand, however, illuminating as it appears and fruitful as the methodic stimulus of Euler proved in the development of the Kantian problem,[1] becomes problematical when considered from the standpoint of modern physics and epistemology. Kant believed that he possessed in Newton's fundamental work, in the *Philosophiae Naturalis Principia Mathematica,* a fixed code of physical "truth"

[1] For more detail concerning Euler and Kant's relation to him, *cf., Erkenntnisproblem* (7), II, 472 ff., 698, 703 f.

and believed that he could definitively ground philosophical knowledge on the *"factum"* of mathematical natural science as he here
found it; but the relation between philosophy and exact science has
since changed fundamentally. Ever more clearly, ever more com-
pellingly do we realize today that the Archimedean point on which
Kant supported himself and from which he undertook to raise the whole
system of knowledge, as if by a lever, no longer offers an unconditionally fixed basis. The *factum* of geometry has lost its unambiguous
definiteness; instead of the one geometry of Euclid, we find ourselves
facing a plurality of equally justified geometrical systems, which
all claim for themselves the same intellectual necessity, and which,
as the example of the general theory of relativity seems to show, can
rival the system of classical geometry in their applications, in their
fruitfulness for physics. And the system of classical mechanics
has undergone an even greater transformation, since in modern
physics the "mechanical" view of the world has been more and more
superseded and replaced by the electro-dynamic view. The laws,
which Newton and Euler regarded as the wholly assured and impregnable possession of physical knowledge, those laws in which they
believed to be defined the concept of the corporeal world, of matter
and motion, in short, of nature itself, appear to us today to be only
abstractions by which, at most, we can master a certain region, a
definitely limited part of being, and describe it theoretically in a
first approximation. And if we turn to contemporary physics with
the old philosophical question as to the "essence" of space and time,
we receive from it precisely the opposite answer to that which Euler
gave the question a hundred and fifty years ago. Newton's concepts of absolute space and absolute time may still count many
adherents among the "philosophers," but they seem definitively
removed from the methodic and empirical foundations of physics.
The general theory of relativity seems herein to be only the ultimate
consequence of an intellectual movement, which receives its decisive
motives equally from epistemological and physical considerations.

The working together of the two points of view has always come
to light with special distinctness at the decisive turning points in the
evolution of theoretical physics. A glance at the history of physics
shows that precisely its most weighty and fundamental achievements
stand in closest connection with considerations of a general epistemological nature. Galileo's *Dialogues on the Two Systems of the*

World are filled with such considerations and his Aristotelian opponents could urge against Galilei that he had devoted more years to the study of philosophy than months to the study of physics. Kepler lays the foundation for his work on the motion of Mars and for his chief work on the harmony of the world in his *Apology for Tycho*, in which he gives a complete methodological account of hypotheses and their various fundamental forms; an account by which he really created the modern concept of physical *theory* and gave it a definite concrete content. Newton also, in the midst of his considerations on the structure of the world, comes back to the most general norms of physical knowledge, to the *regulae philosophandi*. In more recent times, Helmholtz introduces his work, *Uber der Erhaltung der Kraft* (1847), with a consideration of the causal principle as the universal presupposition of all "comprehensibility of nature," and Heinrich Hertz expressly asserts in the preface of his *Prinzipien der Mechanik* (1894), that what is new in the work and what alone he values is "the order and arrangement of the whole, thus the logical, or, if one will, the philosophical side of the subject."[2] But all these great historical examples of the real inner connection between epistemological problems and physical problems are almost outdone by the way in which this connection has been verified in the foundation of the theory of relativity. Einstein himself—especially in the transition from the special to the general theory of relativity—appeals primarily to an epistemological motive, to which he grants, along with the purely empirical and physical grounds, a decisive significance.[3] And even the special theory of relativity is such that its advantage over other explanations, such as Lorentz's hypothesis of contraction, is based not so much on its empirical material as on its pure logical form, not so much on its physical as on its general *systematic* value.[4] In this connection the comparison holds which Planck has drawn between the theory of relativity and the Copernican cosmological reform.[5] The Copernican view could point, when it appeared, to no single new "fact" by which it was absolutely demanded to the exclusion of all earlier astronomical explanations, but its value and real cogency

[2] *Cf.* Helmholtz (29, p. 4); H. Hertz (31, p. XXVII).
[3] *Cf.* Einstein (17, p. 8).
[4] See below, Sect., II.
[5] *Cf.* Planck (68), p. 117 f.

lay in the fundamental and systematic clarity, which it spread over the whole of the knowledge of nature. In the same way, the theory of relativity, taking its start in a criticism of the concept of time, extends into the field of epistemological problems, not merely in its applications and consequences, but from its very beginning. That the sciences, in particular, mathematics and the exact natural sciences furnish the criticism of knowledge with its essential material is scarcely questioned after Kant; but here this material is offered to philosophy in a form, which, even of itself, involves a certain epistemological interpretation and treatment.

Thus, the theory of relativity, as opposed to the classical system of mechanics, offers a new scientific problem by which the critical philosophy must be tested anew. If Kant—as Hermann Cohen's works on Kant urged repeatedly—and proved from all angles—intended to be the philosophical systematizer of the Newtonian natural science, is not his doctrine necessarily entangled in the fate of the Newtonian physics, and must not all changes in the latter react directly on the form of the fundamental doctrines of the critical philosophy? Or do the doctrines of the Transcendental Aesthetic offer a foundation, which is broad enough and strong enough to bear, along with the structure of the Newtonian mechanics, also that of modern physics? The future development of the criticism of knowledge will depend on the answer to these questions. If it is shown that the modern physical views of space and time lead in the end as far beyond Kant as they do beyond Newton, then the time would have come when, on the basis of Kant's presuppositions, we would have to advance beyond Kant. For the purpose of the *Critique of Pure Reason* was not to ground philosophical knowledge once for all in a fixed dogmatic system of concepts, but to open up for it the "continuous development of a science" in which there can be only relative, not absolute, stopping points.

Epistemology, however, closely as its own fate is connected with the progress of exact science, must face the problems which are presented to it by the latter, with complete methodic independence. It stands to physics in precisely the relation, in which, according to the Kantian account, the "understanding" stands to experience and nature: it must approach nature "in order to be taught by it: but not in the character of a pupil, who agrees to everything the master likes, but as an appointed judge, who compels the witnesses to answer

the questions which he himself proposes." Each answer, which
physics imparts concerning the character and the peculiar nature of
its fundamental concepts, assumes inevitably for epistemology the
form of a question. When, for example, Einstein gives as the essential result of his theory that by it "the last remainder of physical
objectivity" is taken from space and time (17, p. 13), this answer of
the physicist contains for the epistemologist the precise formulation
of his real problem. What are we to understand by the physical
objectivity, which is here denied to the concepts of space and time?
To the physicist physical objectivity may appear as a fixed and sure
starting-point and as an entirely definite standard of comparison;
epistemology must ask that its meaning, that what is to be expressed
by it, be exactly defined. For epistemological reflection leads us
everywhere to the insight that what the various sciences call the
"object" is nothing given in itself, fixed once for all, but that it is
first determined by some standpoint of knowledge. According to
the changes of this ideal standpoint, there arise for thought various
classes and various systems of objects. It is thus always necessary
to recognize, in what the individual sciences offer us as their objects
and "things," the specific logical conditions on the ground of which
they were established. Each science *has* its object only by the fact
that it *selects* it from the uniform mass of the given by certain formal
concepts, which are peculiar to it. The object of mathematics is
different from that of mechanics, the object of abstract mechanics
different from that of physics, *etc.*, because there are contained in all
these sciences different questions of knowledge, different ways of
referring the manifold to the unity of a concept and ordering and
mastering the manifold by it. Thus the content of each particular
field of knowledge is determined by the characteristic form of judgment and question from which knowledge proceeds. In the form of
judgment and question the particular special axioms, by which the
sciences are distinguished from each other, are first defined. If we
attempt to gain a definite explanation of the concept of "physical
objectivity" from this standpoint, we are first led to a negative
feature. Whatever this objectivity may mean, in no case can it
coincide with what the naïve view of the world is accustomed to
regard as the reality of things, as the reality of objects of sensuous
perception. For the objects, of which scientific physics treats and
for which it establishes its laws, are distinguished from this reality

by their general fundamental form.   That concepts, such as those of mass and force, the atom or the ether, the magnetic or electrical potential, even concepts, like those of pressure or of temperature, are no simple thing-concepts, no copies of particular contents given in perception: this scarcely needs any further explanation, after all that the epistemology of physics itself has established concerning the meaning and origin of these concepts.   What we possess in them are obviously not reproductions of simple things or sensations, but theoretical assumptions and constructions, which are intended to transform the merely sensible into something measurable, and thus into an "object of physics," that is, into an object *for* physics.   Planck's neat formulation of the physical criterion of objectivity, that everything *that can be measured* exists, may appear completely sufficient from the standpoint of physics; from the standpoint of epistemology, it involves the problem of discovering the fundamental conditions of this measurability and of developing them in systematic completeness.   For any, even the simplest, measurement must rest on certain theoretical presuppositions on certain "principles," "hypotheses," or "axioms," which it does not take from the world of sense, but which it brings to this world as postulates of thought.   In this sense, the reality of the physicist stands over against the reality of immediate perception as something through and through mediated; as a system, not of existing things or properties, but of abstract intellectual symbols, which serve to express certain relations of magnitude and measure, certain functional coördinations and dependencies of phenomena. If we start from this general insight, which within physics itself has been made very clear, especially by Duhem's analysis of the physical construction of concepts, the problem of the theory of relativity gains its full logical definiteness.   That physical objectivity is denied to space and time by this theory must, as is now seen, mean something else and something deeper than the knowledge that the two are not things in the sense of "naïve realism."   For things of this sort, we must have left behind us at the threshold of exact scientific physics, in the formulation of its first judgments and propositions. The property of not being thing-concepts, but pure concepts of measurement, space and time share with all other genuine physical concepts; if, in contrast to these, space and time are also to have a special logical position, it must be shown that they are removed in the same direction as these, a step further from the ordinary thing-concepts, and that they thus represent, to a certain extent, concepts and forms of measurement of an order higher than the first order.

The fact appears even in the first considerations, from which the theory of relativity starts, that the physicist does not have only to hold in mind the measured object itself, but also always the particular conditions of measurement. The theory distinguishes between physical determinations and judgments, which result from measurement from resting and moving systems of reference, and it emphasizes the fact that before determinations, which have been gained from diverse systems of reference, can be compared with each other, a universal methodic principle of transformation and permutation must be given. To each objective measurement, there must be added a certain subjective index, which makes known its particular conditions and only when this has taken place can it be used along with others in the scientific construction of the total picture of reality, in the determination of the laws of nature, and be combined with these others into a unitary result. What is gained by this reflection on the conditions of physical measurement in a pure epistemological regard appears as soon as one remembers the conflicts, which have resulted from the lack of this reflection in the course of the history of philosophy and of exact science. It seems almost the unavoidable fate of the scientific approach to the world that each new and fruitful concept of measurement, which it gains and establishes, should be transformed at once into a thing-concept. Ever does it believe that the truth and the meaning of the physical concepts of magnitude are assured only when it permits certain absolute realities to correspond to them. Each creative epoch of physics discovers and formulates new characteristic measures for the totality of being and natural process, but each stands in danger of taking these preliminary and relative measures, these temporarily ultimate intellectual instruments of measurement, as definitive expressions of the ontologically real. The history of the concept of matter, of the atom, of the concepts of the ether and of energy offer the typical proof and examples of this. All materialism—and there is a materialism not only of "matter" but also of force, of energy, of the ether, *etc.*,—goes back from the standpoint of epistemology, to this one motive. The ultimate constants of physical calculation are not only taken as real, but they are ultimately raised to the rank of that which is alone real. The development of idealistic philosophy itself is not able to escape this tendency. Descartes as an idealistic mathematician was at the same time the founder of the "mechanical view of the world." Since only extension offers us exact and distinct concepts and since all clearly comprehended truth is also the truth of the *existing*, it follows, in his view, that mathematics and nature, the system of measurements and the totality of material existence, must be identified. The manner, in which the same step from the logico-mathematical to the ontological concept has been repeated in the development of modern energetics, is known. Here, after energy has been discovered as a fundamental measure, as a measure which is not limited to the phenomena of motion, but spans equally all physical fields, it was made into an all-inclusive substance, which rivalled "matter" and finally took it up into itself completely. But on the whole, we are here concerned with a metaphysical by-way, which has not seduced science itself from its sure methodic course. For the concept of energy belongs in conception to that general direction of physical thought, which has been called the "physics of principles" in contrast to the physics of pictures and mechanical models. A "principle," however, is never directly related to things and relations of things, but is meant to establish a general rule for complex functional dependencies and their reciprocal connection. This rule proves to be the really permanent and substantial: the epistemological, as well as the physical, value of energetics is not founded on a new pictorial representation to be substituted for the old concepts of "matter" and "force" but on the gaining of *equivalence-numbers*, such as were expressly demanded and discovered by Robert Mayer as the "foundation of exact investigation of nature." (*Cf.* 52, p. 145. 237 ff.)

Even in these two examples we can learn that through the whole history of physics there is a certain intellectual movement, which throughout runs parallel to the movement in epistemology that mediates and passes to and from between the "subject" and the "object" of knowledge. Physical thought is always concerned at first with establishing a characteristic standard of measurement in an objective physical concept, in a certain natural constant. Then it is concerned, in the further development, with understanding more and more clearly the constructive element that is contained in any such original constant, and with becoming conscious of its own conditionality. For, whatever particular properties they may have, no constants are immediately given, but all must be *conceived* and *sought* before they can be found in experience. One of the most pregnant examples of this is found in the history of the concept of the atom. The atoms were postulated by Democritus as ultimate constants of nature long before thought possessed any means of concretely realizing this postulate. Fundamentally, such a realization, such a strictly quantitative meaning of the concept of the atom, was only reached in the beginnings of modern chemistry in the law of multiple proportion. To the extent, however, that we add, to this particular realization of the concept of the atom in the law of multiple proportion, others and still others and the concept of the atom finally comes to characterize and to organize intellectually the most diverse fields, its character as a pure principle, which was originally fused with its thing-character, comes to light. The content of the idea of the atom changes and shifts from place to place in the course of the development of physics and chemistry, but the function of the atom as the temporarily ultimate unit of measurement remains. When we pass from the consideration of "ponderable" matter to the consideration of the ether, when we seek a unity, which comprehends not only the mechanical but also the optical and electrical phenomena, the atom of matter becomes the atom of electricity, the electron. In recent physics, there appears further, with Planck's Quanta Theory the thought of an atomistic structure not only of matter but of energy. It would be in vain were one to attempt to combine all these various applications of the concept of the atom in chemistry, in the kinetic theory of gases, and in the doctrine of light and heat radiation, *etc.*, into a unitary picture. But the unity of its meaning requires no such pictorial unity; it is satisfied, indeed verified in a far stricter logical sense, when it is shown that here a common relation, a peculiar "form" of connection, prevails, which as such can be verified and represented in the most diverse contents. The atom shows itself thereby to be, not an absolute minimum of being, but a relative minimum of measure. It was one of the founders of modern philosophy, Nicholas Cusanus, who, with true speculative profundity, anticipated and announced this as the function of the concept of the atom, which was to be actually realized only in the history of natural science. Cusanus' fundamental doctrine of the infinite and of the unity of opposites in the infinite rested entirely on this insight into the relativity in principle of all determinations of magnitude, on the coincidence of the "greatest" and the "smallest." (*Cf.* 7, I, 40 ff, 265 ff.) Modern criticism of knowledge brings the riddle, with which Cusanus' doctrine of the minimum struggles, to a simple expression. Contradiction only enters when we attempt to unify after the fashion of a thing all the different forms, which the thought of the "smallest" assumes, in the different fields of thought; but it disappears as soon as we reflect that the true unity is never to be sought in things as such, but in intellectual constructions, which we choose according to the peculiarity of the field to be measured, and which are thus in principle possessed of an unlimited variability. It follows from this that, as *what is to be measured* is unlimited in variety, so *what measures* can be represented in infinitely many and infinitely diverse ways. In other words, the unity that we have to seek lies neither in the one nor the other member, but merely in the form of their reciprocal connection, *i.e.*, in the logical conditions of the operation of measurement itself.

This receives new confirmation when we pass from the concept of matter, of energy and of the atom to the real concept of objectivity of modern physics, that of motion. The historical beginnings of the modern theory of motion in Galileo refer directly to the epistemological question, which has received its definitive formulation in the general theory of relativity. What Galileo gained with *his* idea of relativity was the cancelling of the absolute reality of place, and this first step involved for him the most weighty logical consequences, *viz.*, the new concept of the lawfulness of nature and the new interpretation of the particular laws of dynamics. Galileo's doctrine of motion is rooted in nothing less and nothing more than in the choice of a new standpoint from which to estimate and measure the phenomena of motion in the universe. By this choice, there was given him at once the law of inertia and in it the real foundation of the new view of nature. The ancient view saw in place a certain physical property that produced definite physical effects. The "here" and "there," the "above" and "below," were for it no mere relations; but the particular point of space was taken as an independent real, which consequently was provided with particular forces. In the striving of bodies to their "natural places," in the pressure of air and fire upwards and in the sinking of heavy masses downwards, these forces seemed given as immediate empirical realities. Only when one takes account of these fundamental features, not only of ancient astronomy and cosmology, but also of ancient physics, does one under-

stand the boldness of the new intellectual orientation, resulting
from the Copernican system of the world.   One of the most fixed and
certain realities on which Grecian thought had constructed its picture
of the world now became a mere illusion, a purely "subjective" feature.   Even the first adherents of the new doctrine drew the decisive
conclusion with reference to the doctrine of place.   What Gilbert,
e.g., urges against the Aristotelian physics and cosmology is above
all this epistemological feature, *i.e.*, that it confuses the ideal
and the real.   Differences belonging merely to our thought, to
our subjective reflection, are throughout made into objective
oppositions.   But in truth, no place in itself is opposed to any
other, but there are in nature only differences in the mutual positions of bodies and of material masses.   "It is not place which,
in the nature of things, works and produces, which determines the
rest and motion of bodies.   For it is in itself neither a being nor an
effective cause; rather bodies determine their mutual place and position by virtue of the forces which are immanent in them.   The
place is a nothing; it does not exist and exerts no force, but all natural
power is contained and grounded in bodies themselves."   (7, I,
360 f.)   It is implied in this that what we call the "true place" is
never given to us as an immediate sensuous property, but must be
discovered on the basis of calculation and of the "arithmetic of
forces" in the universe.   All determination of place—as Kepler
sharply and clearly expresses this insight which for him resulted
equally from astronomical convictions, physiological optics and
analysis of the general problem of perception—is a work of the mind:
omnis locatio mentis est opus.   (37, II, 55, *cf.* 7, I, 339.)   From this
point the way is open to Galileo's foundation of dynami cs:for since
place has ceased to be something real, the question as to the *ground*
of the place of a body and the ground of its *persistence* in one and
the same place disappears.   Objective physical reality passes from
place to *change of place*, to motion and the factors by which it is
determined as a magnitude.   If such a determination is to be possible
in a definite way, the identity and permanence, which were hitherto
ascribed to mere place, must go over to motion; motion must possess
"being," that is, from the standpoint of the physicist, numerical
constancy.   This demand for the numerical constancy of motion itself
finds its expression and its realization in the law of inertia.   We
recognize here again how closely, in Galileo, the mathematical motive

of his thought was connected with an ontological motive, how his conception of *being* interacted with his conception of *measure*. The new measure, which is found in inertia and in the concept of uniform acceleration, involves also a new determination of reality. In contrast with mere place, which is infinitely ambiguous and differs according to the choice of the system of reference, the inertial movement appears to be a truly intrinsic property of bodies, which belongs to them "in themselves" and without reference to a definite system of comparison and measurement. The velocity of a material system is more than a mere factor for calculation; it not only really belongs to the system but defines its reality since it determines its *vis viva*, *i.e.*, the measure of its dynamic effectiveness. In its measure of motion, in the differential quotient of the space by the time, Galileo's physics claims to have reached the kernel of all physical being, to have defined the intensive reality of motion. By this reality, the dynamic consideration is distinguished from the merely phoronomic. The concept of the "state of motion," not as a mere comparative magnitude, but as an essential element belonging to the moving system intrinsically, now becomes the real mark and characteristic of physical reality. Leibniz, too, in his foundation of dynamics, stands throughout at this standpoint, which becomes for him a starting-point for a new metaphysics of forces. Motion conceived as a mere change of place in the purely phoronomical sense, he explains, remains always something purely relative; it only becomes an expression of a true physical and metaphysical reality when we add to it an inner dynamic principle, a force conceived as an "originally implanted principle of permanence and change," *principium mutationis et perseverantiae.* (42, VI, 100 *cf.* 5, p. 290 ff.) In all these examples, it is evident how sharply, on the one hand the physical thought of modern times has grasped the thought of the relativity of place and of motion, and, on the other hand, how it has shrunk back from following it to its ultimate consequences. If not only place but the velocity of a material system is to signify a magnitude that entirely depends on the choice of the reference body and is thus infinitely variable and infinitely ambiguous, there seems no possibility of an exact determination of magnitude and thus no possibility of an exact objective determination of the state of physical reality. Pure mathematics may be constructed as the ideal doctrine of the comparison and connection of magnitudes, as a system of mere relations and functions and may come to recognize itself as such ever more clearly, but physics seems necessarily to reach an ultimate limit, a *non plus ultra*, if it is not wholly to lose any basis in reality.

The difficulty, which remains in the structure of classical mechanics in the formulation of the principle of inertia, is expressed in an epistemological circle, from which there seems no escape. To understand the meaning of the law of inertia, we need the concept of "equal times" but a practicable physical measure of equal times can, as is discovered, only be gained by presupposing, in its content and validity, the law of inertia. In fact, since Carl Neumann's work, *Uber die Prinzipien der Galilei-Newton'schen Theorie* (57), which set in motion the modern discussion on the law of inertia, it is customary in mechanics to define equal times as times within which a body left to itself traverses equal distances. Maxwell too, in his exposition of the Newtonian mechanics, conceived the law of inertia as a pure definition of measure. The first law of Newton, as he explains clearly and pregnantly, tells under what conditions no external force is present. (51, p. 31.) Thus in the progress of mechanics the principle of inertia is *recognized* with increasing distinctness as what it *meant* fundamentally to Galileo. It is no longer taken as a direct empirical description of given processes of nature, but as the "axiom of the field," the fundamental hypothesis by which the new science of dynamics prescribes to itself a certain form of measurement. Inertia appears, not as an absolute and inherent property of things and of bodies, but as the free establishment of a certain standard and symbol of measurement, by virtue of which we can hope tó reach a systematic conception of the laws of motion. In this alone is rooted its reality, *i.e.*, its objective and physical significance. Thus, within the historical development of physics itself *what measures* is separated with increasing distinctness from *what is measured*, with which it at first seems to coincide; the observable data of experience are separated with increasing distinctness from what must be presupposed and used as a *condition* of observation and of measurement.

And what is here seen in a particular example and within a narrow field is repeated, on closer examination, in all the special fields of physics. Everywhere physical thought must determine for itself its own standards of measurement before it proceeds to observation. There must be established a certain standpoint for the comparison and correlation of magnitudes; certain constants must be established at least hypothetically and in preliminary fashion before a concrete measurement can take place. In this sense, each measurement contains a purely ideal element; it is not so much with the sensuous instruments of measurement that we measure natural processes as with our own thoughts. The instruments of measurement are, as it were, only the visible embodiments of these thoughts, for each of them involves its own theory and offers correct and useful results only in so far as this theory is assumed to be valid. (*Cf.* 8, p. 189 ff.) It is not clocks and physical measuring-rods but principles and postulates that are the real instruments of measurement. For in the multiplicity and mutability of natural phenomena, thought *possesses* a relatively fixed standpoint only by *taking* it. In the choice of this standpoint, however, it is not absolutely determined by the phenomena, but the choice remains its own deed for which ultimately it alone is responsible. The decision is made with reference to experience, *i.e.*, to the connection of observations according to law, but it is not prescribed in a definite way by the mere sum of observations. For these in themselves can always be expressed by a number of intellectual approaches between which a choice is possible only with reference to logical "simplicity," more exactly, to systematic unity and completeness, of scientific exposition. When thought, in accordance with its claims and demands, changes the form of the "simple" fundamental measuring relations, we stand before a new "picture" of the world with regard to content also. The previously gained relations of experience do not indeed lose their validity, but, since they are expressed in a new conceptual language, they enter into a new system of meanings. The fixed Archimedean point of the former view of the world moves; the previous ποῦ στῶ of thought appears transcended. But it is soon seen that thought, by virtue of its peculiar function, can only transcend an earlier construction by replacing it by a more general and more inclusive one; that it only shifts, among phenomena, the constancy and identity, which it cannot cease to demand, to another and deeper place. That every realization, which the demand of thought for ultimate constants can find within the empirical world is always only conditioned and relative, is guaranteed by the unconditionality and radicalism of precisely this demand. The critical theory of knowledge would not only show this connection *in abstracto*, but for it the concrete movement of thought, the continual oscillation between experience and concept, between facts and hypotheses in the history of physics, forms a perpetually new source of instruction. In the midst of the change of particular theoretical instruments of measurement, the critical theory holds fast to the thought of the unity of measurement, which indeed signifies for it no realistic dogma but an ideal goal and a never-to-be-concluded task. Each new physical hypothesis erects, as it were, a new logical system of coördinates, to which we refer phenomena, while nevertheless the doctrine is retained as a regulative idea for investigation that all these systems converge on a certain definite limiting value. In the confusion and continuous flux of phenomena, the understanding seems at first almost arbitrarily to fix and separate out certain points in order to learn through them a definite law of change, but everything which it regards as determined and valid in this sense proves, in the course of further progress, to be a mere approximation. The first construction must be both limited and more exactly defined logically by the second, this again by the third, *etc.* Thus, ever anew does the temporarily chosen theoretical center of thought shift; but in this process, the sphere of being, the sphere of objective knowledge, is more and more penetrated by thought. As often as it seems that thought is overturned by new facts and observations, which are outside its previously formulated laws, it is seen that, in fact, thought has found in them a new point of leverage, around which moves henceforth the totality of empirically provable "facts." The epistemological exposition and evaluation of each new physical theory must always seek to indicate the ideal center and turning-point around which it causes the totality of phenomena, the real and possible observations, to revolve,—whether this point is clearly marked or whether the theory only refers to it indirectly by the intellectual tendency of all its propositions and deductions.

CHAPTER 11

The Empirical and Conceptual Foundations of the Theory of Relativity

If there can be no doubt, according to the opening words of the *Critique of Pure Reason*, that all our knowledge begins with experience, then this holds especially when we are concerned with the origin of a physical theory. The question here can never run as to *whether* the theory has issued from experience but merely as to *how* it is based on experience, and what is the relation of the diverse elements which characterize and make up the concept of experience as such. There is accordingly needed no special epistemological analysis to make clear the relation of the special and general theories of relativity to experience, to the whole of observation and of physical experiment; such an analysis will only have to decide whether the theory in its origin and development is to be taken as an example and witness of the critical or of the sensualistic concept of experience. Does "experience," as it is used here, mean merely the bare sum of particular observations—*experimentorum multorum coacervatio*, as a sensualistic thinker once described it—or is there involved in it an independent intellectual form? Is the construction of the theory merely a matter of joining "fact" to "fact," perception to perception, —or, in this connection of particulars, have there been effective all along certain universal and critical norms, certain methodic presuppositions? No "empiricism" however extreme can ever seek to deny the rôle of thought in establishing and grounding physical theories, and just as little is there, on the other hand, a logical idealism, which would attempt to free "pure thought" from reference to the world of the "factual" and from being bound to it. The question dividing the two views can only be as to whether thought consists in a simple registration of facts, or whether, even in the *establishment*, in gaining an interpretation of "particular facts," thought reveals its characteristic power and function. Is its work completed in arranging particular data, immediately taken from sense perception, like pearls on a thread—or does it face them with its own original measures as independent criteria of judgment?

367

The problem raised here received its first sharp and clear systematic formulation in the Platonic doctrine of ideas.   For Platonic idealism, too, the proposition holds that it is not possible to think save on the basis of some perception: οὐ δυνατὸν ἐννοεῖν ἢ ἔκ τινος αἰσθήσεως. But the function of the "logic in us" consists indeed not in finding the sum of the particular perceptions, not in deriving and deducing the "idea of the equal" from the "equal pieces of wood and stone," but the "logic in us" is revealed in discriminating and judging what is given in perception. This discriminating constitutes the real fundamental character of thought as διάνοια, as *discursus*. Not all perceptions and observations stimulate equally the critical and discriminating activity of thought.   There are some which do not summon the understanding to reflection, since satisfaction is done them by mere sensation, but there are others which in all ways call forth thought, as in their case perception by itself could gain nothing solid.   "Not stimulating, namely, is that which does not pass into an opposite perception; stimulating objects I call those which give opposite perceptions, because here perception gives no more vivid idea of any particular object than of its opposite.   Much in perception is a paraclete of thought (παρακλητικὰ τῆς διανοίας), while other perceptions are not—*such an awakener of thought, namely, is everything, which comes into sense at the same time as its opposite;* but what does not, that also does not arouse thought." (*Republic* 523–524.)   In this Platonic characterization of the relation of thought and sensation, of reason and sensibility, we have, as Cohen has urged, "one of the most fundamental thoughts in the evolution of the critique of cognition." (12, p. 16 ff.) Just as for Plato thought becomes what it is in assertion and contradiction, in dialectic, so only a perception to which this feature corresponds can become the awakener and paraclete of thought.   The dialectic of perception summons that of thought to judgment and decision.   Where the perceptions, as it were, rest peacefully side by side, where there is no inner tension between them, thought rests also; only where they contradict each other, where they threaten to cancel each other does thought's fundamental postulate, its unconditional demand for unity stand forth and demand a transformation, a reshaping of experience itself.

The evolution of the theory of relativity has furnished a new typical proof of this general relation.   It was in fact a fundamental contradiction between physical experiments from which the theory of relativity took its start. On the one side stood the investigation of Fizeau, on the other, that of Michelson, and the two seemed in their results absolutely irreconcilable. Both sought to gain an answer to the question as to how the velocity of light in a moving medium was related to its velocity in a resting medium; and they answered this question in completely opposite ways. The investigation of Fizeau showed that the velocity of light in flowing water was greater than in water at rest; that, however, not the whole velocity of the flowing water, but only a fraction of it was added to the velocity of light in a medium at rest. If we call the velocity of light in the moving medium W and the velocity of light in a medium at rest w and the velocity of the flowing v, it results not simply that $W = w + v$, but rather that $W = w + v \left(1 - \dfrac{1}{n^2}\right)$, in which the magnitude $n = \dfrac{c}{w}$ signifies the exponent of the refraction of the liquid. This result, as interpreted by the theory of Lorentz, spoke directly for the assumption of a motionless ether not carried along by the body in its movement. But the attempt of Michelson, to discover the consequences of the movement of the earth with reference to this motionless ether, failed. In no way could any influence be shown of the motion of the earth on the velocity of the propagation of light; it was rather shown with increasing evidence that all optical phenomena take place as if there were no translation of the earth against the ether.[1] And behind this conflict of "facts" there stood, as one was forced to recognize more and more, a conflict in general principles, to which the theories of mechanical and of optical and electromagnetic phenomena seemed to lead necessarily. Experiments in the latter could finally be summarized in a single proposition, the principle of the constancy of the velocity of light in a vacuum. The validity of the fundamental equations of electrodynamics of Maxwell and Hertz involved the assumption that light in an empty space is always propagated with a definite velocity V independently of the state of motion of the body emitting it. From whatever system one made the observation and from whatever source the light issued there would always be found the same determinate value for its velocity of propagation. But this assumption of the velocity of

[1] For more detail concerning the investigations of Fizeau and Michelson as well as concerning the negative outcome of other investigations on the influence of the movement of the earth on optical and electrical phenomena, *cf.* Laue (40), p. 10 ff.

light as a universal constant the same for all systems, necessarily demanded by the principles of electrodynamics, now comes into opposition with the principle of relativity of the Galileo-Newtonian mechanics. This principle demands that, when any definite Galileian reference body is given —*i.e.*, a body relatively to which a body "left to itself" persists in its state of rest or of uniform motion in a straight line—all the laws, which are valid relatively to this reference body K remain valid when one passes to the system of reference K', which is, with reference to K, in uniform translatory motion. In the transition from K to K', the equations of the "Galileo-transformation" hold, (where v signifies a constant velocity of K' with

$$x' = x - vt, \ y' = y, \ z' = z$$

reference to K parallel to the x and x' axes), to which there is to be added the identical transformation from the time t' = t, which is not especially noticed in classical mechanics. If we seek, however, to apply the principle of relativity of mechanics to electrodynamics, *i.e.*, to recalculate its equations according to the formulae of the Galileo-transformation, it is seen that this cannot be done: the electrodynamic equations, in contrast to the Newtonian equations of motion, alter their form when we insert the coördinates x', y', z', t', into them in place of the coördinates x, y, z, t according to the rules of the Galileo-transformation. The effort to unite mechanics and electro-dynamics by carrying over the principle of relativity of the first into the second thus has to be given up: the Hertzian theory, which represented such an attempt, came into irreconcilable conflict with assured experimental results. Physical investigation stood before the dilemma of giving up a principle which had been verified without exception in all the phenomena of motion and which formed a corner-stone in the structure of classical mechanics—or of retaining it within its field but denying its applicability to optical and electromagnetic phenomena. In both cases, the unity of the explanation of nature, the unity of the very concept of nature, seemed destroyed. Here in fact the condition set up by Plato of the intellectual fruitfulness of experience was fulfilled: here experience stood at a point at which assured observation seemed to pass directly into its opposite. The conflict between the principle of the constancy of the propagation of light and the principle of relativity of mechanics became the "paraclete of thought"—the real awakener of the theory of relativity.

But how did physical thought go about overcoming this conflict, since it was bound to the outcome of observation as such, since it could neither set aside the facts expressed in the principle of the constant velocity of light in a vacuum, nor those expressed in the principle of relativity of mechanics?   If we look back on the historical development of the theory of relativity, we recognize that the latter has followed here a counsel which was once given by Goethe. "The greatest art in theoretical and practical life," wrote Goethe to Zelter, "consists in changing the *problem* into a *postulate; that way one succeeds."   In fact, this was the course which Einstein followed in his fundamental essay. *Zur Elektrodynamik bewegter Systeme* of the year 1905.   The principle of the constancy of the velocity of light was given first place as a postulate, but,—supported by the negative result of all attempts to establish an "absolute" motion with reference to a chosen system of reference, *i.e.*, the "motionless ether,"—the *supposition* was made that there correspond to the concept of absolute rest no properties of phenomena in either mechanics or electrodynamics, but rather that the same electrodynamic and optical laws hold for all systems of coördinates of which the mechanical equations hold.   And this *"supposition"* does not continue such, but is expressly "made a presupposition," *i.e.*, a shaping of theory is demanded which will simultaneously satisfy the conditions of the principle of relativity and those of the principle of the constant propagation of light.   (*Cf.* 16, p. 26).   The two assumptions are indeed not compatible according to the means and habits of thought at the disposal of the kinematics generally accepted before the establishment of the theory of relativity, but they—*ought* no longer to be incompatible.   The demand made of physical theory was that it remove this incompatibility by subjecting precisely these means and habits of thought themselves to a critical examination. By an analysis of the physical concepts of space and time, it now appears that in fact the incompatibility of the principle of relativity with the law of the propagation of light is not to be found; that rather there is only needed a transformation of these concepts in order to reach a logically unobjectionable theory.   The decisive step is taken when it is seen that the measurements, to be gained within a system by definite physical methods of measurement, by the application of fixed measuring-rods and clocks, have no "absolute" meaning fixed once for all, but that they are dependent on the state of motion of the system and must necessarily result differently according to the latter. There now arises the purely mathematical problem of discovering the laws of permutation, according to which the space-time values of an event are changed in going from one reference body to another, which is in uniform translatory motion with regard to the first. This problem is solved, as is known, by the fundamental equations of the "Lorentz-transformation:"

$$x' = \frac{x - v\,t}{\sqrt{1 - \dfrac{v^2}{c^2}}}$$

$$y' = y \qquad\qquad z' = z$$

$$t' = \frac{t - \dfrac{v}{c^2}x}{\sqrt{1 - \dfrac{v^2}{c^2}}}$$

On the basis of these equations, we see that the law of the propagation of light in a vacuum is equally fulfilled for all justified systems K and K'; on the other hand, it is seen that Maxwell's fundamental equations of electrodynamics do not change their form when the formulae of the Lorentz-transformation rather than those of the Galileo-transformation are applied to them. There is thus a universal principle of relativity, which comprehends the totality of physical phenomena; the laws, according to which the states of physical systems change, are independent of whether they are referred to one or the other of two systems of coördinates in uniform translatory motion relative to each other. (*Cf.* 16, p. 29). The principle of relativity of classical mechanics is so little contradicted by this general principle that it is rather contained in it as a special case; the equations of the Galileo-transformation directly issue from those of the Lorentz-transformation when one considers only such velocities v as are very small in comparison with the velocity of light so that the values $\dfrac{v}{c^2}\ \dfrac{v^2}{c^2}$ can practically be left out of account. It follows from this that the principle of relativity of electrodynamics, carried over to mechanics, can come into conflict with no empirical result, while the converse carrying-over of the principle of relativity of mechanics to electrodynamics proves to be impossible, as the collapse of Hertz's theory showed. More closely considered, however,

in the special theory of relativity, the electrodynamic processes are
not used as a key to the mechanical, but a truly universal principle,
a heuristic maxim of investigation in general, is established, which
claims to contain a criterion of the validity and permissibility of all
particular physical fields and theories. Thus it is seen that the
initial contradiction, appearing between the principles of mechanics
and those of electrodynamics, has shown the way to a far more
perfect and deeper unity between them than previously existed.
And this result was not reached entirely by heaping up experiments
by newly instituted investigations, but it rests on a critical transformation of the system of fundamental physical concepts.

On the purely epistemological side, there thus appears with special
distinctness in this intellectual process in which the theory of relativity originates, that peculiar "Copernican revolution." that variation
in the conceptual foundations of the theory of nature, which we have
previously traced in the example of classical mechanics and the older
physics. An essential part of its achievement seems based on the
fact, that it has shifted the previous logical constants of physical
knowledge, that it has set them at another place than before. For
classical mechanics, the fixed and immovable point was the assumption of the identity of the spatial and temporal values gained by
measurement in the various systems. This identity was taken to
be the unquestionable and sure foundation of the concept of objectivity in general: as that which first really constituted the object of
"nature" as a geometrical and mechanical object and distinguished
it from the changeable and relative data of sensation. τὸ μὲν σχμα
καθ’ αὐτό ἐστι, τὸ δὲ γλυκὺ καί ὅλως τὸ αἰσθητὸν πρὸς ἄλλο καὶ ἐν ἄλλοις
—thus runs the proposition, which Democritus brought into
the foundations of atomism, and which in modern times was taken
up by Galileo to support the fundamental distinction between "primary" and "secondary" qualities, and thus the whole "mechanical"
view of the world. Although the principle here established proved
to be very fruitful and has been frequently confirmed in mathematical
physics, the modern evolution of physics shows, with increasing
evidence, that it was conceived too narrowly in a philosophical
and methodological sense. The true goal of science is not mechanism but unity—as Henri Poincaré once formulated the guiding
maxim of modern physics. But concerning this unity the physicist
does not need to ask *whether* it is, but merely *how* it is; *i.e.*, what is

the minimum of presuppositions that are necessary and sufficient
to provide an exact exposition of the totality of experience and its
systematic connection. (72 p. 172 ff.) In order to maintain this
unity, which seemed endangered by the conflict of the principle of
the constancy of the velocity of light and the principle of relativity
of mechanics, and to ground it more deeply and securely, the theory
of relativity renounces the unity of the values of spatial and temporal
magnitudes in different systems. It surrenders the assumption that
the temporal interval between events is a magnitude fixed once for
all independently of the state of motion of the reference body and
that in the same way the spatial distance between two points of a
rigid body is independent of the state of motion of the reference
body. By going back to the method of measuring time and to the
fundamental rôle that the velocity of light plays in all our physical
time measurements, it discovers the *relativity of the simultaneity* of
two processes and further leads to the insight that the magnitude of
the length of a body of its volume, its form, its energy and temperature, *etc.*, are, as results from the formulae of the Lorentz-transformation, to be assumed as different according to the choice of the system
of reference in which measurement takes place. But these "relativi-
zations" are not in contradiction with the doctrine of the constancy
and unity of nature; they are rather demanded and worked out in the
name of this very unity. The variation of the measurements of
space and time constitutes the necessary condition through which the
new invariants of the theory are discovered and grounded. Such
invariants are found in the equal magnitude of the velocity of light
for all systems and further in a series of other magnitudes, such as
the entropy of a body, its electrical charge or the mechanical equivalent of heat, which are unchanged by the Lorentz-transformation and
which thus possess the same value in all justified systems of reference.
But above all it is the general form of natural law which we have to
recognize as the real invariant and thus as the real logical frame-
work of nature in general. While the special theory of relativity
limits itself to regarding all reference bodies K′ which are moving
uniformly in a straight line relatively to a definite justified reference
system K, as equivalent for the formulation of natural laws, the
general theory extends this proposition to the assertion that all
reference bodies KK′, whatever their state of motion may be, are
to be taken as equivalent for the description of nature. (17 p. 9;

18, p. 42.) But the path by which alone this true universality of the concept of nature and of natural law, *i.e.*, a definite and objectively valid description of phenomena independent of the choice of the system of reference, is to be reached, leads, as the theory shows, necessarily through the "relativization" of the spatial and temporal magnitudes, that hold within the individual system; to take these as changeable, as transformable, means to press through to the true invariance of the genuine universal constants of nature and universal laws of nature. The postulate of the constancy of the velocity of light and the postulate of relativity show themselves thus as the two fixed points of the theory, as the fixed intellectual poles around which phenomena revolve; and in this it is seen that the previous logical constants of the theory of nature, *i.e.*, the whole system of conceptual and numerical values, hitherto taken as absolutely determinate and fixed, must be set in flux in order to satisfy the new and more strict demand for unity made by physical thought.

Thus reference to experience, regard for phenomena and their unified exposition, proves to be everywhere the fundamental feature, but at the same time it is seen that, in the words of Goethe, experience is always only half experience; for it is not the mere observational material as such, but the ideal form and the intellectual interpretation, which it is given, that is the basis of the real value of the theory of relativity and of its advantage over other types of explanation. As is known, the investigation of Michelson and Morley, which gave the impetus and starting-point for the development of the theory of relativity, was explained as early as the year 1904 by Lorentz in a manner which fulfilled all purely physical demands. The Lorentzian hypothesis, that each body moving with reference to the motionless ether with a velocity v undergoes a certain shortening in the dimension parallel to the motion, and indeed in the ratio of

$1 : \sqrt{1 - \frac{v^2}{c^2}}$, was sufficient to give a complete explanation of all known

observations. An experimental decision between Lorentz's and Einstein's theories was thus not possible; it was seen that between them there could fundamentally be no *experimentum crucis*.[2] The advocates of the new doctrine accordingly had to appeal—an unusual spectacle in the history of physics—to general philosophical grounds,

---

[2] For more detail *cf. e. g.* Ehrenfest (15a), p. 16 ff.

to the advantages over the assumption of Lorentz which the new doctrine possessed in a systematic and epistemological respect. "A really experimental decision between the theory of Lorentz and the theory of relativity," Laue, *e.g.*, explains in his exposition of the principle of relativity in the year 1911, "is indeed not to be gained, and that the former, in spite of this, has receded into the background, is chiefly due to the fact, that, close as it comes to the theory of relativity, it still lacks the great simple universal principle, the possession of which lends the theory of relativity . . . . an imposing appearance."[3] Lorentz's assumption appeared above all to be epistemologically unsatisfactory because it ascribes to a physical object, the ether, definite effects, while at the same time it results from these effects that ether can never be an object of possible observation. Minkowski too explains in his lecture on space and time that the Lorentzian hypothesis sounds extremely fantastical; for the contraction is not to be conceived as a physical consequence of the resistance of the ether but rather purely as "a gift from above," as an accompaniment of the state of motion. (47 p. 60 f.) What thus, in the last analysis, decided against this assumption was not an empirical but a methodological defect. It conflicted most sharply with a general principle, to which Leibniz has appealed in his struggle against the Newtonian concepts of absolute space and time, and which he formulated as the "principle of observability" (*principe de l'observabilité.*) When Clarke, as the representative of Newton, referred to the possibility that the universe in its motion relatively to absolute space might undergo retardation or acceleration which would not be discoverable by our means of measurement, Leibniz answered that nothing fundamentally outside the sphere of

[3] 40, p. 19 f.; *cf.* 41, p. 106. *Cf.* also the characteristic remark of Lorentz himself in his Haarlem lecture: "The estimation (of the fundamental concepts of Einstein's theory of relativity) *belongs to a very large extent (grösstenteils) to the theory of knowledge,* and one can leave the judgment to the latter in confidence that it will consider the questions mentioned with the necessary thoroughness. But it is certain that it will depend for a great part on the type of thought to which one is accustomed, whether one feels drawn more to the one or the other conception. As far as concerns the lecturer himself, he finds a certain satisfaction in the older conceptions, that ether possesses at least some substantiality, that space and time can be sharply separated, that one can talk of simultaneity without further specification." (46a, p. 23.)

observation possessed "being" in the physical sense: *quand il n'y a point de changement observable, il n'y a point de changement du tout.* (5, p. 247 ff.). It is precisely this principle of "observability," which Einstein applied at an important and decisive place in his theory, at the transition from the special to the general theory of relativity, and which he has attempted to give a necessary connection with the general principle of causality. Any physical explanation of a phenomenon, he urges, is *epistemologically* satisfactory only when there enter into it no non-observable elements; for the law of causality is an assertion concerning the world of experience only when *observable facts* occur as causes and effects. (17, §2). Here we stand before one of the fundamental intellectual motives of the theory of relativity—a motive which not only gives it the advantage over the empirically equivalent hypothesis of Lorentz, but which also produces the advance from the more limited interpretation of the postulate of relativity in the special theory to the completely universal formulation.

The way in which this advance has taken place is especially suited to make clear the conceptual and empirical presuppositions of the theory and their reciprocal connection. The special theory of relativity rests, as has been shown, on two different assumptions, which stand equally justified, side by side: on the postulate of the uniformity of the propagation of light in a vacuum and on the presupposition that all reference systems in rectilinear, uniform and non-rotary motion relatively to a definite justified system K are equally permissible for the formulation of the laws of nature. If one considers these presuppositions, which stand in inseparable connection in the empirical structure of the special theory of relativity, from a purely methodological standpoint, it is seen that in *this* respect they belong to different strata. On the one side, stands the assertion of a general fact, a constant of nature, which results from the experimental findings of optics and electrodynamics; on the other side stands a demand, which we make of the *form* of natural laws. In the first case, it is empirically established that there is a peculiar velocity with a definite finite value, which retains this value in any system independently of the state of motion of the latter. In the second, a general *maxim* is established for the investigation of nature, which is to serve as a "heuristic aid in the *search* for the general laws of nature." In the formal limitation, which is placed

on natural laws by this maxim, lies—as Einstein himself has urged—
the characteristic "penetration" (*Spürkraft*) of the principle of relativity. (18, pp. 28, 67.) But the two principles, the "material"
and the "formal" are not distinguished from each other in the
shaping of the special theory of relativity. The fact that this distinction is made and that the general and "formal" principle is
placed above the particular and "material" principle constitutes, from
the purely epistemological standpoint, the essential step taken by
the general theory of relativity. And this step seems to lead to a
strange and paradoxical consequence; for the particular result is not
taken up into the general, but rather is cancelled by it. From the
standpoint of the general theory of relativity, the law of the constancy
of the velocity of light in a vacuum no longer possesses unlimited
validity. According to the general theory of relativity the velocity
of light is dependent on the gravitation potential and must thus in
general vary with places. The velocity of light must always depend
on the coördinates when a field of gravitation is present; it is only
to be regarded as constant when we have in mind regions with a
constant gravitation potential. This consequence of the general
theory of relativity has often been regarded as a refutation of the
presupposition from which the special theory of relativity took its
start and on which it based all its deductions. But with justice
Einstein rejects any such conclusion. The special theory of relativity, he explains, is not rendered valueless by the fact that one comes
to see that its propositions refer to a definitely limited field, namely,
to the phenomena in an approximately constant field of gravitation.
"Before the establishment of electrodynamics, the laws of electrostatics were regarded as the laws of electricity in general. Today
we know that electrostatics can only describe electrical fields correctly in the case, that is never exactly realized, in which the electric
masses are exactly at rest relatively to each other and to the system
of coördinates. Is electrostatics overthrown by Maxwell's electro-
dynamical equations? Not in the least! Electrostatics is contained
as a limiting case in electrodynamics; the laws of the latter lead
directly to those of the first for the case that the fields are temporarily
unchangeable. The most beautiful fate of a physical theory is to
point the way to the establishment of a more inclusive theory, in
which it lives on as a limiting case." (18, p. 52.) In fact, in the
advance from the special to the general theory of relativity, we

have only a verification of the same principle of the construction of concepts of natural science that is found in the advance from classical mechanics to the special theory of relativity. The constants of measurement and of the theory of nature in general are shifted and magnitudes, which were regarded as absolute from the earlier standpoint, are again, with the gaining of a new theoretical unit of measurement, made into merely relative determinations valid only under definite conditions. While classical mechanics, like the special theory of relativity, distinguishes between certain reference bodies relatively to which the laws of nature were valid and certain relatively to which they were not, this distinction is now cancelled. The expression of the universal physical laws is freed from any connection with a particular system of coördinates or with a certain group of such systems. To be expressed the laws of nature always require some definite system of reference; but their meaning and value is independent of the individuality of this system and remains self-identical whatever change the latter may undergo.

Only with this result do we reach the real center of the general theory of relativity. Now we know where lie its truly ultimate constants, its cardinal points, around which it causes phenomena to revolve. These constants are not to be sought in particular given things, which are selected as chosen systems of reference from all others, such systems as the sun was to Copernicus and as the fixed stars were for Galilei and Newton. No sort of things are truly invariant, but always only certain fundamental relations and functional dependencies retained in the symbolic language of our mathematics and physics, in certain equations. This result of the general theory of relativity, however, is so little a paradox from the standpoint of the criticism of knowledge, that it can rather be regarded as the natural logical conclusion of an intellectual tendency characteristic of all the philosophical and scientific thought of the modern age.[4] To the popular view and its habits of thought the radical resolution of "things" into mere relations remains as ever suspicious and alienating, for this view believes that it would lose with the thing-concept the one sure foundation of all objectivity, of all scientific truth. And thus, from this side not so much the positive as the negative aspect of

[4] Here, indeed, I can only make this assertion in a general way; for its proof I must refer to the more specific explanation in my work *Substance and Function*. (8, pp. 148–310.)

the theory of relativity has been emphasized; what it destroys, not what it constructs has been comprehended. But it is remarkable to find this interpretation not only in popular expositions of the theory of relativity but in investigations of its general "philosophical" significance; and to meet in the latter also the view that it brings an element of subjective arbitrariness into the formulation of the laws of nature and that, along with the unity of space and time, the unity of the concept of nature is destroyed. In truth, as closer consideration shows, the theory of relativity is characterized throughout by the opposite tendency. It teaches that to attain an objective and exact expression of natural process, we cannot take without further consideration the space and time values, gained by measurement within a definite system of reference as the only and universal values, but that we must, in scientifically judging these measurements, take account of the state of motion of the system from which the measurement is made. Only when this is done can we compare measurements which have been made from different systems. Only those relations and particular magnitudes can be called truly objective which endure this critical testing, that is, which maintain themselves not only for one system but for all systems. That not only are there such relations and values, but that there must be such, in so far as a science of nature is to be possible, is precisely the doctrine the theory of relativity sets up as a postulate. If we start, as practically we must do at first, from a definite system of measurement, we must bear in mind that the empirical values, which we gain here, do not signify the final natural values but that, to become such, they must undergo an intellectual correction. What we call the system of nature only arises when we combine the measurements, which are first made from the standpoint of a particular reference body, with those made from other reference bodies, and in principle with those made from all "possible" reference bodies, and bring them ideally into a single result. How there can be found in this assertion any limitation of the "objectivity" of physical knowledge is not evident; obviously it is meant to be nothing but a definition of this very objectivity. "But it is clear," says Kant, "that we have only to do with the manifold of our presentations and that X, which corresponds to them (the object), since it is to be something distinct from all our presentations, is for us nothing; the unity, which makes the object necessary, can be nothing else than the formal unity of consciousness in the synthesis of the manifold of presentations. Thus we say: we know the object when in the manifold of intuition we have produced synthetic unity." The object is thus not gained and known by our going from empirical determinations to what is no longer empirical to the absolute and transcendent, but by our unifying the totality of observations and measurements given in experience into a single complete whole. The theory of relativity shows the whole complexity of this task; but it retains the postulate of the possibility of such a system all the more strenuously and points out a new way to realize it. Classical mechanics believed itself at the goal too soon. It clung to certain reference bodies and believed that it possessed, in connection with them, measures in some way definitive and universal, and thus absolutely "objective." For the new theory, on the contrary, true objectivity never lies in empirical determinations, but only in the manner and way, in the function, of determination itself. The space and time measurements in each particular system are relative; but the truth and universality, which can be gained nevertheless by physical knowledge, consist in the fact that all these measurements correspond mutually and are coördinated with each other according to definite rules. More than this indeed knowledge cannot achieve, but it cannot ask for more, if it understands itself. To wish to know the laws of natural processes independently of all relation to any system of reference, is an impossible and self-contradictory desire; all that can be demanded is that the content of these laws not be dependent on the individuality of the system of reference. It is precisely this independence of the accidental standpoint of the observer that we mean when we speak of the "natural" object and the "laws of nature" as determinate in themselves. Measurements in *one* system, or even in an unlimited number of "justified" systems would in the end give only particularities, but not the true "synthetic unity" of the object. The theory of relativity teaches, first in the equations of the Lorentz-transformation and then in the more far-reaching substitution formulae of the general theory, how we may go from each of these particularities to a definite whole, to a totality of invariant determinations. The anthropomorphism of the natural sensuous picture of the world, the overcoming of which is the real task of physical knowledge,[5]

---

[5] *Cf.* Planck (66) p. 6 ff. and (67) p. 74.

is here again forced a step further back. The mechanical view of the world thought to have conquered it, when it resolved all being and natural process into motion and thus put everywhere pure magnitudes in place of qualitative elements of sensation. But now it is seen that precisely the determination of these values, the measurements, which it applies to motions, are still bound to certain limiting presuppositions. Reflection on the manner in which we make empirical measurements of space and time shows how anthropomorphism reaches into this field that was thought withdrawn from it in principle. It is, as it were, this earthly remainder still belonging to classical mechanics with its assumption of finite fixed reference bodies and motionless inertial systems, from which the theory of relativity seeks to free itself. The conceived unit of connection determined by a system of mathematical equations here takes the place of any sensuously given, and also sensuously conditioned, unit of measurement. As is seen, there is involved here not a cancellation but a critical correction of the empirical concept of objectivity, by which a correction of our empirical spatial and temporal measures and their transformation into the *one* system of natural laws are gained.

We are brought to the same outcome by consideration of the historical problems out of which the theory of relativity has grown. To give the propositions of abstract mechanics, especially the principle of inertia a definite physical meaning had been attempted repeatedly by trying to point out some empirical systems for which they would possess strict validity. But these attempts were all thwarted, in particular, by the discovery of the motion of the solar system and of the fixed stars; to find a fixed and clear empirical meaning for the equations of the Galileo-Newtonian mechanics, nothing remained save to postulate, with Carl Neumann, an absolutely motionless body at some unknown place in space. But such a postulate of the existence of a particular physical object, a body which can never be discovered by observation, remains the strangest anomaly, from the epistemological standpoint. (8, p. 238 ff.) The absolutely motionless ether too, which seemed for a time to offer the lacking physical reference system of the Galileo-Newtonian mechanics, showed itself unsuited to this purpose; since the negative outcome of Michelson's investigation the question seemed to be decided here also. At this point, as has been seen, the theory of relativity begins. It makes a virtue out of the difficulty

into which philosophical thought had fallen in its attempt to find a
particular privileged system of coördinates. Experience had shown
that there is no such system, and the theory, in its most general
interpretation, makes it a postulate that there *cannot* and *must* not
be such. That, for the physical description of the processes of
nature, no particular reference body is to be privileged above any
other is now made a principle. "In classical mechanics, as well as
the special theory of relativity," says Einstein," a distinction is
drawn between reference bodies K relatively to which the laws of
nature are valid and reference bodies K' relatively to which they are
not valid. With this state of affairs no consistently thinking man
can be satisfied. He asks: how is it possible that certain reference
bodies (and their states of motion) are privileged over other reference
bodies (and their states of motion)? In vain, I seek in classical
mechanics for something real to which I might trace the difference
in the behavior of the body with reference to the systems of reference K and K'." (18, p. 49.) In this argument from the principle of
insufficient reason, the physicist seems to move on slippery ground.
One is inevitably reminded of the argument of Euler, who thought
that he *proved* the principle of inertia of classical mechanics by
explaining that, if a body changed its state of motion without the
influence of external forces, there would be no reason why it should
choose any particular change of magnitude and direction of its
velocity. (23.) The circle involved here, namely, that "the state
of motion" of a body is assumed to be a determinate magnitude,
while it is only defined as such by the law of inertia itself, is easily
seen. In Einstein's appeal to the "principle of reason," there is
doubtless involved a more general and deeper epistemological motive.
If we assume that the final objective determinations, which our
physical knowledge can reach, *i.e.*, the laws of nature, are provable
and valid only for certain chosen systems of reference, but not for
others, then, since experience offers no certain criterion that we have
before us such a privileged reference system, we can never reach a
truly universal and determinate description of natural processes.
This is only possible if some determinations can be pointed out, which
are indifferent to every change in the system of reference taken as a
basis. Only those relations can we call laws of nature, *i.e.*, ascribe
to them objective universality, whose form is independent of the
particularity of our empirical measurements of the special choice

of the four variables x_1, x_2, x_3, x_4, which express the space and time parameters. In this sense, one could conceive the principle of the universal theory of relativity, that the universal laws of nature are not changed in form by arbitrary changes of the space-time variables, as an analytic assertion; as an explanation of what is meant by a "universal" law of nature. But the demand, that there must in general be such ultimate invariants, is synthetic.

In fact, it can be shown that the general doctrine of the invariability and determinateness of certain values, which is given first place by the theory of relativity, must recur in some form in any theory of nature, because it belongs to the logical and epistemological nature of such a theory. To start from the picture of the world of general energetics—Leibniz, in establishing the law of the "conservation of *vis viva*" as a universal law of nature, referred to this logical element in it. He first defines the *vis viva* of a physical system as a quantity of work; he determines that forces are to be called equal, when they are able to perform equal mechanical work, no matter what their properties may be in detail; thus if they produce an equal degree of tension in an equal number of elastic springs, raise an equal weight to the same height, communicate to an equal number of bodies the same amount of velocity, *etc.*, they are equal. In this definition it is assumed that measurement of the *vis viva* by different systems of measurement will give results equivalent to each other, and thus that force which, when measured by a certain effect, prove to be equal or in a definite relation of greater or smaller, will retain this same relation if we measure them by any other effect. If this were not the case, and did there result a different relation of forces according to the different effect one uses as a measure, nature would be without laws; the whole science of dynamics would be superfluous; and it would not be possible to measure forces, for forces would have become something indeterminate and contradictory, *quiddam vagum et absonum*. (42, III, 208 ff.; VI, 209 f.; *cf.* 5, p. 305 ff.) The same process of thought has been repeated on broader physical lines in the discovery and grounding of the modern principle of energy. Here, too, the energy of a material system in a certain state was defined—*e.g.*, by W. Thompson—first as the amount of all the effects, expressed in mechanical units of work, called forth outside the system when the system passes in any way from its state into a definite but arbitrarily defined state of nullity. This explanation at first leaves it entirely undecided as to whether there exists a determinate value of what is here called "energy," *i.e.*, whether the results of the measurement of the amount of work of a system turn out the same or differently according to the method of bringing the system from the given state into a definite state of nullity. But that this determinateness in fact exists, that there always results the same amount of energy no matter what effect we use as the measure of work and what type of transition we choose, is precisely what the principle of the conservation of energy affirms. This affirms nothing else and has no other physically comprehensible meaning than that the amount of all the effects, measured in units of mechanical work, which a material system calls forth in its external environment, when it passes from a definite state in any arbitrary manner to an arbitrarily defined state of nullity, has a determinate value, and is thus independent of the type of transition. If this independence did not exist—and that it exists only experience can teach us—it would follow that what we called "energy" is not an exact physical determination; energy would not be a universal constant of measurement. We would have to seek for other empirical values to satisfy the fundamental postulate of determinateness. But it holds, conversely, that if energy is once established as a constant of measurement, it thus becomes a constant of nature also, a "concept of a definite object." Now from a physical standpoint a "substantial" conception of energy can be carried through without arousing suspicion; energy can be regarded as a sort of "reserve supply" of the physical system, the quantity of which is completely[6] determined by the totality of the magnitudes of the states, which belong to the system involved. From the epistemological standpoint, it must be remembered that such an interpretation is nothing more than a convenient expression of the relations of measurement, that alone are known, an expression which adds to them nothing essential. The unity and determinateness of measurement can be immediately understood and expressed as the unity and determinateness of the object, precisely because the empirical object means nothing but a totality of relations according to law. It follows from this analogy from a new angle that the advance in "relativization" which takes place in the theory of relativity, represents no contrast to the general

[6] In more detail in Planck (63) p. 92 ff.

task of objectification, but rather signifies one step in it, since, by the
nature of physical thought, all its knowledge of objects *can* consist
in nothing save knowledge of objective relations. "Whatever we
may know of matter," here, too, we can cite the *Critique of Pure
Reason*, "is nothing but relations, some of which are independent and
permanent and by which a certain object is given us." (34, p. 341;
cf. Müller's Trans. p. 232.) The general theory of relativity has
shifted these "independent and permanent relations" to another
place by breaking up both the concept of matter of classical mechanics
and the concept of the ether of electrodynamics; but it has not contested them as such, but has rather most explicitly affirmed them in
its own invariants, which are independent of every change in the
system of reference. The criticism made by the theory of relativity
of the physical concepts of objects springs thus from the same method
of scientific thought, which led to the establishment of these concepts, and only carries this method a step further by freeing it still
more from the presuppositions of the naïvely sensuous and "substantialistic" view of the world. To grasp this state of affairs in its
full import we must go back to the general epistemological questions
offered to us by the theory of relativity; we must go back to the
transformation of the *physical concept of truth* involved in it by which
it comes into direct contact with the fundamental problem of logic.

CHAPTER III

The general principle of the relativity of knowledge received its
first complete systematic working out in the history of ancient
skepticism. Here it possessed, according to the fundamental character of skepticism, an exclusively negative meaning; it signified the
limit in principle which is set to all knowledge and by which it is
separated once for all from the definitive apprehension of the truth
as "absolute." Among the skeptical "tropes" intended to show the
uncertainty of sensuous and conceptual knowledge, the "trope"
of πρός τι stands in the first place. To know the object, our knowledge would, above all, have to be in a position to grasp it in its pure
"in itself" and to separate it from all the determinations, which only
belong to it relatively to us and other things. But this separation
is impossible, not only actually, but in principle. For what is
actually given to us only under certain definite conditions can never
be made out logically as what it is in itself and under abstraction from
precisely these conditions. In what we call the perception of a
thing, we can never separate what belongs to the objective thing from
what belongs to the subjective perception and contrast the two as
independent factors. The form of subjective organization enters as a
necessary element into all our so-called objective knowledge of things
and properties. The "thing" appears, accordingly, not only differently to the various senses but it is limitlessly variable for the same
organ according to the time and varying conditions of perception.
For its whole character depends on the relations under which it is
presented to us. No content is given us in experience unmixed with
others in a purely self-identical character, but what is given us is
always only a general combination of impressions. It is not one or
the other, "this" or "that" definite quality, but only the reciprocal
relation of the one to the other and the other to the one that is here
known, indeed that is alone knowable.

Modern science has overcome the objections of skepticism to the
possibility of knowledge, not by contesting their content, but by
drawing from them a wholly different, indeed, opposite logical

consequence. Modern science also assumes the reduction of what is taken in the naïve view of the world, as fixed and absolute "properties" of things to a system of mere relations. "With regard to the properties of the objects of the outer world," we read in, *e.g.*, Helmholtz's *Handbuch der physiologischen Optik*, "it is easy to see that all the properties we can ascribe to them, signify only the effects they produce either on our senses or on other natural objects. Color, sound, taste, smell, temperature, smoothness, solidity belong to the first class; they signify effects on our sense organs. The chemical properties are likewise related to reactions, *i.e.*, effects, which the natural body in question exerts on others. It is thus with the other physical properties of bodies, the optical, the electrical, the magnetic. Everywhere we are concerned with the mutual relations of bodies to each other, with effects which depend on the forces different bodies exert on each other. . . . . From this it follows that in fact, the properties of the objects of nature do not signify, in spite of their name, anything proper to the particular objects in and for themselves, but always a relation to a second object (including our sense organs). The type of effect must naturally always depend on the peculiarities of the effecting body as well as on those of the body on which the effect is exerted. . . . . To question whether cinnabar is really red as we see it, or whether this is only an illusion of the senses, is therefore meaningless. The sensation of red is the normal reaction of normally constituted eyes to the light reflected from cinnabar. A color-blind person will see the cinnabar black or dark grey; this also is the correct reaction of his peculiarly constituted eye. . . . . In itself, the one sensation is not more correct and not more false than the other. . . . ." (30, p. 588 f.) The old skeptical "trope," the argument of the πρός τι here stands before us again in all distinctness. But renunciation of the absoluteness of things involves no longer renunciation of the objectivity of knowledge. For the truly objective element in modern knowledge of nature is not so much things as laws. Change in the elements of experience and the fact that no one of them is given in itself, but is always given with reference to something else, constitute no objection to the possibility of objectively real knowledge in so far as the laws establish precisely these relations themselves. The constancy and absoluteness of the elements is sacrificed to gain the permanency and necessity of laws. If we have gained the latter, we no longer need the former. For the objection of skepticism, that we can never know the absolute properties of things, is met by science in that it defines the concept of property in such a way that the latter involves in itself the concept of relation. Doubt is overcome by being outdone. When it is seen that "blue" can mean absolutely nothing save a relation to a seeing eye, that "heavy" means nothing save a relation of reciprocal acceleration and that in general all "having" of properties can be resolved purely and simply into a "being-related" of the elements of experience, then the longing to grasp the ultimate absolute qualities of things, secretly at the basis of skepticism, loses its meaning. Skepticism is refuted, not by showing a way to a possible fulfillment of its demands, but by understanding and thus rendering ineffective the dogmatic import of these demands themselves.

In this transformation of the general ideal of knowledge, modern science and modern logic are both involved; the development of the one is in closest connection with that of the other. Ancient logic is entirely founded on the relation of "subject" and "predicate," on the relation of the given concept to its also given and final properties. It seeks finally to grasp the absolute and essential properties of absolute self-existent substances. Modern logic, on the contrary, in the course of its development, comes more and more to abandon this ideal and to be made into a pure doctrine of form and relation. The possibility of all determinate character of the content of knowledge is grounded, for it, in the laws of these forms, which are not reducible to mere relations of subsumption but include equally all the different possible types of relational construction and connection of elements of thought. But here doubt must begin in a new and deeper sense. If knowledge of things is understood as knowledge of laws and if an attempt is made to ground the former in the latter and to protect it from the attacks of skepticism, then what guarantees the objectivity, the truth and universality of the knowledge of laws? Do we have, in the strict sense, knowledge of laws or does all that we can gain resolve itself in the most favorable case into knowledge of particular cases? Here as we see, the problem of skepticism is reversed on the basis of the modern conception of law. What perplexed the ancient skeptic, who sought the substance of things, was the limitless relativity of all phenomena; it was the fact that phenomena would not remain fixed individual data, but were reduced for knowledge ever again into mere relations and relations of relations.

But for the modern skeptic, to whom the objective truth, in so far as it is attainable, means the one all-inclusive and necessary law of all process, the basis of doubt lies in the fact that reality is never given us in this universal intellectual form, but is always divided and broken up into mere punctual particularities. We grasp only a here and a now, only a particularity isolated in space and time, and it is not to be seen how we could ever pass from this perception of the individual to a view of the objective form of the whole. No more than the continuum can be built up and generated by the summation of mere unextended points can a truly objective and necessary law be gained and deduced by the simple aggregation of however many particular cases. This is the form of Hume's skepticism, which is characteristically distinguished from the ancient. While the ancient skeptic could not reach the absolute substance because of the relativities in which the phenomenal world involved him, the modern skeptic fails to reach laws as universal relations because of the absolute particularities of sensation. While in the former it is the certainty of things that is questionable, in the latter it is the certainty of causal connections. The connections of processes become an illusion; what remains is only their particular atoms, the immediate data of sensation, in which all knowledge of "facts," of "matter of fact" ultimately consists.

If it is possible to overcome this essentially more radical form of skepticism also, it can only be by there being shown in it too a concealed dogmatic assumption, which lies implicitly at its basis. And this assumption consists in fact in its concept of empirical "givenness" itself. This givenness of "bare" impressions in which abstraction is made in principle from all elements of form and connection, proves to sharper analysis to be a fiction. When this is understood, doubt is directed, not on the possibility of knowledge, but on the possibility of the logical measuring-rod with which knowledge is measured here. Instead of the criterion of the "impression" making the universal formal relations of knowledge and its axioms questionable, the validity of this criterion must be contested on the basis of these relations. The only refuge from radical doubt lies in its being not set aside but intensified, in our learning to question, as ultimate elements of knowledge known in themselves, not only "things" and "laws" but especially sensations. The skepticism of Hume left the "simple" sensation as a completely unproblematic certainty, as a simple and unquestionable expression of "reality." While antique skepticism rested completely on the tacit assumption of absolute things, that of Hume rests on the assumption of absolute sensations. The hypostasization in the one case concerns "outer" being, in the other, "inner" being, but its general form is the same. And only by this hypostasization does the doctrine of the relativity of knowledge gain its skeptical character. Doubt does not result directly from the content of this doctrine, but, on the contrary, it depends on the fact that the doctrine is not truly and consistently thought through. As long as thought contents itself with developing, with reference to phenomena and according to demands of its own form, its logical axioms, and truth as a system of pure relations, it moves within its own circle with complete certainty. But when it affirms an absolute, whether of outer or inner experience, it is forced skeptically to annihilate itself with reference to this absolute. It strikes this absolute of things or of sensations again and again as if against the wall of the cell in which it is enclosed. Relativity, which is, fundamentally, its immanent force, becomes its immanent limit. It is no longer the principle, which renders possible and governs the positive advance of knowledge, but is merely a necessary instrument of thought, which by that fact confesses itself not adequate to being the absolute object and the absolute truth.

This relation is indeed changed when we contrast to both the dogmatic and the skeptical concept of truth, which are united by a common root, the idealistic concept of truth. For the latter does not measure the truth of fundamental cognitions by transcendent objects, but it grounds conversely the meaning of the concept of the object on the meaning of the concept of truth. Only the idealistic concept of truth overcomes finally the conception which makes knowledge a copying, whether of absolute things or of immediately given "impressions." The "truth" of knowledge changes from a mere pictorial to a pure functional expression. In the history of modern philosophy and logic, this change is first represented in complete clarity by Leibniz, although in his case, the new thought appears in the setting of a metaphysical system, in the language of the monadological scheme of the world. Each monad is, with all its contents, a completely enclosed world, which copies or mirrors no outer being but merely includes and governs by its own law the whole of its presentations; but these different individual worlds express, nevertheless, a common universe and a common truth.

This community, however, does not come about by these different pictures of the world being related to each other as copies of a common "original" but by the fact that they correspond functionally to each other in their inner relations and in the general form of their structure. For one fact, according to Leibniz, expresses another when there exists between what can be said of the one and of the other a constant and regular relation. Thus a perspective projection expresses its appropriate geometrical figure, an algebraic equation expresses a definite figure, a drawn model a machine; not as if there existed between them any sort of factual likeness or similarity, but in the sense that the relations of the one structure correspond to those of the other in a definite conceptual fashion. (43, VII, 263 f, 44, II, 233; cf. 7, II, 167.) This Leibnizian concept of truth was taken up and developed by Kant who sought to free it from all the unproved metaphysical assumptions that were contained in it. In this way he gained his own interpretation of the critical concept of the object, in which the relativity of knowledge was affirmed in a far more inclusive meaning than in ancient or modern skepticism, but in which also this relativity was given a new positive interpretation. The theory of relativity of modern physics can be brought without difficulty under this interpretation, for, in a general epistemological regard, it is characterized by the fact that in it, more clearly and more consciously than ever before, the advance is made from the copy theory of knowledge to the functional theory. As long as physics retained the postulate of absolute space, the question still had a definite meaning as to which of the various paths of a moving body that result when we regard it from different systems of reference, represents the real and "true" motion; thus a higher objective truth had to be claimed for certain spatial and temporal values, obtained from the standpoint of certain selected systems, than for others. The theory of relativity ceases to make this exception; not that it would abandon the determinateness of natural process, but because it has at its disposal new intellectual means of satisfying this demand. The infinite multiplicity of possible systems is not identical with the infinite indeterminateness of the values to be gained in them—in so far as all these systems are to be related and connected with each other by a common rule. In this respect, the principle of relativity of physics has scarcely more in common with "relativistic positivism," to which it has been compared, than the name. When there is seen in the former a renewal of ancient sophistical doctrines, a confirmation of the Protagorean doctrine that man is the "measure of all things," its essential achievement is mistaken.[1] The physical theory of relativity teaches not that what appears to each person is true to him, but, on the contrary, it warns against taking appearances, which hold only from a particular system, as the truth in the sense of science, *i.e.*, as an expression of an inclusive and final law of experience. The latter is gained neither by the observations and measurements of a particular system nor by those of however many such systems, but only by the reciprocal coördination of the results of all possible systems. The general theory of relativity purports to show how we can gain assertions concerning all of these, how we can rise above the fragmentariness of the individual views to a total view of natural processes. (*Cf.* above.) It abandons the attempt to characterize the "object" of physics by any sort of pictorial properties, such as can be revealed in presentation, and characterizes it exclusively by the unity of the laws of nature. When, for example, it teaches that a body regarded from one system possesses spherical form and, regarded from another system, in motion relatively to the first, appears as an ellipsoid of rotation, the question can no longer be raised as to which of the two optical images here given is like the absolute form of the object, but it can and must be demanded that the multiplicity and diversity of the sensuous data here appearing can be united into a universal concept of experience. Nothing more is demanded by the critical concept of truth and the object. According to the critical view, the object is no absolute model to which our sensuous presentations more or less correspond as copies, but it is a "concept, with reference to which presentations have synthetic unity." This concept the theory of relativity no longer represents in the form of a picture but as a physical theory, in the form of equations and systems of equations, which are covariant with reference to arbitrary substitutions. The "relativization," which is thus accomplished, is itself of a purely logical and mathematical sort. By it the object of physics is indeed determined as the "object in the phenomenal world;" but this phenomenal world no longer possesses any subjective arbitrariness and contingency. For the ideality of the forms and conditions of knowledge, on which physics rests as a science, both assures and grounds the empirical reality of all that is established by it as a "fact" and in the name of objective validity.

[1] *Cf.* Petzoldt (61).

CHAPTER IV

MATTER, ETHER AND SPACE

In the structure of physics we must, it seems, distinguish two different classes of concepts from each other. One group of concepts concerns only the form of order as such, the other the content that enters into this form; the first determines the fundamental schema which physics uses, the other concerns the particular properties of the real by which the physical object is characterized. With regard to the pure formal concepts, they appear to persist as relatively fixed unities in spite of all changes of physical ideals in detail. In all the diversity and conflict of the systematic concepts of physics, space and time are distinguished as the ultimate, agreeing unities. They seem, in this sense, also, to constitute the real *a priori* for any physics and the presupposition of its possibility as a science. But the first step from these bare possibilities to reality, which is a matter not of the spatio-temporal form, but of the *somewhat* that is thought to be somehow "given" *in* space and *in* time, seems to force us beyond the circle of the *a priori*. Kant indeed, in the *Metaphysischen Anfangsgründen der Naturwissenschaft*, attempted an *a priori* deduction and construction of the concept of "matter" as a necessary concept of physics; but it is easy to see that this deduction does not stand on the same plane and cannot claim the same force as the Transcendental Aesthetic or the Analytic of the Pure Understanding. He himself believed that he possessed in these deductions a philosophical grounding of the presuppositions of the science of Newton; today we recognize to an increasing extent that what he so regarded was in fact nothing but a philosophical circumlocution for precisely these presuppositions. As a fundamental definition of the physical concept of the object, the classical system of mechanics is only one structure, by the side of which there are others. Heinrich Hertz, in his new grounding of the mechanical principles, distinguished three such structures: the first is given in the Newtonian system, which is founded on the concepts of space, of time, of force, and of mass, as given presentations; the second leaves the presuppositions of space, time and mass unchanged, but substitutes for the concept of force as the mechanical "cause of acceleration" the universal concept of energy, which is divided into two different forms, potential and kinetic energy. Here, too, we have four mutually independent concepts, whose relations to each other are to constitute the content of mechanics. Hertz's own formulation of mechanics offers a third structure in which the concept of force or of energy as an independent idea is set aside and the construction of mechanics is accomplished by only three independent fundamental ideas, space, time and mass. The circle of possibilities would thus have seemed completely surveyed—had not the theory of relativity once more given a new interpretation to the mutual relation between the pure formal concepts and the physical concepts of the object and substance, and thus transformed the problem not only in content but in principle.

The concept of "nature," the gaining of which is the real methodic problem of physics, leaves room, as the history of physical thought shows, for a dualism of presuppositions, which as such seems necessary and unavoidable. Even in the first logical beginnings of genuine natural science, which are found in Greek thought, this dualism appears in full distinctness and clarity. Antique atomism, which is the first classical example of a conceptual and scientific picture of the world, can only describe and unify the "being" of nature by building it up out of two heterogeneous elements. Its view of nature is founded on the opposition of the "full" and the "void." The two, the full and the void, prove necessary elements for the constitution of the object of physics. To the being of the atom and matter as the παμπλῆρες ὄν, there is opposed by Democritus the not-being, the μὴ ὄν of empty space; both this being and this not-being possessed for him, however, uncontested physical truth and thus indubitable physical reality. The reality of motion was only intelligible by virtue of this dual presupposition; motion would disappear if we did not both distinguish empty space from the material filling of space and conceive the two as in inseparable mutual relation, as fundamental elements in all natural processes. At the beginning of modern times, Descartes attempted philosophically to overcome this duality in the foundations of physical thought. Proceeding from the thought of the unity of *consciousness*, he postulated also a new unity of *nature*. And this seemed to him only attainable by abandoning the opposition of the "full" and the "void," of matter and extension. The physical being of the body and the geometrical being of extension constitute one and the same object:

the "substance" of a body is reduced to its spatial and geometrical determinations. Thus a new approach to physics, methodologically deeper and more fruitful, was found, the concrete realization of which, however, could not be accomplished by Descartes' physics. When Newton fought the hypothetical and speculative premises of the Cartesian physics, he also abandoned this approach. His picture of the world was rooted in the dualistic view, which was even intensified in it and which set its seal on his universal law of nature and the cosmos. On the one side, there stands space as a universal receptacle and vessel; on the other, bodies, inert and heavy masses, which enter into it and determine their reciprocal position in it on the basis of a universal dynamic law. The "quantity of matter," on the one hand, the purely spatial "distance" of the particular masses from each other, on the other, give the universal physical law of action, according to which the cosmos is constructed. Newton as a physicist always declined to ask for a further "why," for a reason for this rule. It was for him the unitary mathematical formula, which included all empirical process under it and thus perfectly satisfied the task of the exact knowledge of nature. That this formula concealed—in the expression for the cosmic *masses* and in the expression for their *distance*—two wholly different elements seemed a circumstance that no longer concerned the physicist but only the metaphysician and the speculative philosopher of nature. The proposition *"hypotheses non fingo"* cuts off any further investigation in this direction. For Newton as for Democritus, matter and space, the full and the void, form for us the ultimate but mutually irreducible elements of the physical world, the fundamental building-stones of all reality, because as equally justified and equally necessary factors, they enter into the highest law of motion taught us by experience.

If we contrast this view with the picture of the world of modern and very recent physics, there results the surprising fact that the latter seems to be again on the road to Descartes, not indeed in content, but certainly in method. It too strives from various sides toward a view in which the dualism of "space" and "matter" is cancelled, in which the two no longer occur as different classes of physical object-concepts. There now appears in the concept of the "field" a new mediating concept between "matter" and "empty space;" and this it is which henceforth appears with increasing definiteness as the genuine expression of the physically real since it is the perfect expression of the physical law of action. In this concept of the field, the typical manner of thought of modern physics has gained, from the epistemological standpoint, its sharpest and most distinct expression. There now takes place, starting from electrodynamics, a progressive transformation of the concept of matter. Already with Faraday, who constructed matter out of "lines of force," there is expressed the view that the field of force cannot depend on matter, but that on the contrary, what we call matter is nothing else than specially distinguished places of this field.[1] In the progress of electrodynamics, this view is confirmed and assumes ever more radical expression. The doctrine is carried through more and more of a pure "field-physics," which recognizes neither bare undifferentiated space by itself nor matter by itself subsequently entering into this finished space, but which takes as a basis the intuition of a spatial manifold determined by a certain law and qualified and differentiated according to it. Thus, e.g., there was established by Mie a more general form of electrodynamics on the basis of which it seemed possible to construct matter out of the field. The concept of a substance existing along with the electromagnetic field seemed unnecessary in this approach; according to the new conception, the field no longer requires for its existence matter as its bearer, but matter is considered and treated, on the contrary, as an "outgrowth of the field." It is the last consequence of this type of thought that is drawn by the theory of relativity. For it, too, the real difference finally disappears between an "empty" space and a space-filling substance, whether one calls this matter or ether, since it includes both moments in one and the same act of *methodic determination.* The "riddle of weight" is revealed to us, according to the fundamental thought of Einstein's theory of gravitation, in the consideration and analysis of the inner relations of measurement of the four dimensional space-time manifold. For the ten functions $g_{\mu\nu}$, which occur in the determination of the linear elements of the general theory of relativity $ds^2 = \sum_1^4 g_{\mu\nu} d_{x\mu} d_{x\nu}$ (μ, $\nu = 1, 2, 3, 4$), represent also the ten components of the gravitation potential of Einstein's theory. It is thus the same determinations, which, on the one hand, designate and express the metrical properties

---

[1] On Faraday, *cf.* Buek (4, esp. p. 41 ff.); *cf.* also Weyl (83, p. 142).

of the four-dimensional space and, on the other, the physical properties of the field of gravitation. The spatio-temporal variability of
the magnitudes $g_{\mu\nu}$ and the occurrence of such a field prove to be
equivalent assumptions differing only in expression. Thus it is
shown most distinctly that the new physical view proceeds neither
from the assumption of a "space in itself," nor of "matter" nor of
"force in itself"—that it no longer recognizes space, force and matter
as physical objects separated from each other, but that for it exists
only the unity of certain *functional relations*, which are differently
designated according to the system of reference in which we express
them. All dynamics tends more and more to be resolved into pure
metrics, a process in which indeed the concept of metrics undergoes,
in contrast with classical geometry, an extraordinary broadening and
generalization whereby the measurements of Euclidean geometry
appear as only a special case within the total system of possible
measurements in general. "The world," as is said by Weyl, in
whose account of the general theory of relativity one can trace and
survey this development most clearly," is a $(3 + 1) =$ dimensional
metrical manifold; all physical phenomena are expressions of world
metrics. . . . . The dream of Descartes of a purely geometrical
physics seems to be about to be fulfilled in a wonderful way, which
could not have been foreseen by him." (83, p. 244; *cf*. p. 85 ff.,
170 ff).

Just as the dualism of matter and space is superseded here by a
unitary physical conception, so the opposition between "matter"
and "force" is to be overcome by the principle and law of the new
physics. Since Newton, as a physicist, established this opposition
between the "inert masses" and the forces that affect them in the
Philosophiae Naturalis Principia Mathematica, attempts, indeed,
have not been lacking to overcome it from the philosophical and
speculative side. Leibniz led the way here; but although, in his
metaphysics, he wholly resolved substance into force he retained in
the construction of his mechanics, the duality of an "active" and a
"passive" force, whereby matter is subsumed under the concept of
the latter. The essence of matter consists in the dynamic principle
immanent in it; but this expresses itself, on the one hand, in activity and striving for change, on the other hand, in the resistance which
a body opposes, according to its nature, to change coming upon it from

without.[2] As for Newton, the opposition in fundamental concepts, which he assumes, threatens finally to destroy the unity of the physical structure of his world; he can only retain this unity by introducing at a certain place a *metaphysical* factor. The principle of the conservation of *vis viva* is disputed by him because all bodies consist of "absolutely hard" atoms, and in the rebounding of such atoms, mechanical energy must be lost; the sum total of force is in a continuous decrease, so that for its preservation the world needs from time to time a new divine impulse. (58, p. 322 ff.) Kant attempted in a youthful work, the *Monadologia Physica* of the year 1756, a reconciliation and mediation between the principles of the Leibnizian philosophy and those of Newtonian mechanics; and in the *Metaphysischen Anfangsgründen der Naturwissenschaft* he returns to the attempted purely dynamic deduction and construction of matter. The "essense" of matter *i.e.*, its pure concept for experience, according to which it is nothing else than a totality of external relations, is resolved into a pure interaction of forces acting at a distance; but since these forces themselves occur in a double form, as attracting and repelling forces, the dualism is not fundamentally overcome, but is only shifted back into the concept of force itself.

Modern physics has sought, from essentially different standpoints and motives, to overcome the old opposition between matter and force, which seemed sanctioned and made eternal in the classical system of mechanics. Heinrich Hertz's *Prinzipien der Mechanik* takes the opposite course to that of previous philosophical speculation by placing the sought unity in the concept of mass, instead of in the concept of force. Along with the fundamental concepts of space and time, only the concept of mass enters into the systematic construction of mechanics. The carrying out of this view presupposes, indeed, that we do not remain with gross perceptible mass and gross perceptible motion but supplement the sensuously given elements, which by themselves do not constitute a lawful world, by assuming certain "concealed" masses and "concealed" motions. This supplementation takes place when it is shown to be necessary for the description and calculation of phenomena, and without arousing suspicion since mass is conceived by Hertz from the beginning merely as a definite factor of calculation. It is intended to express

[2] *Cf.* (44), I, 204, 267 ff., 332 II, 290 ff., 303.

nothing but certain coördination of space and time values: "a particle of mass," as Hertz defines it, "is a property by which we coördinate unambiguously a certain point of space to a certain point of time (and)[3] a certain point of space to every other time." (31, p. 29 ff., 54.) Another attempt was made by general energetics to reach a unified foundation for physics and with it for mechanics. Inert mass appears here merely as a definite factor of energy, as the capacity-factor of the energy of motion, which with certain other capacity-factors shares with the different types of energy, e.g., electricity, the empirical property of quantitative conservation. Energetics refuses to grant this law of conservation a special place and to recognize matter as a particular substance along with energy. (Cf. 60, p, 282 ff.) But precisely in this we see very distinctly what is logically unsatisfactory, which consists in that the principle of conservation refers to wholly different moments between which an inner connection is not to be seen.

The theory of relativity brings important clarification here too in that it combines the two principles of conservation, that of the conservation of energy and that of the conservation of mass into a single principle. This result it gains by applying its characteristic manner of thought; it is led to this result by general considerations on the conditions of measurement. The demand of the theory of relativity (at first of the special theory) is that the law of the conservation of energy be valid not only with reference to any system of coördinates K but also with reference to any other in uniform rectilinear motion relatively to it; it results from this presupposition, however, combined with the fundamental equations of Maxwell's electrodynamics that when a body in motion takes up energy E_0 in the form of radiation its inert mass increases by a definite amount $(\frac{E}{c^2})$. The mass of a body is thus a measure of its content of energy; if the energy content alters a definite amount then its mass alters proportionately.[4] Its independent constancy is thus only an appearance; it holds good only in so far as the system takes up and gives off no energy. In the modern electron theory, it follows from the well-known investigation of Kaufmann that the "mass" of an electron is not unchangeable, but that it rapidly increases with the velocity of the electron as soon as the latter approaches the velocity of light.

[3] Trans.
[4] Einstein (16a) and Planck (64 and 65).

While previously a distinction had been made between a "real" and a "fictitious" mass of electrons, *i.e.*, between an inertia, which came from its ponderable mass, and another, which they possessed solely because of their motion and their electric charge, in so far as this opposed a certain resistance to every change of velocity, it now turns out that the alleged ponderable mass of the electrons is to be taken as strictly = 0.

The inertia of matter thus seems completely replaced by the inertia of energy; the electron—and thus the material atom as a system of electrons—possesses no material but only "electromagnetic" mass. What was previously regarded as the truly fundamental property of matter, as its substantial kernel, is resolved into the equations of the electro-magnetic field. The theory of relativity goes further in the same direction; but it reveals in this too its peculiar *nuance* and character. This comes out especially in the process by which it gains one of its fundamental propositions: in the establishment of the equivalence of phenomena of inertia and weight. Here it is at first merely a calculation, a consideration of the same phenomena from different systems of reference, which points the way. We can, as it shows, regard one and the same phenomenon now as a pure inertial movement and now as a movement under the influence of a field of gravitation according to the standpoint we choose. The equivalence of *judgment,* here indicated, grounds for Einstein the *physical identity* of phenomena of inertia and weight. If certain accelerated motions occur for an observer within his sphere of observation, he can interpret them either by ascribing them to the effects of a field of gravitation or conceive the system of reference from which he makes his measurements as in a certain acceleration. The two assumptions accomplish precisely the same in the description of the facts and can thus be applied without distinction. We can— as Einstein expresses it—produce a field of gravitation by a mere change of the system of coördinates. (17, p. 10; *cf.* 18, p. 45 ff.) Hence, it follows that to attain a universal theory of gravitation we need only assume such a shift of the system of reference and establish its consequences by calculation. It suffices that in purely ideal fashion we place ourselves at another standpoint to be able to deduce certain physical consequences from this change of standpoints. What was previously done in the Newtonian theory of gravitation by the dynamics of forces is done by pure kinematics in Einstein's theory, *i.e.*, by the consideration of different systems of reference moving relatively to each other.

In emphasizing this ideal element in Einstein's theory of gravitation, the *empirical* assumption on which it rests must naturally not be forgotten. That we change in thought, by the mere introduction of a new system of reference, a field of inertia into a field of gravitation, and a field of gravitation of special structure into a field of inertia, rests on the empirical equality of inert and gravitating masses of bodies, as was established with extraordinary exactitude by the investigation of Eötvös to which Einstein refers. Only the fact that gravitation imparts to all bodies found at the same place in the field of gravitation, the same amount of acceleration, and that thus it is for any definite body the same constant, *i.e.*, mass, which determines its inertial effects and its gravitational effects, renders possible that transformation of the one into the other, from which the Einstein theory starts.[5] But it is especially interesting and important from a general methodological standpoint that this fundamental fact is given a completely different interpretation than in the Newtonian mechanics. What Einstein urges against the latter is that it *registered* the phenomenon of the equivalence of gravitating and inert masses, but did not interpret it. (18, p. 44.) What was established as a fact by Newton is now to be understood from principles. In this problem one can trace how gradually the question as to the "essence" of matter and of gravitation is superseded by another epistemological formulation of the question, which finds the "essence" of a physical process expressed wholly in its quantitative relations and its numerical constants. Newton never ceased to reject the question as to essence, which met him ever again, and the phrase that physics has to do merely with the "description of phenomena" was first formulated in his school and is an expression of his method.[6] But so little was he able to escape this question that he expressly urged that universal attraction was not itself grounded in the essence of body, but that it came to it as something new and alien. Weight is, as he emphasizes, indeed a universal but not an essential property of matter. (59, Vol. III, p. 4.) What this distinction between the universal and the essential means from the standpoint of the

---

[5] For more detail, *cf.* Freundlich (24), pp. 28 and 60 f. and Schlick (79) p. 27 ff; *cf.* Einstein (18), p. 45 ff.

[6] Keill, *Introductio ad veram Physicam* (1702), (36); *cf.* 7, II, 404 ff.

physicist, who has to do merely with the laws of phenomena, and thus with the universality of the rule to which they are subjected, is here left in the dark. Here lies a difficulty, which has been felt again and again in the tedious controversy of physicists and philosophers on the actuality and possibility of force acting at a distance. Kant, in his *Metaphysischen Anfangsgründen der Naturwissenschaft*, urges against Newton that, without the assumption that all matter merely by virtue of its essential properties exercises the action we call gravitation, the proposition that the universal attraction of bodies is proportional to their inert mass, would be a totally contingent and mysterious fact. (35, IV, p. 421.) In its solution of this problem the general theory of relativity has followed the path prescribed by the peculiarity of the physical method. The numerical *proportion* which is universally found between inert and heavy masses becomes the expression of physical *equivalence*, of the essential likeness of the two. The theory of relativity concludes that it is the *same* quality of the body, which is expressed according to circumstances as "inertia" or as "weight." We have here in principle the same procedure before us, which, *e.g.*, in the electromagnetic theory of light led to insight into the "identity" of light waves and electrical waves. For this identity too means nothing else and nothing more mysterious than that we can represent and master the phenomena of light and the phenomena of dielectric polarization by the same *equations* and that the same numerical value results for the velocity of light and for that of dielectric polarization. This equality of values means to the physicist likeness in essence—since for him essence is defined in terms of exact determinations of measure and magnitude. In the advance to this insight, there may be traced historically a definite series of steps, a culmination of physical theories. The physics of the eighteenth century was in general rooted in a substantialistic view. In the fundamental investigations of Sadi Carnot on thermodynamics heat was still regarded as a material, and the assumption seemed unavoidable, in understanding electricity and magnetism, of a particular electric and magnetic "matter." Since the middle of the nineteenth century, however, there appears in place of this "physics of materials," ever more definitely and distinctly the physics that has been called the "physics of principles." Here a start is not made from the hypothetical existence of certain materials and agents, but from certain universal relations, which are regarded as the criteria for the interpretation of particular phenomena. The general theory of relativity stands methodologically at the end of this series, since it collects all particular systematic principles into the unity of a supreme postulate, in the postulate not of the constancy of things, but of the invariance of certain magnitudes and laws with regard to all transformations of the system of reference.

The same evolution, that is characteristic of physical conceptual construction in general, is seen when we go from the concept of matter to the second fundamental concept of modern physics, to the concept of the ether.[7] The idea of the ether, as the bearer of optical and magnetic effects was at first conceived in the greatest possible analogy and affinity with our presentations of empirically given materials and things. A sensuous description of its fundamental properties was sought by comparing it now with a perfectly incompressible fluid, now with a perfectly elastic body. But the more one attempted to work these pictures out in detail, the more distinctly was it seen that they demanded the impossible of our faculty of presentation, that they demanded the unification of absolutely conflicting properties. Thus modern physics was more and more forced to abandon in principle this sort of sensuous description and illustration. But the difficulty was unchanged also when one asked, not concerning any concrete properties of the ether, but merely concerning the abstract laws of its motion. The attempt to construct a mechanics of the ether led little by little to the sacrifice of all the fundamental principles of classical mechanics; it was seen that, really to carry it through, one would have to give up not only the principle of the equality of action and reaction, but the principle of impenetrability in which, e.g., Euler saw the kernel and inclusive expression of all mechanical laws. Ether was and remained accordingly, in an expression of Planck, the "child of sorrow of the mechanical theory;" the assumption of the exact validity of the Maxwell-Hertzian differential equations for electrodynamic processes in the pure ether excludes the possibility of their mechanical explanation.[8] An escape from

[7] Here I do not go into details in the development of the hypothesis of the ether; they have been expounded from the standpoint of epistemology by e. g., Aloys Müller (55, p. 90 ff.) and Erich Becher (2, p. 232 ff.). On the following cf. Substance and Function (8, p. 215 ff.).

[8] Cf. Planck (67), p. 64 ff. Lenard (45a and b), especially declares for the possibility and necessity of a "mechanics of the ether."

this antinomy could only be reached by reversing the treatment. Instead of asking about the properties or constitution of the ether as a real thing, the question must be raised as to by what right here in general one seeks for a particular substance with particular material properties and a definite mechanical constitution. What if all the difficulties of the answer are based on the question itself, there being in it no clear and definite physical meaning? That is, in fact, the new position which the theory of relativity takes to the question of the ether. According to the outcome of Michelson's investigation and the principle of the constancy of the propagation of light, each observer has the right to regard his system as "motionless in the ether;" one must thus ascribe to the ether simultaneous rest with reference to wholly different systems of coördinates K, K', K'', which are in uniform translatory motion relatively to each other. That, however, is an obvious contradiction and it forces us to abandon the thought of the ether as a somehow moving or motionless "substance," as a thing with a certain "state of motion." Physics, instead of imagining some sort of hypothetical substratum of phenomena and losing itself in consideration of the nature of this substratum, is satisfied, as it becomes a pure "physics of fields," with the body of field-equations themselves and their experimentally verifiable validity. "One cannot define," says e.g., Lucien Poincaré, "ether by material properties without committing a real fallacy, and to characterize it by other properties than those, the direct and exact knowledge of which is produced for us by experiment, is an entirely useless labor condemned to sterility from the beginning. The ether is defined when we know the two fields, which can exist in it, the electric and magnetic fields, in their magnitude and direction at each point. The two fields can change; by custom we speak of a motion propagated in the ether; but the phenomenon accessible to experiment is the propagation of these changes." (75, p. 251.) Here we again face one of those triumphs of the critical and functional concept over the naïve notion of things and substances, such as are found more and more in the history of exact science. The physical rôle of the ether is ended as soon as a type of exposition is found for the electrodynamic laws into which it does not enter as a condition. "The theory of relativity," remarks one of its representatives, "rests on an entirely new understanding of the propagation of electromagnetic effects in empty space; they are not carried by a medium, but neither do they take place by unmediated action at a distance. But the electromagnetic field in empty space is a thing possessing self-existent physical reality independently of all substance. Indeed, one must first accustom himself to this idea; but perhaps this habituation will be made easier by the remark that the physical properties of this field, which are given most adequate expression in Maxwell's equations, are much more perfectly and exactly known than the properties of any substance." (Laue, 41, p. 112.) Habituation with regard to a "thing independent of any substance" can indeed be as little attributed to common human understanding as to the epistemologically trained understanding; for precisely to the latter does substance mean the *category* on the application of which rests all possibility of positing "things." But it is obvious that we have here only an inexactitude of expression and that the "independent physical reality" of the electromagnetic field can mean nothing but the reality of the relations holding within it which are expressed in the equations of Maxwell and Hertz. Since they are for us the ultimate attainable object of physical knowledge, they are set up as the ultimate attainable reality for us. The idea of the ether as an inexperienceable substance is excluded by the theory of relativity in order to give conceptual expression merely to the pure properties of empirical knowledge.

For this purpose, however, according to the theory of relativity, we do not need the fixed and rigid reference body, to which classical mechanics was ultimately referred. The general theory of relativity no longer measures with the rigid bodies of Euclidean geometry and classical mechanics, but it proceeds from a new and more inclusive standpoint in its determination of the universal linear element *ds*. In place of the rigid rod which is assumed to retain the same unchanging length for all times and places and under all particular conditions of measurement there now appear the curved coördinates of Gauss. If any point P of the space-time continuum is determined by the four parameters x_1, x_2, x_3, x_4, then for it and an infinitely close point P' there is a certain "distance" *ds*, which is expressed by the formula:

$$ds^2 = g_{11}dx_1^2 + g_{22}dx_2^2 + g_{33}dx_3^2 + g_{44}dx_4^2 + 2g_{12}dx_1dx_2 + 2g_{13}dx_1dx_3 + \ldots$$

in which the magnitudes g_{11}, $g_{22} \ldots g_{44}$ have values, which vary with the place in the continuum. In this general expression, the formula for the linear element of the Euclidean continuum is contained as a special case. We need not here go into details of this determination;[9]—its essential result, however, is that measurements in general different from each other result for each place in the space-time continuum. Each point is referred, not to a rigid and fixed system of reference outside of it, but to a certain extent only to itself and to infinitely close points. Thus all measurements become infinitely fluid as compared with the rigid straight lines of Euclidean geometry, which are freely movable in space without change of form; and yet, on the other hand, all these infinitely various determinations are collected into a truly universal and unitary system. We now apply, instead of given and finite reference bodies, only "reference mollusks" as Einstein calls them; but the conceptual system of all these "mollusks" satisfies the demand for an exact description of natural processes. For the universal principle of relativity demands that all these systems can be applied as reference bodies with equal right and with the same consequences in the formulation of the universal laws of nature; the form of the law is to be completely independent of the choice of the mollusk. (18, p. 67.) Here is expressed again the characteristic procedure of the general theory of relativity; while it destroys the *thing-form* of the finite and rigid reference body it would thereby only press forward to a higher form of object, to the true *systematic form* of nature and its laws. Only by heightening and outdoing the difficulties which resulted even for classical mechanics from the fact of the relativity of all motions, does it hope to find an escape in principle from these difficulties. "The clearer our concepts of space and time become," as was said in the outline of mechanics, which Maxwell has given in his short work, *Matter and Motion*, "the more do we see that everything to which our dynamic doctrines refer, belongs in a single system. At first we might think that we, as conscious beings, must have as necessary elements of our knowledge, an absolute knowledge of the place, in which we find ourselves, and of the direction in which we move. But this opinion, which was undoubtedly that of many sages of antiquity, disappears more and more from the idea of the physicist. In space, there are no milestones; one part of space is precisely like any other part, so that we cannot know where we are. We find ourselves in a

---

[9] *Cf.* Einstein (17 and 18, pp. 59 ff.); *cf.* below VI.

waveless sea without stars, without compass and sun, without wind and tide, and cannot say in what direction we move. We have no log that we can cast out to make a calculation; we can indeed determine the degree of our motion in comparison with neighboring bodies, but we do not know what the motion in space of these bodies is." (51, p. 92 f.)   From this mood of *"ignorabimus,"* into which physics was sinking more and more, only a theory could free it which grasped the problem at its root; and, instead of modifying the previous solutions, transformed fundamentally the formulation of the question. The question of absolute space and absolute motion could receive only the solution which had been given to the problem of the perpetual *mobile* and the squaring of the circle. It had to be made over from a mere negative expression into a positive expression, to be changed from a limitation of physical knowledge to a principle of such knowledge, if the true philosophic import, which was concealed in it, was to be revealed.

CHAPTER V

We have hitherto sought primarily to understand the special and general theory of relativity on its physical side. In fact, this is the standpoint from which it must be judged and one does it poor service if one seeks precipitately to interpret its results in purely "philosophical" or indeed in speculative and metaphysical terms. The theory contains not one concept, which is not deducible from the intellectual means of mathematics and physics and perfectly representable in them. It only seeks to gain full consciousness of precisely these intellectual means by seeking not only to represent the result of physical measurement, but to gain fundamental clarity concerning the form of any physical measurement and its conditions.

Thereby it seems indeed to come into the immediate neighborhood of the critical and transcendental theory, which is directed on the "possibility of experience;" but it is nevertheless different from it in its general tendency. For, in the language of this transcendental criticism, the doctrine of space and time developed by the theory of relativity is a doctrine of empirical space and empirical time, not of pure space and pure time. As far as concerns this point, there is scarcely possible a difference of opinion; and, in fact, all critics, who have compared the Kantian and the Einstein-Minkowski theories of space and time seem to have reached essentially the same result.[1] From the standpoint of a strict empiricism, one could attempt to dispute the possibility of a doctrine of "pure space" and of "pure time;" but the conclusion cannot be avoided that in so far as such a doctrine is justified, it must be independent of all results of concrete measurement and of the particular conditions, which prevail in the latter. If the concepts of pure space and pure time have in general any definite justified meaning, to use a phrase of the theory of relativity, then this meaning must be invariant with regard to all transformations of the doctrine of the empirical measurement of space and time. The only thing that such transformations

---

[1] *Cf,* esp. Natorp (56, p. 392 ff.). Hönigswald (33, p. 88 ff.). Frischeisen-Köhler (26, p. 323 ff.) and more recently Sellien (81, p. 14 ff.).

can and will accomplish is that they teach us to draw the line more sharply between what belongs to the purely philosophical, "transcendental," criticism of the concepts of space and time and what belongs merely to the particular applications of these concepts. Here, in fact, the theory of relativity can perform an important indirect service for the general criticism of knowledge,—if we resist the temptation to translate its propositions directly into propositions of the criticism of knowledge.

Kant's doctrine of space and time developed to a large extent on the basis of physical problems, and the conflict carried on in the natural science of the eighteenth century on the existence of absolute time and absolute space affected him keenly from the beginning. Before he approached the problems of space and time as a critical philosopher, he had himself lived through the various and opposite solutions by which contemporary physics sought to master these problems. Here, at first, contrary to the dominant scholastic opinion, he took his stand throughout on the basis of the relativistic view. In his *Neuen Lehrbegriff der Bewegung und der Ruhe* of the year 1758, the thirty-four year old Kant set up the principle of the relativity of all motion with all decisiveness and from it attacked the traditional formulation of the principle of inertia. "Now I begin to see," he says after he has illustrated the difficulties of the concept of "absolute motion" with well-known examples, "that I lack something in the expression of motion and rest. I should never say: a body rests without adding with regard to what thing it rests, and never say that it moves without at the same time naming the objects with regard to which it changes its relation. If I wish to imagine also a mathematical space free from all creatures as a receptable of bodies, this would still not help me. For by what should I distinguish the parts of the same and the different places, which are occupied by nothing corporeal?" (35, II, 19.) But Kant, in his further development did not at first remain true to the norm, which he here set up so decisively and of which a modern physicist has said that it deserves to be set up in iron letters over each physical lecture hall.[2] He ventured to abandon the concept of inertial force, of *vis inertiae;* he refused to pour his thoughts on the principles of mechanics "into the mill of the Wolffian or of any

[2] Streintz (82), p. 42.

other famous system of doctrine." But while he opposed in this way the authority of the leading philosophers, he could not permanently withdraw himself from the authority of the great mathematical physicists of his time. In his *Versuch, den Begriff der negativen Grössen in die Weltweisheit einzuführen* of the year 1763, he took his place at the side of Euler to defend with him the validity of the Newtonian concepts of absolute space and absolute time, and six years later, in his essay on the first grounds of the difference of regions in space (1769), he sought to support the proof, that Euler had attempted, of the existence of absolute space from the principles of mechanics, by another, purely geometrical consideration, which "would give practical geometricians a conclusive reason to be able to affirm the reality of their absolute space with the "evidence" which is customary to them." (35, II, 394.) But this is indeed only an episode in Kant's evolution; for only a year later the decisive critical turn in the question of space and time had taken place in his Inaugural Dissertation of the year 1770. By it the problem receives an entirely new form; it is removed from the field of physics to that of "transcendental philosophy" and must be considered and solved according to the general principles of the latter.

But the transcendental philosophy does not have to do primarily with the reality of space or of time, whether these are taken in a metaphysical or in a physical sense, but it investigates the objective *significance* of the two concepts in the total structure of our empirical knowledge. It no longer regards space and time as things, but as "sources of knowledge." It sees in them no independent objects, which are somehow present and which we can master by experiment and observation, but "conditions of the possibility of experience," conditions of experiment and observation themselves, which again for their part are not to be viewed as things.

What—like time and space—makes possible the positing of objects can itself never be given to us as a particular object in distinction from others. For the *"forms"* of possible experience, the forms of intuition as well as the pure concepts of the understanding, are not met again as *contents* of real experience. Rather the only possible manner in which we can ascribe any sort of "objectivity" to these forms must consist in that they lead to certain *judgments* to which we must ascribe the values of necessity and universality. The meaning is thus indicated, in which one can henceforth inquire as to the objec-

tivity of space or time. Whoever demands absolute thing-like
correlates for them strains after shadows. For their whole "being"
consists in the meaning and function they possess for the complexes
of judgments, which we call science, whether geometry or arithmetic,
mathematical or empirical physics. What they can accomplish
as presuppositions in this connection can be exactly determined by
transcendental criticism; what they are as things in themselves is a
vain and fundamentally unintelligible question. This basic view
comes out clearly even in the Inaugural Dissertation. Even here
absolute space and time possessing an *existence* separate from empirical bodies and from empirical events, are rejected as nonentities, as
mere conceptual fictions (*inane rationis commentum.*) The two,
space and time, signify only a fixed law of the mind, a schema of
connection by which what is sensuously perceived is set in certain
relations of coexistence and sequence. Thus the two have, in
spite of their "transcendental ideality," "empirical reality," but this
reality means always only their validity for all experience, which
however must never be confused with their existence as isolated objective contents of this experience itself. "Space is merely the form
of external intuition (formal intuition) and not a real object that can
be perceived by external intuition. Space, as prior to all things
which determine it (fill or limit it), or rather which give an empirical
intuition determined by its form, is, under the name of absolute
space, nothing but a mere possibility of external phenomena.
If we try to separate one from the other, and to place space
outside all phenomena, we arrive at a number of empty determinations of external intuition, which, however, can never be possible
perceptions; for instance, motion or rest of the world in an infinite
empty space, *i.e.*, a determination of the mutual relation of the two,
which can never be perceived, and is therefore nothing but the predicate of a mere idea." (34, p. 457; Müller trans., p. 347.)

Accordingly, when Einstein characterizes as a fundamental feature
of the theory of relativity that it takes from space and time "the
last remainder of *physical objectivity*," it is clear that the theory only
accomplishes the most definite application and carrying through of
the standpoint of critical idealism within empirical science itself.
Space and time in the critical doctrine are indeed distinguished in
their validity as types of order from the contents, which are ordered
in them; but these forms possess for Kant a separate existence neither

in the subjective nor in the objective sense. The conception, that space and time as subjective forms into which sensations enter "lie ready in the mind" before all experience, not as "physical" but as "psychical" realities, today scarcely needs refutation. This conception indeed seems to be indestructible, although Fichte poured upon it his severe but appropriate scorn; but it disappears of itself for everyone who has made clear to himself even the first conditions of the transcendental formulation of the question in opposition to the psychological. The meaning of the principle of order can in general be comprehended only in and with what is ordered; in particular, it is urged in the case of the measurement of time that the determination of the temporal positions of particular empirical objects and processes cannot be derived from the relations of the phenomena to absolute time, but that conversely the phenomena must determine and make necessary their positions in time for each other. "This unity in the determination of time is dynamical only, that is, time is not looked upon as that in which experience assigns immediately its place to every existence, for this would be impossible; because absolute time is no object of perception by which phenomena could be held together; but the rule of the understanding through which alone the existence of phenomena can receive synthetical unity in time determines the place of each of them in time, therefore *a priori* and as valid for all time." (34, p. 245 and 262; *cf.* 56, p. 332; *cf.* Müller trans., p. 175.)

It is such a "rule of the understanding," in which is expressed the synthetic unity of phenomena and their reciprocal dynamical relation, on which rests all empirical spatial order, all objective relations of spatial "community" in the corporeal world. The *"communio spatii,"* i.e., that *a priori* form of coexistence, which in Kant's language is characterized as "pure intuition" is, as he expressly urges, only empirically knowable for us by the *commercium* of substances in space, *i.e.*, by a whole of physical effects, that can be pointed out in experience. We read, in a passage of the *Critique of Pure Reason*, which appears especially significant and weighty in connection with the development of the modern theory of relativity: "The word *communion* (*Gemeinschaft*), may be used in two senses, meaning either *communio* or *commercium*. We use it here in the latter sense: as a dynamical communion, without which even the local *communio spatii* could never be known empirically. We can easily perceive in our experience, *that continuous influences only can lead our senses in all parts of space from one object to another; that the light which plays between our eyes and celestial bodies produces a mediate communion between us and them*, and proves the coexistence of the latter; that we cannot change any place empirically (perceive such a change) unless matter itself renders the perception of our own place possible to us, and that by means of its reciprocal influence only matter can evidence its simultaneous existence, and thus (though mediately only) its coexistence, even to the most distant objects." (34, p. 260; *cf.*, Müller trans., p. 173 f.)   The spatial order of the corporeal world, in other words, is never given to us directly and sensuously, but is the result of an intellectual construction, which takes its start from certain empirical laws of phenomena and from that point seeks to advance to increasingly general laws, in which finally is grounded what we call the unity of experience as a spatio-temporal unity.

But is there not found in this last expression the characteristic and decisive opposition between the theory of space and time of critical idealism and the theory of relativity? Is not the essential result of this theory precisely the destruction of the unity of space and time demanded by Kant?   If all measurement of time is dependent on the state of motion of the system from which it is made there seem to result only infinitely many and infinitely diverse "place-times," which, however, never combine into the unity of "the" time.   We have already seen, however, that this view is erroneous, that the destruction of the substantialistic unity of space and time does not destroy their functional unity but rather truly grounds and confirms it. (*Cf.* above, p. 33 ff. 54 ff.)   In fact, this state of affairs is not only granted by the representatives of the theory of relativity among the physicists, but is expressly emphasized by them. "The boldness and the high philosophical significance of Einstein's doctrine consists," we read, *e.g.*, in the work of Laue, "in that it clears away the traditional prejudice of one time valid for all systems.   Great as the change is, which it forces upon our whole thought, there is found in it not the slightest epistemological difficulty.   For in Kant's manner of expression time is, like space, a pure form of our intuition; a schema in which we must arrange events, so that in opposition to subjective and highly contingent perceptions they may gain objective meaning. This arranging can only take place on the basis of empirical knowledge of natural laws. The place and time of the observed change of a heavenly body can only be established on the basis of optical laws. That two differently moving observers, each one regarding himself at rest, should make this arrangement differently on the basis of the same laws of nature, contains no logical impossibility. Both arrangements have, nevertheless, objective meaning since there may be deduced exactly from each of them by the derivative transformation formulae that arrangement valid for the other moving observer." (40, p. 36 f.) This one-to-one correlation and not the oneness of the values gained in the different systems, is what remains of the notion of the "unity of time"; but precisely in it is expressed all the more sharply the fundamental view that this unity is not to be represented in the form of a particular objective content, but exclusively in the form of a system of valid relations. The "dynamic unity of temporal determinations" is retained as a postulate; but it is seen that we cannot satisfy this postulate if we hold to the laws of the Newtonian mechanics, but that we are necessarily driven to a new and more universal and concrete form of physics. The "objective" determination shows itself thus to be essentially more complex than the classical mechanics assumed, which believed it could literally grasp with its hands the objective determination in its privileged systems of reference. That a step is thereby taken beyond Kant is incontestible; for he shaped his "Analogies of Experience" essentially on the three fundamental Newtonian laws: the law of inertia, the law of the proportionality of force and acceleration, and the law of the equality of action and reaction. But in this very advance the doctrine that it is the "rule of the understanding," that forms the pattern of all our temporal and spatial determinations, is verified anew. In the special theory of relativity, the principle of the constancy of the velocity of light serves as such a rule; in the general theory of relativity this principle is replaced by the more inclusive doctrine that all Gaussian coördinate systems are of equal value for the formulation of the universal natural laws. It is obvious that we are not concerned here with the expression of an empirically observed fact, but with a principle which the understanding uses hypothetically as a norm of investigation in the interpretation of experience, for how could an infinite totality be "observed"? And the meaning and justification of this norm rest precisely on the fact that only by its application could we hope to regain the lost unity of the object, namely, the "synthetic unity of phenomena according to temporal relations." The physicist now depends neither on the constancy of those objects with which the naïve sensuous view of the world rests nor on the constancy of particular spatial and temporal measurements gained from a particular system, but he affirms, as a condition of his science, the existence of "universal constants" and universal laws, which retain the same values for all systems of measurement.

In his *Metaphysischen Anfangsgründen der Naturwissenschaft*, Kant, returning to the problem of absolute space and time, formulates a happy terminological distinction, which is suited to characterize more sharply the relation of critical idealism to the theory of relativity. Absolute space, he urges here too, is *in itself nothing and indeed no object;* it signifies only a *space relative to every other* which I can think outside of any given space. To make it a real thing means to confuse the *logical universality* of any space with which I can compare any empirical space as included in it with the *physical universality* of real extension and to misunderstand the Idea of reason. The true logical universality of the Idea of space thus not only does not include its physical universality, as an all-inclusive container of things, but it is precisely of a sort to exclude it. We should, in fact, *conceive* an absolute space, *i.e.*, an ultimate unity of all spatial determinations; but not in order to know the absolute movements of empirical bodies, but to represent in the same "all movements of the material as merely relative to each other, as alternatively reciprocal, but not as absolute motion or rest." "Absolute space is thus necessary not as a concept of a real object, but as an Idea, which should serve as a rule for considering all motions in it as merely relative, and all motion and rest must be reduced to the absolute space, if the phenomena of the same are to be made into a definite concept of experience that unifies phenomena." (35, IV, 383 f., 472 f.) The logical universality of such an idea does not conflict with the theory of relativity; it starts by regarding all motions in space as merely relative because only in this way can it combine them into a definite concept of experience, that unifies all phenomena. On the basis of the demand for the totality of determinations it negates every attempt to make a definite particular system of reference the norm for all the others. The one valid norm is merely the idea of the unity of nature, of exact determination itself. The mechanical view of the world is overcome from this standpoint.

The "unity of nature" is grounded by the general theory of relativity in a new sense, since it includes under a supreme principle of knowledge along with the phenomena of gravitation, which form the real classical field of the older mechanics, the electrodynamic phenomena. That in order to advance to this "logical universality of the Idea," many trusted presentational pictures must be sacrificed need not disturb us; this can affect the "pure intuition" of Kant only in so far as it is misunderstood as a mere picture and not conceived and estimated as a constructive method.

In fact, the point at which the general theory of relativity must implicitly recognize the methodic presupposition, which Kant calls "pure intuition" can be pointed out exactly. It lies, in fact, in the concept of "coincidence" to which the general theory of relativity ultimately reduces the content and the form of all laws of nature. If we characterize events by their space-time coördinates x_1, x_2, x_3, x_4, x'_1, x'_2, x'_3, x'_4, etc., then, as it emphasizes, everything that physics can teach us of the "essence" of natural processes consists merely in assertions concerning the coincidences or meetings of such points. We reach the construction of physical time and of physical space merely in this way; for the whole of the space-time manifold is nothing else than the whole of such coördinations.[3]   Here is the point at which the ways of the physicist and of the philosopher definitely part, without their being thereby forced into conflict. What the physicist calls "space" and "time" is for him a concrete measurable manifold, which he gains as the *result* of coördination, according to law, of the particular points; for the philosopher, on the contrary, space and time signify nothing else than the forms and *modi*, and thus the presuppositions, of this coördination itself. They do not result for him from the coördination, but they *are* precisely this coördination and its fundamental directions. It is coördination from the standpoint of coexistence and adjacency or from the standpoint of succession, which he understands by space and time as "forms of intuition." In this sense, both are expressly defined in the Kantian Inaugural Dissertation. *"Tempus non est objectivum aliquid et reale . . . . sed subjectiva conditio, per naturam mentis humanae necessaria, quaelibet sensibilia certa lege sibi coordinandi et intuitus purus. . . . . Spatium est . . . .*

---

[3] Einstein (17), p. 13 f.; (18), p. 64.

subjectivum et ideale et e natura mentis stabili lege profisiscens velut
schema omnia omnino externe sensa sibi coordinandi.' '(35; II, 416,
420.) Whoever recognizes this law and this schema, this possibility
of relating point to point and connecting them with each other, has
recognized space and time in their "transcendental meaning," for we
can abstract here from any psychological by-meaning of the concept
of form of intuition. We can thus conceive the "world-points"
x_1 x_2 x_3 x_4 and the world-lines, which result from them, so abstractly
that we understand under the values x_1 x_2 x_3 x_4 nothing but certain
mathematical parameters; the "meeting" of such world-points
involves a comprehensible meaning only if we take as a basis that
"possibility of succession," which we call time. A coincidence,
which is not to mean identity, a unification, which is still a
separation, since the same point is conceived as belonging to
different lines: all this finally demands that synthesis of the
manifold, for which the term "pure intuition" was formulated.
The most general meaning of this term, which indeed was not always
grasped by Kant with equal sharpness, since more special meanings
and applications were substituted involuntarily in his case, is merely
that of the serial form of coexistence and of succession. Nothing is
thereby presupposed concerning special relations of measurement in
the two, and in so far as these depend in particular on the relations of
the physical in space, we must guard against seeking to find an exhaustive determination in the mere "forms of possibility" of the relations
of the "real." (*Cf.* below VI.) When *e.g.*, in the mathematical
foundations of the theory of relativity the formula is deduced for the
"distance" of the two infinitely close points x_1 x_2 x_3 x_4, and $x_1 +$
dx_1, $x_2 + dx_2$, $x_3 + dx_3$, $x_4 + dx_4$, this cannot indeed be conceived
as a rigid Euclidean distance in the ordinary sense, since there is
involved in it, by the addition of time as a fourth dimension, not a
magnitude of space but rather one of motion; but the fundamental
form of coexistence and succession and their reciprocal relation and
"union" is unmistakably contained in this expression of the general
linear element. Not that the theory, as has been occasionally
objected, presupposes space and time as something already given, for
it must be declared free of this epistemological circle, but in the sense
that it cannot lack the form and function of spatiality and temporality
in general.

What seems to render understanding difficult at this point between the physicist and the philosopher is the fact that a common problem is found here, which both approach from entirely different sides. The process of measurement interests the critic of knowledge only in so far as he seeks to survey in systematic completeness the concepts, which are used in this process, and to define them in the utmost sharpness. But any such definition is unsatisfying and fundamentally unfruitful to the physicist as long as it is not connected with any definite indication as to how the measurement is to be made in the concrete particular case. "The concept exists for the physicist," says Einstein in one place neatly and characteristically, "only when the possibility is given of finding out in the concrete case whether the concept applies or not." (18, p. 14.) Thus the concept of simultaneity, for example, only receives a definite meaning, when a method is given by which the temporal coincidence of two events is determined by certain measurements, by the application of optical signals; and the difference which is found in the results of this measurement seems to have as a consequence the ambiguity of the concept. The philosopher has to recognize unconditionally this longing of the physicist for concrete determinateness of concepts; but he is ever again brought to the fact that there are ultimate ideal determinations without which the concrete cannot be conceived and made intelligible. To make clear the opposition in formulation of the question which is here fundamental, one can contrast to Einstein's expression one of Leibniz. "On peut dire," we read in Leibniz' Nouveaux Essais, "qu'il ne faut point s'imaginer deux étendues, l'une abstraite, de l'espace, l'autre concrète, du corps; le concret n'étant tel que par l'abstrait." (43, V, 115.) As we see, it is the unity of the abstract and the concrete, of the ideal and the empirical in which the demands of the physicist and the philosopher agree; but while the one goes from experience to the idea, the other goes from the idea to experience. The theory of relativity holds fast to the "pre-established harmony between pure mathematics and physics;" Minkowski, in the well-known concluding words of his lecture, "Space and Time," has expressly taken up again and brought to honor this Leibnizian term. But this harmony is for the physicist the incontestable premise from which he strives to reach the particular consequences and applications, while for the critic of knowledge the "possibility" of this harmony constitutes the real problem. The basis of this possibility he finds ultimately in the fact that any physical assertion, even the simplest determination of magnitude established by experiment and concrete measurement, is connected with universal conditions, which gain separate treatment in pure mathematics, *i.e.*, that any physical assertion involves certain logico-mathematical constants. If we desire to bring all of these constants into a short formula, we can point out the concept of number, the concept of space, the concept of time, and the concept of function as the fundamental elements, which enter as presuppositions into every question which physics can raise. None of these concepts can be spared or be reduced to another so that, from the standpoint of the critique of cognition, each represents a specific and characteristic motive of thought; but on the other hand, each of them possesses an actual empirical use only along with the others and in systematic connection with them. The theory of relativity shows with especial distinctness how, in particular, the thought of function is effective as a necessary motive in each spatio-temporal determination. Thus physics knows its fundamental concepts never as logical "things in themselves," but only in their reciprocal combination; it must, however, be open to epistemology to analyze this product into its particular factors. It thus cannot admit the proposition that the meaning of a concept is identical with its concrete application, but it will conversely insist that this meaning must be already established before any application can be made. Accordingly, the thought of space and time in their meaning as connecting forms of order is not first created by measurement but is only more closely defined and given a definite content. We must have grasped the concept of the "event" as something spatio-temporal, we must have understood the meaning expressed in it, before we can ask as to the coincidence of events and seek to establish it by special methods of measurement.

In general, physics sees itself placed by its fundamental problem from the beginning between two realms, which it has to recognize and between which it has to mediate without asking further as to their "origin." On the one side, stands the manifold of data of sensation, on the other a manifold of pure functions of form and order. Physics, as an empirical science, is equally bound to the "material" content, which sense perception offers it, and to these formal principles in which is expressed the universal conditions of the "possibility of experience." It has to "invent" or to derive deductively the one as little as the other, *i.e.*, neither the whole of empirical contents nor the whole of characteristic scientific forms of thought, but its task consists in progressively relating the realm of "forms" to the data of empirical observation and, conversely, the latter to the former.   In this way, the sensuous manifold increasingly loses its "contingent" anthropomorphic character and assumes the imprint of thought, the imprint of systematic unity of form.   Indeed "form," just because it represents the active and shaping, the genuinely creative element, must not be conceived as rigid, but as living and moving.   Thought comprehends more and more that form in its peculiar character cannot be given to it at one stroke, but that the existence of form is only revealed to it in the becoming of form and in the law of this becoming.   In this way, the history of physics represents not a history of the discovery of a simple series of "facts," but the discovery of ever new and more special means of thought. But in all change of these means of thought there is nevertheless revealed, as surely as physics follows the "sure course of a science," the unity of those methodic principles upon which rests the formulation of its question.   In the system of these principles, space and time take their fixed place, although they are not to be conceived as fixed things or contents of presentation.   The ancient view believed that it possessed and encompassed the spatio-temporal unity of being directly in presentation.   To Parmenides and fundamentally the whole ancient world being was given "like the mass of a well-rounded sphere."   With the reform of Copernicus, the security of this possession was gone once for all.   Modern science knows that there is a definite spatio-temporal order of phenomena for knowledge only in so far as knowledge progressively establishes it, and that the only means of establishing it consists in the scientific concept of law.   But the problem of such a general orientation remains for thought and becomes the more urgent the more thought knows it as a problem never to be solved definitively.   Precisely because the unity of space and time of empirical knowledge seems to flee eternally before all our empirical measurements, thought comprehends that it must seek it eternally and that it must avail itself of new and ever sharper instruments.   It is the merit of the theory of relativity not only to have proved this in a new way but also to have established a principle, *i.e.*, the principle of the co-variancy of the universal laws of nature with regard to all arbitrary substitutions, by which thought can master, out of itself, the relativity which it calls forth.

In the analysis of spatial and temporal measurements, made by the theory of relativity this fundamental relation can be traced in detail. This analysis does not begin by accepting the concept of the "simultaneity" of two processes as a self-evident and immediately known datum, but by demanding an explanation of it—an explanation, which, as a physical explanation, cannot consist in a general conceptual definition, but only in the indication of the concrete methods of measurement, by which "simultaneity" can be empirically pointed out. The simultaneity of such processes as take place practically in "the same" point of space or in immediate spatial adjacency is at first presupposed; we assume, as Einstein explains, the determinability of "simultaneity" for events, which are immediately adjacent spatially, or, more exactly, for events in immediate spatio-temporal adjacency (coincidence), without defining this concept (17, §3.) In fact, recourse here to a mediating physical method of measurement seems neither desirable nor possible; for any such method would always presuppose the possibility of making a temporal coördination between diverse events, thus, *e.g.*, of establishing "the simultaneity" of a definite event with a certain position of the hands of a clock found at the "same" place. The real problem of the theory of relativity begins only when we are no longer concerned with temporally connecting spatially adjacent series of events with each other, but rather series of events spatially removed from each other. If we assume that there is established for the two points of space A and B a certain "place-time," then we possess only an "A-time" and "B-time" but no time *common* to A and B. It is seen that every attempt to establish such a common time as an empirically measurable time, is bound to a definite empirical presupposition concerning the velocity of light. The assumption of the uniform velocity of light enters implicitly into all our assertions concerning the simultaneity of what is spatially distant. A time common to A and B is gained when one establishes *by definition* that the "time," which light takes in going from A to B is equal to the "time," which it takes in going from B to A. Let us assume that a ray of light is sent at A-time t_A from a clock found in A to B, and then at B-time, t_B, the ray of light is reflected to A and reaches A again at A-time, t'_A; then we establish by definition that the two clocks of A and B are to be called "synchronous" if $t_B - t'_A = t'_A - t_B$. Thus for the first time an exact determination is made of what we are to under-

stand by the "time" of an event and by the "simultaneity" of two
processes; "the time" of an event is what is told us by a motionless
clock found at the place of the event simultaneously with the event,
a clock which runs synchronously with a certain motionless clock
and indeed synchronously with the latter at all times." (16, p. 28 f.)

That the "forms" of space and time as definite forms of the coordination of different contents already enter into the concrete
determinations, which are here made for the procedure of the physical
measurement of time, scarcely needs special explanation.   The two
are immediately assumed in the concept of the "place-time;" for the
possibility is involved in it of grasping a definitely distinguished
"now" in a definitely distinguished "here."   This "here" and "now"
does not signify indeed the whole of space and time, to say nothing of
all the concrete relations within the two to be established by measurement; but it represents the first foundation, the unavoidable basis
of the two.   The first primitive difference, which is expressed in the
mere positing of a "here" and a "now" remains thus, for the theory
of relativity, too, an indefinable on which it grounds its complex
physical definitions of space and time values.   And while for these
definitions it appeals to a definite assumption concerning the law of
the propagation of light, this, too, involves the presupposition that a
certain condition that we call "light" occurs *in succession* at different
places and according to a definite rule, in which what space and
time mean as mere schemata of coördination, is obviously contained.
The epistemological problem seems indeed to be heightened when we
reflect on the reciprocal relation of space and time values in the
fundamental equations of physics.   What is given in these equations
is the four-dimensional "world," the continuum of events in general,
the temporal determinations in this continuum not being separated
from the spatial.   The intuitive difference between a spatial distance and a temporal duration, which we believe ourselves to grasp
immediately, plays no rôle in this purely mathematical determination.   According to the temporal equation of the Lorentz-
transformation:

$$t' = \frac{t - \frac{v}{c^2}x}{\sqrt{1 - \frac{v^2}{c^2}}}$$

the time differential Δt' between two events with reference to K
does not disappear when the time differential Δt of the same disappears with reference to K; the purely spatial distance of two events
with reference to K has as a consequence in general the temporal
sequence of the same with reference to K'. This leveling of space
and time values is developed even further in the general theory of
relativity. Here it is seen to be impossible to construct a reference
system out of fixed bodies and clocks of such a sort that place and
time are directly indicated by a fixed arrangement of measuring rods
and clocks relatively to each other; but each point of the continuous
series of events is correlated with four numbers, x_1, x_2, x_3, x_4, which
possesses no direct physical meaning, but only serve to enumerate
the points of the continuum in a definite but arbitrary way. This
correlation need not have such properties that a certain group of
values x_1 x_2 x_3 must be understood as the spatial coördinates and
opposed to the "temporal" coördinate x_4. (18, p. 38, 64.) The
demand of Minkowski that "space for itself and time for itself be
completely degraded to shadows" and that only "a sort of union of
the two shall retain independence" seems thus now to be strictly
realized. Now at any rate, this demand contains nothing
terrible for the critical idealist, who has ceased to conceive space and
time as things in themselves or as given empirical objects. For the
realm of ideas is for him a "realm of shadows," as Schiller called it,
since no pure idea corresponds directly to a concrete real object,
but rather the ideas can always only be pointed out in their systematic community, as fundamental moments of concrete objective
knowledge. If it thus appears that physical space and time measurements can be assumed only as taking place in common, the difference
in the fundamental character of space and time, or order in coexistence and succession is not thereby destroyed. Even if it is true that,
as Minkowski urges, no one has *perceived* a place save at a time and
a time save at a place, there remains a difference between what is to
be *understood* by spatial and by temporal discrimination. The
factual interpenetration of space and time in all empirical physical
measurements does not prevent the two from being different in principle, not as objects, but as types of objective discrimination.
Although two observers in different systems K and K' can assume
the arrangement of the series of events in the orders of space and
time to be different, it is still always a *series of events* and thus a

continuum both spatial and temporal, which they construct in their measurements. Each observer distinguishes from his standpoint of measurement a continuum, which he calls "space," from another, which he calls "time;" but he can, as the theory of relativity shows, not assume without further consideration that the arrangement of phenomena in these two schemata must be similar from each system of reference. There may thus, according to Minkowski's "world postulate," be given only the four-dimensional world in space and time, and "the projection into space and time" may be possible "with a certain freedom;" this only affects the different spatio-temporal interpretations of phenomena, while the difference of the form of space from that of time is unaffected.

For the rest, here too the transformation-equation reëstablishes objectivity and unity, since it permits us to translate again the results found in one system into those of the other. Also, if one seeks to clarify the proposition of Minkowski that only the insepa-rable union of space and time possesses independence, by saying that this union itself, according to the results of the general theory of relativity, becomes a shadow and an abstraction, and that only the unity of space, time and things possesses independent reality,[4] then this classification only leads us back again to our first epistemo-logical insight. For that neither "pure space" nor "pure time" nor the reciprocal connection of the two, but only their realization in some empirical material gives what we call "reality," i.e., the physical being of things and of events, belongs to the fundamental doctrines of critical idealism. Kant himself did not weary of referring re-peatedly to this indissoluble connection, this reciprocal correlation of the spatio-temporal form and the empirical content in the exist-ence and structure of the *world of experience*. "To give an object," we read, "if this is not meant again as mediate only, but if it means to represent something immediately in intuition, is nothing else but to refer the representation of the object to experience. . . . . Even space and time, however pure these concepts may be of all that is empirical, and however certain it is that they are represented in the mind entirely *a priori*, would lack nevertheless all objective validity, all sense and meaning, if we could not show the necessity of their use with reference to all objects of experience. Nay, their

---

[4] See Schlick (79), p. 51; *cf*. p. 22.

representation is a pure schema, always referring to that reproductive imagination, which calls up the objects of experience, without which objects would be meaningless." (34, p. 195; *cf.* Müller trans., p. 127 f.) The "*ideal*" meaning, that space and time possess "in the mind" thus does not involve any sort of *particular* existence, which they would possess prior to things and independently of them, but it rather expressly denies it—the ideal separation of pure space and pure time from things (more exactly, from empirical phenomena), not only permits but demands precisely their empirical "union." This union the general theory of relativity has verified and proved in a new way, since it recognizes more deeply than all preceding physical theories the dependency belonging to all empirical measurement, to all determination of concrete spatio-temporal relations.[5] The relation of experience and thought that is established in the critical doctrine does not contradict this result in any way, but rather it confirms it and brings it to its sharpest expression. It is indeed at first glance strange and paradoxical that the most diverse epistemological standpoints, that radical empiricism and positivism as well as critical idealism have all appealed to the theory of relativity in support of their fundamental views. But this is satisfactorily explained by the facts that empiricism and idealism meet in certain presuppositions with regard to the doctrine of empirical space and of empirical time, and that the theory of relativity sets up just such a doctrine. Both here grant to experience the decisive rôle, and both teach that every exact measurement presupposes universal empirical *laws.*[6] But the question becomes all the more urgent as to how we reach these laws, on which rests the possibility of all empirical measurement, and what sort of validity, of logical "dignity" we grant to them. Strict positivism has only one answer to this question: for it all knowledge of laws, like all knowledge of objects, is grounded in the simple elements of sensation and can never go beyond their realm. The knowledge of laws possesses accordingly in principle the same purely passive character that belongs to our knowledge of any particular sensuous qualities. Laws are treated like things whose properties one can read off by immediate perception. Mach attempts, quite consistently with his standpoint, to extend this

[5] On the "relativization" of the difference of space and time, *cf.* also below, VII.

[6] (8), p. 191 ff.; *cf.* Sellien (81), p. 14 ff.

manner of consideration to pure mathematics also and the deduction of its fundamental relations. The way in which we gain the differential quotient of a certain function, as he explains, is not distinguished in principle from the way in which we establish any sort of properties or changes of physical things. As in the one case we subject the thing, so in the other case we subject the function to certain operations and simply observe how it "reacts" to them. The reaction of the function $y = x^m$ to the operation of differentiation out of which the equation $\frac{dy}{dx} = mx^{m-1}$ results "is a distinguishing mark of x^m just as much as the blue-green color in the solution of copper in sulphuric acid." (49, p. 75.) Here we find clearly before us the sharp line of distinction between critical idealism and positivism of Mach's type. That the equations governing larger or smaller fields are to be regarded as what is truly permanent and substantial, since they make possible the gaining of a stable picture of the world,[7] that they thus constitute the kernel of physical objectivity: this is the fundamental view in which the two theories combine. The question concerns only the manner of establishing, only the exact grounding, of these equations. Idealism urges against the standpoint of "pure experience" as the standpoint of mere sensation, that all equations are results of measurement; all measurement, however, presupposes certain theoretical principles and in the latter certain universal functions of connection, of shaping and coördination. We never measure mere sensations, and we never measure with mere sensations, but in general to gain any sort of relations of measurement we must transcend the "given" of perception and replace it by a conceptual symbol, which possesses no copy in what is immediately sensed. If there is anything that can serve as a typical example of this state of affairs, it is the development of modern physics in the theory of relativity. It is verified again that every physical theory, to gain conceptual expression and understanding of the facts of experience, must free itself from the form in which at first these facts are immediately given to perception.[8] That the theory of relativity is founded on experience and observation is, of course,

[7] See Mach (49), p. 429.

[8] *Cf.* Duhem (15, p. 322): "*Les faits d'expérience, pris dans leur brutalité native, ne sauraient servir au raisonnement mathématique; pour alimenter ce raisonnement, ils doivent être transformés et mis sous forme symbolique.*"

beyond question. But, on the other hand, its essential achievement consists in the new interpretation that it gives to the observed facts, in the conceptual interpretation by which it is progressively led to subject the most important intellectual instruments of classical mechanics and the older physics to a critical revision. It has been pointed out with justice that it has been precisely the oldest empirical fact of mechanics, the equality of inert and heavy masses, which, in the new interpretation it has received from Einstein, has become the fulcrum of the general theory of relativity. (24a.) The way in which the principle of equivalence and with it the foundations of the new theory of gravitation have been deduced from this fact can serve as a logical example of the meaning of the pure "thought-experiment" in physics. We conceive ourselves in the position of an observer, who, experimenting in a closed box, establishes the fact that all bodies left to themselves move, always with constant acceleration, toward the floor of the box. This fact can be represented conceptually by the observer in a double manner: in the first place, by the assumption that he is in a temporarily constant field of gravity in which the box is hung up motionless, or, in the second place, by the assumption that the box moves upward with a constant acceleration whereby the fall of bodies in it would represent a movement of inertia. The two: the inertial movement and the effect of gravitation, are thus in truth a single phenomenon seen and judged from different sides. It follows that the fundamental law that we establish for the movement of bodies must be such that it includes equally the phenomena of inertia and those of gravitation. As is seen, we have here no empirical proposition abstracted from particular observations, but a rule for our construction of physical concepts: a demand that we make, not directly of experience, but rather of our manner of intellectually representing it. "Thought-experiments" of such force and fruitfulness cannot be explained and justified by the purely empiristic theory of physical knowledge. It is not in contradiction with this that Einstein refers gratefully to the decisive stimulus, which he received from Mach (20); for a sharp distinction must be made between what Mach has accomplished as a physicist in his criticism of Newton's fundamental concepts, and the general philosophical consequences he has drawn from this achievement. Mach himself has, as is known, granted wide scope to the pure "thought-experiment" in his own logic of physics;

but, more closely considered, he has thereby already left the ground of a purely sensatory founding of the fundamental concepts of physics.[9] That there is no necessary connection between the theory of relativity and Mach's philosophy may be concluded from the fact, among other things, that it is precisely one of the first advocates of this theory, Max Planck, who among all modern physicists has most sharply criticized and fought against the presuppositions of this philosophy. (69.) Even if one take the theory of relativity as an achievement and outcome of purely empirical thought, it is thereby a proof and confirmation of the constructive force immanent in this thought by which the system of physical knowledge is distinguished from a mere "rhapsody of perceptions."

[9] See Mach (50, p. 180 ff.); cf. (8), p. 316 ff. and (39), p. 86 f.

CHAPTER VI

Euclidean and Non-Euclidean Geometry

In the preceding considerations, however, we have taken up only incidentally an achievement of the general theory of relativity, which, above all others, seems to involve a "revolution of thought." In the working out of the theory, it is seen that the previous Euclidean measurements are not sufficient; the development of the theory can only take place by our going from the Euclidean continuum, which was still taken as a basis by the special theory of relativity, to a non-Euclidean four-dimensional space-time continuum and seeking to express all relations of phenomena in it. Thus a question seems answered physically which had concerned the epistemology of the last decades most vitally and which had been answered most diversely within it. Physics now proves not only the possibility, but the reality of non-Euclidean geometry; it shows that we can only understand and represent theoretically the relations, which hold in "real" space, by reproducing them in the language of a four-dimensional non-Euclidean manifold.

The solution of this problem from the side of physics was, on the one hand, for a long time hoped for as keenly, as, on the other hand, its possibility was vigorously denied. Even the first founders and representatives of the doctrine of non-Euclidean geometry sought to adduce experiment and concrete measurement in confirmation of their view. If we can establish, they inferred, by exact terrestrial or astronomical measurements, that in triangles with sides of very great length the sum of the angles differs from two right angles, then empirical proof would be gained that in "our" empirical space the propositions not of Euclidean geometry, but of one of the others were valid. Thus, *e.g.*, Lobatschefsky, as is known, used a triangle $E_1\ E_2\ S$, whose base $E_1\ E_2$ was formed by the diameter of the orbit of the earth and whose apex S was formed by Sirius and believed that he could, in this way, prove empirically a possible constant curvature of our space. (48.) The fallacy in method of any such attempt must be obvious, however, to any sharper epistemological analysis of the problem and it has been pointed out from the side of the mathematicians with special emphasis by H. Poincaré. No measurement, as Poincaré objects with justice, is concerned with space itself but always only with the empirically given and physical objects in space. No experiment therefore can teach us anything about the *ideal* structures, about the straight line and the circle, that pure geometry takes as a basis; what it gives us is always only knowledge of the relations of material things and processes. The propositions of geometry are therefore neither to be confirmed nor refuted by experience. No experiment will ever come into conflict with the postulates of Euclid; but, on the other hand, no experiment will ever contradict the postulates of Lobatschefsky. For granted, that some experiment could show us a variation in the sums of the angles of certain very great triangles, then the conceptual representation of this fact would never need to consist in, and methodologically could not consist in, changing the axioms of geometry, but rather in changing certain hypotheses concerning physical things. What we would have *experienced*, in fact, would not be another structure of space, but a new law of optics, which would teach us that the propagation of light does not take place in strictly rectilinear fashion. "However, we turn and twist," Poincaré therefore concludes, "it is impossible to attach a rational meaning to empiricism in geometry." (72, p. 92 ff.) If this decision holds and if it can be proved, on the other hand, that among all possible self-consistent geometries the Euclidean possesses a certain advantage of "simplicity" since it defines the minimum of those conditions under which experience is possible in general, there would then be established for it an exceptional position from the standpoint of the critique of knowledge. It would be seen that the different geometries, which are equivalent to each other from a purely, formal standpoint, as regards their logical conceivability, are yet distinguished in their fruitfulness in the founding of empirical science. "The geometries are distinguished from each other in principle," one can conclude, "only by reference to their epistemological relation to the concept of experience; for this relation is positive only in the case of the Euclidean geometry."[1]

In connection, however, with the new development of physics in the general theory of relativity, this epistemological answer seems to become definitely untenable. Again and again the fact has been

---

[1] *Cf.* Hönigswald (32); on the following *cf.* Bauch (1), p. 126 ff.

appealed to in the controversy concerning the epistemological justification of the different geometries that what determines value must
not be sought in formal but in transcendental logic; that the compatibility of a geometry with experience is not involved but rather its
"positive fruitfulness," *i.e.*, the "founding of experience," that it can
give. And this latter was thought to be found in Euclidean
geometry. The latter appeared as the real and unique "foundation
of possibility of knowledge of reality," the others, on the contrary,
always as only the foundations of the possible. But with regard to
the extraordinary rôle that the concepts and propositions of Riemannian geometry played in the grounding and construction of Einstein's
theory of gravitation, this judgment cannot be supported. Supported
by the same logical criterion of value, one now seems forced rather
to the opposite conclusion: non-Euclidean space is alone "real,"
while Euclidean space represents a mere abstract possibility. In
any event, the logic of the exact sciences now finds itself placed before
a new problem. The fact of the fruitfulness of non-Euclidean geometry for physics can no longer be contested, since it has been
verified, not only in particular applications, but in the structure of a
complete new system of physics; what is in question is the explanation to be given to this fact. And here we are first forced to a negative decision, which is demanded by the first principles of the theory
of relativity. Whatever meaning we may ascribe to the idea of
non-Euclidean geometry for physics, for purely empirical thought,
the assertion has lost all meaning for us that any space, whether
Euclidean or non-Euclidean, is the "real" space. Precisely this was
the result of the general principle of relativity, that by it "the last
remainder of physical objectivity" was to be taken from space.
Only the various relations of measurement within the physical manifold, within that inseparable correlation of space, time, and the
physically real object, which the theory of relativity takes as ultimate, are pointed out; and it is affirmed that these relations of
measurement find their simplest exact mathematical expression in
the language of non-Euclidean geometry. This language, however,
is and remains purely ideal and symbolic, precisely as, rightly understood, the language of Euclidean geometry could alone be. The
reality which alone it can express is not that of things, but that of
laws and relations. And now we can ask, epistemologically, only
one question: whether there can be established an exact relation

and coördination between the symbols of non-Euclidean geometry and the empirical manifold of spatio-temporal "events." If physics answers this question affirmatively, then epistemology has no ground for answering it negatively. For the "a priori" of space that it affirms as the condition of every physical theory involves, as has been seen, no assertion concerning any definite particular structure of space in itself, but is concerned only with that function of "spatiality" in general, that is expressed even in the general concept of the linear element ds as such, quite without regard to its character in detail.

If it is seen thus, that the determination of this element as is done in Euclidean geometry, does not suffice for the mastery of certain problems of knowledge of nature then nothing can prevent us, from a methodological standpoint, from replacing it by another measure, in so far as the latter proves to be necessary and fruitful physically. But in either case one must guard against taking the "preëstablished harmony between pure mathematics and physics," that is revealed to us in increasing fullness and depth in the progress of scientific knowledge, as a naïve copy theory. The structures of geometry, whether Euclidean or non-Euclidean, possess no immediate correlate in the world of *existence*. They exist as little physically in things as they do psychically in our "presentations" but all their "being," *i.e.*, their validity and truth, consists in their ideal *meaning*. The existence, that belongs to them by virtue of their definition, by virtue of a pure logical act of assumption is, in principle, not to be interchanged with any sort of empirical "reality." Thus also the applicability, which we grant to any propositions of pure geometry, can never rest on any direct coinciding between the elements of the ideal geometrical manifold and those of the empirical manifold. In place of such a sensuous congruence we must substitute a more complex and more thoroughly mediate relational system. There can be no copy or correlate in the world of sensation and presentation for what the points, the straight lines and the planes of pure geometry signify. Indeed, we cannot in strictness speak of any degree of similarity, of greater or less difference of the "empirical" from the ideal, for the two belong to fundamentally different species. The theoretical relation, which science nevertheless establishes between the two, consists merely in the fact, that it, while granting and holding fast to the difference in content of the two series, seeks to establish a more exact and perfect correlation between them. All verification, which the propositions of geometry can find in physics, is possible only in this way. The particular geometrical truths or particular axioms, such as the principle of parallels, can never be compared with particular experiences, but we can always only compare with the whole of physical experience the whole of a definite system of axioms. What Kant says of the concepts of the understanding in general, that they only serve "to make letters out of phenomena so that we can read them as experiences" holds in particular of the concepts of space. They are only the letters, which we must make into words and propositions, if we would use them as expressions of the laws of experience. If the goal of harmony is not reached in this indirect way, if it appears that the physical laws to which observation and measurement lead us cannot be represented and expressed with sufficient exactitude and simplicity by a given system of axioms, then we are free to determine which of the two factors we shall subject to a transformation to reëstablish the lost harmony between them. Before thought advances to a change of one of its "simple" geometrical laws it will first make the complex physical conditions that enter into the measurement responsible for the lack of agreement; it will change the "physical" factors before the "geometrical." If this does not lead to the goal and if it is seen on the other hand, that surprising unity and systematic completeness can be reached in the formulation of the "laws of nature" by accepting an altered conception of geometrical methods, then in principle there is nothing to prevent such a change. For if we conceive the geometrical axioms, not as copies of a given reality, but as purely ideal and constructive structures, then they are subjected to no other law than is given them by the *system* of thought and knowledge. If the latter proves to be realizable in a purer and more perfect form by our advancing from a relatively simpler geometrical system to a relatively more complex, then the criticism of knowledge can raise no objection from its standpoint. It will be obliged to affirm only this: that here too "no intelligible meaning can be gained" for empiricism in geometry. For here, too, experience does not *ground* the geometrical axioms, but it only selects from among them, as various logically possible systems, of which each one is derived strictly rationally, certain ones with regard to their concrete use in the interpretation of phenomena.[2] Here, too,

---

[2] On this relation of the problem of metageometry to the problem of "experience," *cf.* esp. Albert Görland (28, p. 324 ff.)

Platonically speaking, phenomena are measured by Ideas, by the foundations of geometry, and these latter are not directly read out of the sensuous phenomena.

But when one grants to non-Euclidean geometry in this sense meaning and fruitfulness for physical experience, the general methodic difference can and must be urged, that still remains between it and Euclidean geometry. This difference can no longer be taken from their relation to experience, but it must be recognized as based on certain "inner" moments, *i.e.*, on general considerations of the *theory of relations*. A special and exceptional logical position, a fundamental simplicity of ideal structure, can be recognized in Euclidean geometry even if it must abandon its previous sovereignty within physics. And here it is precisely the fundamental doctrine of the general theory of relativity, that, translated back into the language of logic and general methodology, can establish and render intelligible this special position. Euclidean geometry rests on a definite axiom of relativity, which is peculiar to it. As the geometry of space of a constant curvature 0, it is characterized by the thorough-going relativity of all places and magnitudes. Its formal determinations are in principle independent of any absolute determinations of magnitude. While, *e.g.*, in the geometry of Lobatschefsky, the sum of the angles of a rectilinear triangle is different from 180° and indeed the more so, the more the surface area of the triangle increases, the absolute magnitude of the lines enters into none of the propositions of Euclidean geometry. Here for every given figure a "similar" can be constructed; the particular structures are grasped in their pure "quality," without any definite "quantum," any absolute value of number and magnitude, coming into consideration in their definition. This indifference of Euclidean structures to all absolute determinations of magnitude and the freedom resulting here of the particular points in Euclidean space from all determinations and properties, form a logically positive characteristic of the latter. For the proposition, *omnis determinatio est negatio*, holds here too. The assumption of the indeterminate serves as the foundation for the more complex assumptions and determinations which can be joined to it. In this sense, Euclidean geometry is and remains the "simplest," not in any practical, but in a strictly logical meaning; Euclidean space is, as Poincaré expresses it, "simpler not merely in consequence of our mental habits or in consequence of any direct intuition, which we possess of it, but it is in itself simpler, just as a polynomial of the first degree is simpler than a polynomial of the second degree." (72, p. 67.)   This logical simplicity belonging to Euclidean space in the system of our intellectual meanings wholly independently of its relations to experience, is shown, *e.g.*, in the fact that we can make any "given" space, that possesses any definite curvature, into Euclidean by regarding sufficiently small fields of it from which the difference conditioned by the curvature disappears.   Euclidean geometry shows itself herein as the real geometry of infinitely small areas, and thus as the expression of certain elementary relations, which we take as a basis in thought, although we advance from them in certain cases to more complex forms.

The development of the general theory of relativity leaves this methodic advantage of Euclidean geometry unaffected.   For Euclidean measurements do not indeed hold in it absolutely but they hold for certain "elementary" areas, which are distinguished by a certain simplicity of physical conditions.   The Euclidean expression of the linear element shows itself to be unsatisfactory for the working out of the fundamental thought of the general theory of relativity, since it does not fulfill the fundamental demand of retaining its form in every arbitrary alteration of the system of reference.   It must be replaced by the *general* linear element $(ds^2 = \sum_1^4 g_{\mu\nu} dx_\mu dx_\nu)$, which satisfies this demand.   If, however, we consider infinitely small four-dimensional fields, it is expressly demanded that the presuppositions of the special theory of relativity, and thus its Euclidean measurements shall remain adequate for them.   The form of the universal linear element here passes over into the Euclidean element of the special theory when the ten magnitudes g, which occur in this as functions of the coördinates of particular points assume definite constant values.   The physical explanation of this relation, however consists in that the magnitudes $g_{\mu\nu}$ are recognized as those which describe the gravitational field with reference to the chosen system of reference.   The condition, under which we can pass from the presuppositions of the general theory of relativity to the special theory, can accordingly be expressed in the form that we only consider regions within which abstraction can be made from the effects of fields of gravitation.   This is always possible for an infinitely small field and it holds further for finite fields in which, with appropriate choice of the system of reference, the body considered undergoes no noticeable acceleration. As we see, the variability of the magnitudes $g_{\mu\nu}$, which expresses the variation from the homogeneous Euclidean form of space, is recognized as based on a definite *physical* circumstance. If we consider fields in which this circumstance is absent or if we cancel it in thought, we again stand within the Euclidean world. Thus the assertion of Poincaré that all physical theory and physical measurement can prove absolutely nothing about the Euclidean or non-Euclidean character of *space*, since it is never concerned with the latter but only with the properties of *physical reality in space* remains entirely in force. The abstraction (or, better expressed, the pure function) of homogeneous Euclidean space is not destroyed by the theory of relativity, but is only known as such through it more sharply than before.

In fact, the pure meaning of geometrical concepts is not limited by what this theory teaches us about the conditions of measurement. These concepts are indeed, as is seen now anew, neither an empirical datum nor an empirical *dabile*, but their ideal certainty and meaning is not in the least affected thereby. It is shown that in fields where we have to reckon with gravitational effects of a definite magnitude, the preconditions of the ordinary methods of measurement fall short, that here we can no longer use "rigid bodies" as measures of length, nor ordinary "clocks" as measures of time. But this change of relations of measurement does not affect the calculation of space, but the calculation of the physical relation between the measuring rods and rays of light determined by the field of gravitation. (*Cf.* 83, p. 85 ff.) The truths of Euclidean geometry would only be also affected if one supposed that these propositions themselves are nothing but generalizations of empirical observation, which we have established in connection with fixed bodies. Such a supposition, however, epistemologically regarded, would amount to a *petitio principii*. Even Helmholtz, who greatly emphasizes the empirical origin of the geometrical axioms occasionally refers to another view, which might save their purely ideal and "transcendental" character. The Euclidean concept of the straight line might be conceived not as a generalization from certain physical observations, but as a purely ideal concept, to be confirmed or refuted by no experience, since we would have to decide by it whether any bodies of nature were to be regarded as fixed bodies. But, as he objects, the geometrical axioms would then cease to be synthetical propositions in Kant's sense, as they would only affirm something that would follow analytically from the concepts of the fixed geometrical structures necessary to measurement. (30a, II, 30.) It is, however, overlooked by this objection that there are *fundamentally synthetic forms of unity* besides the form of analytic identity, which Helmholtz has here in mind and which he contrasts with the empirical concept as if the form of analytic identity were unique, and that the axioms of geometry belong precisely to the former. Assumptions of this sort refer to the object in so far as in their totality they "constitute" the object and render possible knowledge of it; but none of them, taken for itself, can be understood as an assertion concerning things or relations of things. Whether they fulfill their task as moments of empirical knowledge can be decided always only in the indicated indirect way: by using them as building-stones in a theoretical and constructive system, and then comparing the consequences, which follow from the latter, with the results of observation and measurement. That the elements, to which we must ascribe, methodologically, a certain "simplicity," must be adequate for the interpretation of the laws of nature, can not be demanded *a priori*. But even so, thought does not simply give itself over passively to the mere *material* of experience, but it develops out of itself new and more complex *forms* to satisfy the demands of the empirical manifold.

If we retain this general view, then one of the strangest and, at first appearance, most objectionable results of the general theory of relativity receives a new light. It is a necessary consequence of this theory that in it one can no longer speak of an immutably given geometry of measurement, which holds once for all for the whole world. Since the relations of measurement of space are determined by the gravitational potential and since this is to be regarded as in general changeable from place to place, we cannot avoid the conclusion that there is in general no unitary "geometry" for the totality of space and reality, but that, according to the specific properties of the field of gravitation at different places, there must be found different forms of geometrical structure. This seems, in fact, the greatest conceivable departure from the idealistic and Platonic conception of geometry, according to which it is the "science of the eternally existent," knowledge of what always "is in the same state" (ἀεὶ κατὰ ταὐτὰ ὡσαύτως ἔχον). Relativism seems here to pass over directly into the field of logic; the relativity of places involves that of geometrical truth. And yet this view is, on the other hand, only the sharpest expression of the fact that the problem of space has lost all ontological meaning in the theory of relativity. The purely methodological question has been substituted for the question of being. We are no longer concerned with what space "is" and with whether any definite character, whether Euclidean, Lobatschefskian or Riemannian, is to be ascribed to it, but rather with what use is to be made of the different systems of geometrical presuppositions in the interpretation of the phenomena of nature and their dependencies according to law. If we call any such system a particular "space," then indeed we can no longer attempt to grasp all of these spaces as intuitive parts to be united into an intuitive whole. But this impossibility rests fundamentally on the fact that we have here to do with a problem, which as such stands outside the limits of intuitive representation in general. The space of pure intuition is always only *ideal*, being only the space constructed according to the laws of this intuition, while here we are not concerned with such ideal syntheses and their unity, but with the relations of measurement of the empirical and the physical. These relations of measurement can only be gained on the basis of natural laws, *i.e.*, by proceeding from the dynamic dependency of phenomena upon each other, and by permitting phenomena to determine their positions reciprocally in the space-time manifold by virtue of this dependency. Kant too decisively urged that this form of dynamic determination did not belong to intuition as such, but that it is the "rules of the understanding" which alone give the existence of phenomena synthetic unity and enable them to be collected into a definite concept of experience. (*Cf.* above p. 79.) The step beyond him, that we have now to make on the basis of the results of the general theory of relativity, consists in the insight that geometrical axioms and laws of other than Euclidean form can enter into this determination of the understanding, in which the empirical and physical world arises for us, and that the admission of such axioms not only does not destroy the unity of the world, *i.e.*, the unity of our experiential concept of a total order of phenomena, but first truly grounds it from a new angle, since in this way the particular laws of nature, with which we have to calculate in space-time determination, are ultimately brought to the unity of a supreme principle,—that of the universal postulate of relativity. The renunciation of intuitive simplicity in the picture of the world thus contains the guarantee of its greater intellectual and systematic completeness. This advance, however, can not surprise us from the epistemological point of view; for it expresses only a general law of scientific and in particular of physical thought. Instead of speaking ontologically of the being or indeed of the coexistence of diversely constituted "spaces," which results in a tangible contradiction, the theory of relativity speaks purely methodologically of the possibility or necessity of applying different measurements, *i.e.*, different geometrical conceptual languages in the interpretation of certain physical manifolds. This possible application tells us nothing concerning the "existence" of space, but merely indicates that by an appropriate choice of geometrical presuppositions certain physical relations, such as the field of gravitation or the electromagnetic field, can be described.

The connection between the purely conceptual thought, involved in the working out of the general doctrine of the manifold and order, and physical empiricism (*Empirie*) here receives a surprising confirmation. A doctrine, which originally grew up merely in the immanent progress of pure mathematical speculation, in the ideal transformation of the hypotheses that lie at the basis of geometry, now serves directly as the form into which the laws of nature are poured. The same functions, that were previously established as expressing the metrical properties of non-Euclidean space, give the equations of the field of gravitation. These equations thus do not need for their establishment the introduction of new unknown forces acting at a distance, but are derived from the determination and specialization of the general presuppositions of measurement. Instead of a new complex of things, the theory is satisfied here by the consideration of a new general complex of conditions. Riemann, in setting up his theory, referred to its future physical meaning in prophetic words of which one is often reminded in the discussion of the general theory of relativity. In the "question as to the inner ground of the relations of measurement of space," he urges, "the remark can be applied that in a discrete manifold the principle of measurement is already contained in the concept of this manifold, but in the case of a continuous manifold it must come from elsewhere. Either the real lying at the basis of space must be a discrete manifold or the basis of measurement must be sought outside it in binding forces working upon it. The answer to this question can only be found by proceeding from the conception of phenomena, founded by Newton and hitherto verified by experience and gradually reshaping this by facts that cannot be explained from it; investigations, which, like the one made here, proceed from universal concepts, can only serve to the effect that these works are not hindered by limitations of concepts and the progress in knowledge of the connection of things not hindered by traditional prejudices." (77). What is here demanded is thus full freedom for the construction of geometrical concepts and hypotheses because only thereby can physical thought also attain full effectiveness, and face all future problems resulting from experience with an assured and systematically perfected instrument. But this connection is expressed, in the case of Riemann, in the language of Herbartian *realism*. At the basis of the pure form of geometrical space a real is to be found in which is to be sought the ultimate cause for the inner relations of measurement of this space. If we carry out, however, with reference to this formulation of the problem, the critical, "Copernican," revolution and thus conceive the question so that a real does not appear as a ground of space but so that space appears as an ideal ground in the construction and progress of knowledge of reality, there results for us at once a characteristic transformation. Instead of regarding "space" as a self-existent real, which must be explained and deduced from "binding forces" like other realities, we ask now rather whether the *a priori* function, the universal ideal relation, that we call "space" involves possible formulations and among them such as are proper to offer an exact and exhaustive account of certain physical relations, of certain "fields of force." The development of the general theory of relativity has answered this question in the affirmative; it has shown what appeared to Riemann as a geometrical hypothesis, as a mere possibility of thought, to be an organ for the knowledge of reality. The Newtonian dynamics is here resolved into pure kinematics and this kinematics ultimately into geometry. The content of the latter must indeed by broadened and the "simple" Euclidean type of geometrical axioms must be replaced by a more complex type; but in compensation we advance a step further into the realm of being, *i.e.*, into the realm of empirical knowledge, without leaving the sphere of geometrical consideration. By abandoning the form of Euclidean space as an undivided whole and breaking it up analytically and by investigating the place of the particular axioms and their reciprocal dependence or independence, we are led to a system of pure *a priori* manifolds, whose laws thought lays down constructively, and in this construction we possess also the fundamental means for representing the relation of the real structures of the empirical manifold.

The realistic view that the relations of measurement of space must be grounded on certain physical determinations, on "binding forces" of matter, expresses this peculiar double relation one-sidedly and thus, epistemologically regarded, inexactly and unsatisfactorily. For this *metaphysical* use of the category of ground would destroy the *methodological* unity, which should be brought out. What relativistic physics, which has developed strictly and consistently from a theory of space and time measurement, offers us is in fact only the combination, the reciprocal determination, of the metrical and physical elements. In this, however, there is found no one-sided relation of ground and consequent, but rather a purely reciprocal relation, a correlation of the "ideal" and "real" moments, of "matter" and "form," of the geometrical and the physical. In so far as we assume any division at all in this reciprocal relation and take one element as "prior" and fundamental, the other as "later" and derivative, this distinction can be meant only in a logical, not in a real sense. In this sense, we must conceive the pure space-time manifold as the logical *prius;* not as if it existed and were given in some sense outside of and before the empirical and physical, but because it constitutes a principle and a fundamental condition of all knowledge of empirical and physical relations. The physicist as such need not reflect on this state of affairs; for in all the concrete measurements, which he makes, the spatio-temporal and the empirical manifold is given always only in the unitary operation of measurement itself, not in the abstract isolation of its particular conceptual elements and conditions.

From these considerations the relation between Euclidean and non-Euclidean geometry appears in a new light. The real superiority of Euclidean geometry seems at first glance to consist in its concrete and intuitive determinateness in the face of which all "pseudo-geometries" fade into logical "possibilities." These possibilities exist only for thought, not for "being;" they seem analytic plays with concepts, which can be left unconsidered when we are concerned with experience and with "nature," with the synthetic unity of objective knowledge. When we look back over our earlier considerations, this view must undergo a peculiar and paradoxical reversal. Pure Euclidean space stands, as is now seen, not closer to the demands of empirical and physical knowledge than the non-Euclidean manifolds but rather more removed. For precisely because it represents the logically simplest form of spatial construction it is not wholly adequate to the complexity of content and the material determinateness of the empirical. Its fundamental property of homogeneity, its axiom of the equivalence in principle of all points, now marks it as an abstract space; for, in the concrete and empirical manifold, there never is such uniformity, but rather thorough-going differentiation reigns in it. If we would create a conceptual expression for this fact of differentiation in the sphere of geometrical relations themselves, then nothing remains but to develop further the geometrical conceptual language with reference to the problem of the "heterogeneous." We find this development in the construction of metageometry. When the concept of the special three-dimensional manifold with a curvature 0 is broadened here to the thought of a system of manifolds with different constant or variable curvatures, a new ideal means is discovered for the mastery of complex manifolds; new conceptual symbols are created, not as expressions of things, but of possible relations according to law. Whether these relations are realized within phenomena at any place only experience can decide. But it is not experience that grounds the content of the geometrical concepts; rather these concepts foreshadow it as methodological anticipations, just as the form of the ellipse was anticipated as a conic section long before it attained concrete application and significance in the courses of the planets. When they first appeared, the systems of non-Euclidean geometry seemed lacking in all empirical meaning, but there was expressed in them the intellectual preparation for problems and tasks, to which experience was to lead later. Since the "absolute differential calculus," which was grounded on purely mathematical considerations by Gauss, Riemann and Christoffel, gains a surprising application in Einstein's theory of gravitation, the possibility of such an application must be held open for all, even the most remote constructions of pure mathematics and especially of non-Euclidean geometry. For it has always been shown in the history of mathematics that its complete freedom contains the guarantee and condition of its fruitfulness. Thought does not advance in the field of the concrete by dealing with the particular phenomena like pictures to be united into a single mosaic, but by sharpening and refining its own means of determination while guided by reference to the empirical and by the postulate of its determinateness according to law. If a proof were needed for this logical state of affairs, the development of the theory of relativity would furnish it. It has been said of the special theory of relativity that it "substituted mathematical constructions for the apparently most tangible reality and resolved the latter into the former." (38, p. 13). The advance to the general theory of relativity has brought this constructive feature more distinctly to light; but, at the same time, it has shown how precisely this resolution of the "tangible" realities has verified and established the connection of theory and experience in an entirely new way. The further physical thought advances and the higher universality of conception it reaches the more does it seem to lose sight of the immediate data, to which the naïve view of the world clings, so that finally there seems no return to these data. And yet the physicist abandons himself to these last and highest abstractions in the certainty and confidence of finding in them reality, *his* reality in a new and richer sense. In the progress of knowledge the deep words of Heraclitus hold that the way upward and the way downward are one and the same: ὁδὸς ἄνω κάτω μίη. Here, too, ascent and descent necessarily belong together: the direction of thought to the universal principles and grounds of knowledge finally proves not only compatible with its direction to the particularity of phenomena and facts, but the correlate and condition of the latter.

CHAPTER VII

The Theory of Relativity and the Problem of Reality

We have attempted to show how the new concept of nature and of the object, which the theory of relativity establishes, is grounded in the *form* of physical thought and only brings this form to a final conclusion and clarity. Physical thought strives to determine and to express in pure objectivity merely the natural object, but thereby necessarily expresses itself, its own law and its own principle. Here is revealed again that "anthropomorphism" of all our concepts of nature to which Goethe, in the wisdom of old age, loved to point. "All philosophy of nature is still only anthropomorphism, *i.e.*, man, at unity with himself, imparts to everything that he is not, this unity, draws it into his unity, makes it one with himself. . . . . We can observe, measure, calculate, weigh, *etc.*, nature as much as we will, it is still only our measure and weight, as man is the measure of all things." Only, after all our preceding considerations, this "anthropomorphism" itself is not to be understood in a limited psychological way but in a universal, critical and transcendental sense. Planck points out, as the characteristic of the evolution of the system of theoretical physics, a progressive emancipation from anthropomorphic elements, which has as its goal the greatest possible separation of the system of physics from the individual personality of the physicist. (68, p. 7.) But into this "objective" system, free from all the accidents of individual standpoint and individual personality, there enter those universal conditions of system, on which depends the peculiarity of the physical way of formulating problems. The sensuous immediacy and particularity of the particular perceptual qualities are excluded, but this exclusion is possible only through the concepts of space and time, number and magnitude. In them physics determines the most general content of reality, since they specify the direction of physical thought as such, as it were the form of the original physical apperception. In the formulation of the theory of relativity this reciprocal relation has been confirmed throughout. The principle of relativity has at once an objective and a subjective, or methodological meaning. The "postulate of the absolute world," which it involves according to an expression of

445

Minkowski, is ultimately a postulate of absolute method. The general relativity of all places, times and measuring rods must be the last word of physics, because "relativization," the resolution of the natural object into pure relations of measurement constitutes the kernel of physical *procedure*, the fundamental cognitive function of physics.

If we understand, however, how, in this sense, the affirmation of relativity develops with inner consequence and necessity out of the very form of physics, a certain critical limitation of this affirmation also appears. The postulate of relativity may be the purest, most universal and sharpest expression of the physical concept of objectivity, but this concept of the *physical* object does not coincide, from the standpoint of the general criticism of knowledge, with reality absolutely. The progress of epistemological analysis is shown in that the assumption of the simplicity and oneness of the concepts of reality is recognized more and more as an illusion. Each of the original directions of knowledge, each interpretation, which it makes of phenomena to combine them into the unity of a theoretical connection or into a definite unity of meaning, involves a special understanding and formulation of the concept of reality. There result here not only the characteristic differences of meaning in the objects of science, the distinction of the "mathematical" object from the "physical" object, the "physical" from the "chemical," the "chemical" from the "biological," but there occur also, over against the whole of *theoretical* scientific knowledge, other forms and meanings of independent type and laws, such as the ethical, the aesthetic "form." It appears as the task of a truly universal criticism of knowledge not to level this manifold, this wealth and variety of forms of knowledge and understanding of the world and compress them into a purely abstract unity, but to leave them standing as such. Only when we resist the temptation to compress the totality of forms, which here result, into an *ultimate* metaphysical unity, into the unity and simplicity of an absolute "world ground" and to deduce it from the latter, do we grasp its true concrete import and fullness. No individual form can indeed claim to grasp absolute "reality" as such and to give it complete and adequate expression. Rather if the thought of such an ultimate definite reality is conceivable at all, it is so only as an Idea, as the problem of a totality of determination in which each particular function of knowledge and consciousness must coöperate according to its character and within its definite limits. If one holds fast to this general view, there results even within the pure concepts of nature a possible diversity of approaches of which each one can lay claim to a certain right and characteristic validity. The "nature" of Goethe is not the same as that of Newton, because there prevail, in the original *shaping* of the two, different principles of form, types of synthesis, of the spiritual and intellectual combination of the phenomena. Where there exist such diversities in fundamental *direction* of consideration, the *results* of consideration cannot be directly compared and measured with each other. The naïve realism of the ordinary view of the world, like the realism of dogmatic metaphysics, falls into this error, ever again. It separates out of the totality of possible concepts of reality a single one and sets it up as a norm and pattern for all the others. Thus certain necessary formal points of view, from which we seek to judge and understand the world of phenomena, are made into things, into absolute beings. Whether we characterize this ultimate being as "matter" or "life," nature" or "history," there always results for us in the end confusion in our view of the world, because certain spiritual functions, that coöperate in its construction, are excluded and others are over-emphasized.

It is the task of systematic philosophy, which extends far beyond the theory of knowledge, to free the idea of the world from this one-sidedness. It has to grasp the *whole system* of symbolic forms, the application of which produces for us the concept of an ordered reality, and by virtue of which subject and object, ego and world are separated and opposed to each other in definite form, and it must refer each individual in this totality to its fixed place. If we assume this problem solved, then the rights would be assured, and the limits fixed, of each of the particular forms of the concept and of knowledge as well as of the general forms of the theoretical, ethical, aesthetic and religious understanding of the world. Each particular form would be "relativized" with regard to the others, but since this "relativization" is throughout reciprocal and since no single form but only the systematic totality can serve as the expression of "truth" and "reality," the limit that results appears as a thoroughly immanent limit, as one that is removed as soon as we again relate the individual to the system of the whole.

We trace the general problem, which opens up here, no further but use it merely to designate the limits, that belong to any, even the most universal, *physical* formulation of problems, because these limits are necessarily grounded in the concept and essence of this way of formulating the question. All physics considers phenomena under the standpoint and presupposition of their measurability. It seeks to resolve the structure of being and process ultimately into a pure structure or order of numbers. The theory of relativity has brought this fundamental tendency of physical thought to its sharpest expression. According to it the procedure of every physical "explanation" of natural process consists in coördinating, to each point of the space-time continuum, four numbers, x_1, x_2, x_3, x_4, which possess absolutely no direct physical meaning but only serve to enumerate the points of the continuum "in a definite, but arbitrary way." (18, p. 64). The ideal, with which scientific physics began with Pythagoras and the Pythagoreans, finds here its conclusion; all qualities, including those of pure space and time, are translated into pure numerical values. The logical postulate contained in the concept of number, which gives this concept its characteristic form, seems now fulfilled in a degree not to be surpassed; all sensuous and intuitive heterogeneity has passed into pure homogeneity. The classical mechanics and physics seeks to reach this immanent goal of conceptual construction by relating the manifold of the sensuously given to the homogeneous and absolutely uniform time. All difference of sensation is hereby reduced to a difference of motions; all possible variety of content is resolved into a mere variety of spatial and temporal positions. But the ideal of strict homogeneity is not reached here since there are still always two fundamental forms of the homogeneous itself that are opposed to each other as pure space and pure time. The theory of relativity in its development advances beyond this opposition also; it seeks to resolve not only the differences of sensation but also those between spatial and temporal determinations into the unity of numerical determinations. The particularity of each "event" is expressed by the four numbers x_1, x_2, x_3, x_4, whereby these numbers among themselves have reference to no differences, so that some of them x_1, x_2, x_3, cannot be brought into a special group of "spatial" coördinates and contrasted with the time coördinate" x_4. Thus all differences belonging to spatial and temporal apprehension in subjective consciousness seem to be consistently set aside in the same way that nothing of the subjective visual sensation enters into the physical concept of light and color.[1] Not only are all spatial and temporal values exchangeable with each other, but all inner differences of the temporal itself, unavoidable for the subjective consciousness, all differences of *direction*, which we designate by the words "past" and "future," are cancelled. The direction into the past and that into the future are distinguished from each other in this form of the concept of the world by nothing more than are the $+$ and $-$ directions in space, which we can determine by arbitrary definition. There remains only the "absolute world" of Minkowski; the world of physics changes from a *process* in a three-dimensional world into a *being* in this four-dimensional world, in which time is replaced as a variable magnitude by the imaginary "ray of light" (*Lichtweg*) $(x_4 = \sqrt{-1}\, c\, t)$.[2]

This transformation of the time-value into an imaginary numerical value seems to annihilate all the "reality" and qualitative determinateness, which time possesses as the "form of the inner sense," as the form of immediate experience. The "stream of process," which, psychologically, constitutes consciousness and distinguishes it as such, stands still; it has passed into the absolute rigidity of a mathematical cosmic formula. There remains in this formula nothing of that form of time, which belongs to all our experience as such and enters as an inseparable and necessary factor into all its content.[3] But, paradoxical as this result seems from the standpoint of this experience, it expresses only the course of mathematical and physical objectification, for, to estimate it correctly from the epistemological standpoint, we must understand it not in its mere result, but as a process, a method. In the resolution of subjectively experienced qualities into pure objective numerical determinations, mathematical physics is bound to no fixed limit. It must go its way to the end; it can stop before no form of consciousness no matter how original and fundamental; for it is precisely its specific cognitive task to translate everything enumerable into pure number, all quality into quantity, all particular forms into a universal order and it only "conceives" them scientifically by virtue of this transformation. Philosophy would seek in vain to bid this tendency halt at any point and to declare *ne plus ultra*. The task of philosophy must rather be limited to

[1] On this latter point *cf.* now Planck, *Das Wesen des Lichts* (71).
[2] *Cf.* Minkowski (54, p. 62 ff.); Einstein (18, p. 82 f.).
[3] *Cf.*, e. g., J. Cohn (14, p. 228 ff.).

recognizing fully the logical *meaning* of the mathematical and physical concept of objectivity and thereby conceiving this meaning in its logical limitedness. All particular physical theories including the theory of relativity receive their definite meaning and import only through the unitary cognitive will of physics, which stands back of them. The moment that we transcend the field of physics and change not the means but the very goal of knowledge, all particular concepts assume a new aspect and form. Each of these concepts means something different, depending on the general "modality" of consciousness and knowledge with which it is connected and from which it is considered. Myth and scientific knowledge, the logical and the aesthetic consciousness, are examples of such diverse modalities. Occasionally concepts of the same name, but by no means of the same meaning, meet us in these different fields. The conceptual relation, which we generally call "cause" and "effect" is not lacking to mythical thought, but here its meaning is specifically distinct from the meaning that it receives in scientific, and in particular, in mathematical and physical thought. In a similar way, all the fundamental concepts undergo a characteristic intellectual change of meaning when we trace them through the different fields of intellectual consideration. Where the copy theory of knowledge seeks a simple identity, the functional theory of knowledge sees complete diversity, but, indeed, at the same time complete correlation of the individual forms.[4]

If we apply these considerations to the concepts of space and time, then it is obvious what the transformation of these concepts in modern physics means, in its philosophical import, and what it cannot mean. The content of physical deductions cannot, without falling into the logical error of a $\mu\epsilon\tau\dot{\alpha}\beta\alpha\sigma\iota\varsigma$ $\epsilon\iota\varsigma$ $\alpha\lambda\lambda o$ $\gamma\dot{\epsilon}\nu o\varsigma$ be simply carried over into the language of fields whose structure rests on a totally different principle. Thus, what space and time are as immediate contents of experience and as they offer themselves to our psychological and phenomenological analysis is unaffected by the use we make of them in the determination of the object, in the course of objective conceptual knowledge. The distance between these two types of consideration and conception is only augmented by the theory of

---

[4] I am aware of the fragmentary character of these suggestions: for their supplementation and more exact proof I must refer to some subsequent more exhaustive treatment. *Cf.* also the essay *Goethe und die mathematische Physik* (11).

relativity and thus only made known more distinctly, but is not first produced by it. Rather it is clear that even to attain the first elements of mathematical and physical knowledge and of the mathematical and physical object we assume that characteristic transformation of "subjective" phenomenal space and of "subjective" phenomenal time, which leads, in its ultimate consequences, to the results of the general theory of relativity. From the standpoint of strict sensualism too, it is customary to admit this transformation, this opposition between the "physiological" space of our sensation and presentation and the purely "metrical" space, which we make the basis of geometry. The latter rests on the assumption of the equivalence of all places and directions, while for the former the distinction of places and directions and the marking out of the one above the others is essential. The space of touch, like that of vision, is anistropic and inhomogeneous, while metrical Euclidean space is distinguished by the postulate of isotropism and homogeneity. Compared with "metrical" time, physiological time shows the same characteristic variations and differences of meaning; one must, as Mach himself urges, as clearly distinguish between the immediate sensation of duration and the measuring number as between the sensation of warmth and temperature.[5]

---

[5] Mach (50, p. 331 ff., 415 ff.). If, with Schlick (79, p. 51ff.), one would call the psychological space of sensation and presentation the space of intuition, and contrast with it physical space as a conceptual construction, no objection could be made against this as a purely terminological determination; but one must guard against confusing this use of the word "intuition" with the Kantian, which rests on entirely different presuppositions. When Schlick sees in the insight that objective physical time has just as little to do with the intuitive experience of duration as the three-dimensional order of objective space with optical or "haptical" extension, "the kernel of truth in the Kantian doctrine of the subjectivity of time and space," and when he on the other hand combats, on the basis of this distinction, the Kantian concept of "pure intuition," this rests on a psychological misunderstanding of the meaning of the Kantian concepts. The space and time of pure intuition are for Kant never sensed or perceived space or time, but the "mathematical" space and time of Newton; they are themselves constructively generated, just as they form the presupposition and foundation of all further mathematical and physical construction. In Kant's thought, "pure intuition" plays the rôle of a definite fundamental *method of objectification;* it coincides in no way with "subjective," *i.e.*, psychologically experienceable time and space. When Kant speaks of the subjectivity of space and time, we must never understand

This contrast between subjective, "phenomenal" space and time, on the one hand, and objective and mathematical space and time, on the other, comes to light with special distinctness, when one considers a property which seems at first glance to be common to them. Of both we are accustomed to predicate the property of *continuity*, but we understand thereby, more closely considered, something wholly different in the two cases. The continuity, which we ascribe to time and processes in it on the basis of the form of our experience, and that which we define in mathematical concepts by certain constructive methods of analysis, not only do not coincide but differ in their essential moments and conditions. The experiential continuity affirms that each temporal content is given to us only in the way of certain characteristic *"wholes,"* which can not be resolved into ultimate simple *"elements;"* analytic continuity demands reduction to such elements. The first takes time and duration as "organic" unities in which according to the Aristotelian definition, "the whole precedes the parts;" the second sees in them only an infinite *totality* of parts, of particular sharply differentiated *temporal points.* In the one case, the continuity of becoming signifies that living flux, that is given to our consciousness only as a flux, as a transition, but not as separated and broken up into discrete parts; in the other, it is demanded that we continue our analysis beyond all limits of empirical apprehension; it is demanded that we do not allow the division of elements to cease where sensuous perception, which is bound to definite but accidental limits in its capacity for discrimination, allows it to end, but that we follow it purely intellectually *ad infinitum.* What the mathematician calls the "continuum" is thus never the purely experiential quality of "continuity," of which there is no longer possible any further "objective" definition, but it is a purely conceptual construction, which he puts in the place of the latter. Here too he must follow his universal method; he must reduce the quality of continuity to mere number, *i.e.*, precisely to the fundamental form of all intellectual *discreteness.* (Cf. 6, p. 21). The only continuum he knows and the one to which he reduces all others, is always the continuum of *real numbers* which modern analysis and

experiential subjectivity but their "transcendental" subjectivity as conditions of the possibility of "objective," *i.e.*, of objectifying empirical knowledge. (*Cf.* also the significant remarks of Selliens against Schlick; 81, p. 19, 39).

theory of groups seek, as is known, to construct strictly conceptually with renunciation in principle of any appeal to the "intuition" of space and time. The continuum thus considered, as Henri Poincaré especially has urged with all emphasis is nothing but a totality of individuals, which are conceived in a definite order and are given indeed in infinite number, of which each one is opposed to the others as something separate and external. We are here no longer concerned with the ordinary view, according to which there exists between the elements a sort of "inner bond" by which they are connected into a whole, so that, *e.g.*, the point does not precede the line, but the line the point. "Of the famous formula, that the continuum is the unity of the manifold," concludes Poincaré, "there remains only the manifold,— the unity has disappeared. The analysts are nevertheless right when they define continuity as they do, for in all their inferences they are concerned, in so far as they claim rigor, only with this concept of the continuous. But this circumstance suffices to make us attentive to the fact that the true mathematical continuum is something totally different from that of the physicist and the metaphysician." (72, p. 30.) In so far as physics is an objectifying science working with the conceptual instruments of mathematics, the physical continuum is conceived by it as related to and exactly correlated with the mathematical continuum of pure numbers. But the "metaphysical" continuum of the pure and original "subjective" form of experience can never be comprehended in this way, for the very *direction* of mathematical consideration is such that, instead of leading to this form, it continually leads away from it. The critical theory of knowledge, which does not have to select from among the different sorts of knowledge, but merely to establish what each of them "is" and means, can make no normative decision as to the opposite aspects under which the continuum here appears, but its task consists in defining the two with reference to each other in utmost distinctness and clarity. Only by such a delimitation can be reached, on the one hand, the goal of phenomenological analysis of the temporal and spatial consciousness, and, on the other hand, the goal of the exact foundation of mathematical analysis and its concepts of space and time. "With regard to the objection," says a modern mathematical author in concluding his investigation of the continuum, "that nothing is contained in the intuition of the continuum of the logical principles that we must adduce in the exact definition of the concept of the real number, we have taken account of the fact that what can be found in the intuitive continuum and in the mathematical world of concepts are so alien to each other, that the demand that the two coincide must be rejected as absurd. In spite of this, those abstract schemata, which mathematics offers us, are helpful in rendering possible an exact science of fields of objects in which continua play a rôle. The exact temporal or spatial point does not lie in the given (phenomenal) duration or extension as an ultimate indivisible element, but only reason reaching through this can grasp these ideas and they crystallize into full determinateness only in connection with the purely formal arithmetical and analytical concept of the real number."[6]

If we bear in mind this state of affairs, the deductions of the theory of relativity in its determination of the four-dimensional space and time continuum lose the appearance of paradox, for it is seen that they are only the final consequence and working out of the fundamental methodic idea on which rests mathematical analysis in general. But the question as to which of the two forms of space and time, the psychological or the physical, the space and time of immediate experience or of mediate conception and knowledge, expresses the *true* reality has lost fundamentally for us all definite meaning. In the complex that we call our "world," that we call the being of our ego and of things, the two enter as equally unavoidable and necessary moments. We can cancel neither of them in favor of the other and exclude it from this complex, but we can refer each to its definite *place* in the whole. If the physicist, whose problem consists in objectification, affirms the superiority of "objective" space and time over "subjective" space and time; if the psychologist and the metaphysician, who are directed upon the totality and immediacy of experience draw the opposite conclusion; then the two judgments express only a false "absolutization" of the norm of knowledge by which each of them determines and measures "reality." In which direction this "absolutization" takes place and whether it is directed on the "outer" or the "inner" is a matter of indifference from the standpoint of pure epistemology. For Newton it was certain that the absolute and mathematical time, which by its nature flowed uniformly, was the "true" time of which all empirically given tem-

⁶ Weyl, 84, p. 83, 71.

poral determination can offer us only a more or less imperfect copy;
for Bergson, this "true" time of Newton is a conceptual fiction and
abstraction, a barrier, which intervenes between our apprehension
and the original meaning and import of reality. But it is forgotten
that what is here called absolute reality, *dureé réelle*, is itself no absolute but only signifies a standpoint of consciousness opposed to that
of mathematics and physics. In the one case, we seek to gain a
unitary and exact measure for all objective process, in the other we
are concerned in retaining this process itself in its pure qualitative
character, in its concrete fullness and subjective inwardness and
"contentuality." The two standpoints can be understood in their
meaning and necessity; neither suffices to include the actual whole
of being in the idealistic sense of "being for us." The symbols that
the mathematician and physicist take as a basis in their view of the
outer and the psychologist in his view of the inner, must both be
understood as *symbols*. Until this has come about the true philosophical view, the view of the *whole*, is not reached, but a partial experience is hypostasized into the whole. From the standpoint of mathematical physics, the total content of the immediate qualities, not
only the differences of sensation, but those of spatial and temporal
consciousness, is threatened with complete annihilation; for the
metaphysical psychologist, conversely, all reality is reduced to this
immediacy, while every mediate conceptual cognition is given only
the value of an arbitrary convention produced for the purposes of our
action. But both views prove, in their absoluteness, rather perver-
sions of the full import of being, *i.e.*, of the full import of the *forms*
of knowledge of the self and the world. While the mathematician
and the mathematical physicist stand in danger of permitting the
real world to be identified with the world of their *measures*, the metaphysical view, in seeking to narrow mathematics to practical goals,
loses the sense of its purest and deepest *ideal* import. It violently
closes the door against what, according to Plato, constitutes the real
meaning and the real value of mathematics; that, namely, "by each
of these cognitions an *organ of the soul* is purified and strengthened,
which under other occupations is lost and blinded; for its preservation is more important than that of a thousand eyes; for by this
alone is the truth seen." And between the two poles of consideration,
which we find here, there stand the manifold concepts of truth of the
different concrete sciences—and therewith their concepts of space

and time. *History*, to set up its temporal measure, cannot do without the methods of the objectifying sciences: chronology is founded on astronomy and through this on mathematics. But the time of the historian is nevertheless not identical with that of the mathematician and physicist, but possesses in contrast to it a peculiar concrete form. In the concept of time of history, the "objective" content of knowledge and the "subjective" experiential content enter into a new characteristic reciprocal relation. An analogous relation is presented, when we survey the aesthetic meaning and shaping of the forms of space and time. Painting presupposes the objective laws of perspective, architecture the laws of statics, but the two serve here only as material out of which develops the unity of the picture and of the architectural spatial form, on the basis of the original artistic laws of form. For music, too, the Pythagoreans sought a connection with pure mathematics, with pure number; but the unity and rythmical division of a melody rests on wholly different structural principles than those on which we construct time in the sense of the unity of objective physical processes of nature. What space and time truly *are* in the philosophical sense would be determined if we succeeded in surveying completely this wealth of nuances of intellectual meaning and in assuring ourselves of the underlying formal law under which they stand and which they obey. The theory of relativity cannot claim to bring this philosophical problem to its solution; for, by its development and scientific tendency from the beginning, it is limited to a definite particular motive of the concepts of space and time. As a physical theory it merely develops the meaning that space and time possess in the system of our empirical and physical measurements. In this sense, final judgment on it belongs exclusively to physics. In the course of its history, physics will have to decide whether the world-picture of the theory of relativity is securely founded theoretically and whether it finds complete experimental verification. Its decision on this, epistemology cannot anticipate; but even now it can thankfully receive the new impetus which this theory has given the general doctrine of the principles of physics.

BIBLIOGRAPHY

(1) BAUCH, BRUNO: *Studien zur Philosophie der exakten Wissenschaften.* Heidelberg, 1911.

(2) BECHER, ERICH: *Philosophische Voraussetzungen der exakten Naturwissenshaften.* Leipzig, 1906.

(3) BERG, OTTO: *Das Relativitätsprinzip der Elektrodynamik.* (*Abt. der Friesschen Schule, N. F.*, Göttingen, 1912.)

(3a) BLOCH, W.: *Einführung in die Relativitätstheorie,* Lpz. und Berl., 1918. (*Aus Natur. u. Geisteswelt,* Bd. 618.)

(4) BUEK, OTTO: *Mich. Faradays System der Natur und seine begrifflichen Grundlagen. Philos. Abhandlungen zu H. Cohens 70. Geburtstag.* Berlin, 1912.

(5) CASSIRER, ERNST: *Leibniz' System in seinen wissenschaftlichen Grundlagen.* Marburg, 1902.

(6) CASSIRER, ERNST: *Kant und die moderne Mathematik. Kant-Studien* XII (1901).

(7) CASSIRER, ERNST: *Das Erkenntnisproblem in der Philosophie und Wissenschaft der neueren Zeit. Bd. I und II, 2. Aufl.,* Berlin, 1911; *Bd. III,* Berlin, 1920.

(8) CASSIRER, ERNST: *Substance and Function.*

(9) CASSIRER, ERNST: *Erkenntnistheorie nebst den Grenzfragen der Logik. In Jahrb. für Philosophie, hg. von Max Frischeisen-Köhler,* Berlin, 1913.

(10) CASSIRER, ERNST: *Kants Leben und Lehre,* Berlin, 1918.

(11) CASSIRER, ERNST: *Goethe und die mathematische Physik* (In: *Idee und Gestalt, Fünf Aufsätze,* Berlin, 1920).

(12) COHEN, HERMANN: *Kants Theorie der Erfahrung, 3. Aufl.,* Berlin, 1918.

(13) COHEN, HERMANN: *Logik der reinen Erkenntnis,* Berlin, 1902.

(13a) COHN, EMIL: *Physikalisches über Raum und Zeit, 4. Aufl.,* Berlin, 1920.

(14) COHN, JONAS: *Relativität und Idealismus, Kant-Studien* (XXI), 1916, p. 222 ff.

(15) DUHEM, PIERRE: *La Théorie Physique, son objet et sa structure.* Paris, 1906.

(15a) EHRENFEST, P.: *Zur Krise der Lichtaether-Hypothese,* Berlin, 1913.

(16) EINSTEIN, ALBERT: *Zur Elektrodynamik bewegter Systeme. Annalen der Physik, 4. F., XVII,* p. 891 (1905), (Cited from No. 47).

(16a) EINSTEIN, ALBERT: *Ist die Trägheit eines Körpers von seinem Energiegehalt abhängig? Annal. der Physik* (17), 1905.

(17) EINSTEIN, ALBERT: *Die Grundlagen der allgemeinen Relativitätstheorie,* Lpz., 1916.

(18) EINSTEIN, ALBERT: *Über die spezielle und die allgemeine Relativitätstheorie (Sammlung Vieweg, Heft 38) 2. Aufl.,* Braunschweig, 1917.

(19) EINSTEIN, ALBERT: *Die formalen Grundlagen der allgemeinen Relativitätstheorie. Sitzungsber. der Berliner Akad. d. Wiss.,* XLI, 1916.

(20) EINSTEIN, ALBERT: *Ernst Mach, Physikalische Zeitschrift,* XVII (1916), p. 101 ff.

(21) ERDMANN, BENNO: *Die Axiome der Geometrie. Eine philosophische Untersuchung der Riemann-Helmholtzschen Raumtheorie* (1877).

(22) EULER, LEONHARD: *Réflexions sur l'espace et le temps. Hist. de l'Acad. des Sciences et Belles Lettres*, Berlin, 1748.

(23) EULER, LEONHARD: *Briefe an eine deutsche Prinzessin* (1768).

(24) FREUNDLICH, ERWIN: *Die Grundlagen der Einsteinschen Gravitationstheorie.* Berlin, 1916.

(24a) FREUNDLICH, ERWIN: *Die Entwicklung des physikalischen Weltbildes bis zur allgemeinen Relativitätstheorie, Weisse Blätter*, 1920, p. 174 ff.

(25) FRISCHEISEN-KÖHLER, MAX: *Wissenschaft u. Wirklichkeit*, Leipzig and Berlin, 1912.

(26) FRISCHEISEN-KÖHLER, MAX: *Das Zeitproblem. In: Jahrb. f. Philosophie* [s. Nr. 9]. Berlin, 1913.

(27) GALILEI: *Opere*, ed. Alberi, Firenze, 1843 ff.

(28) GÖRLAD, ALBERT: *Aristoteles und Kant bezüglich der Idee der theoretischen Erkenntnis untersucht.* [*Philos. Arbeiten, hg. von H. Cohen u. P. Natorp, Band* II.] Giessen, 1909.

(29) HELMHOLTZ, HERMANN: *Über die Erhaltung der Kraft* (1847). *Ostwalds Klassiker der exakten Wissenschaft*, H. 1, 1889).

(30) HELMHOLTZ, HERMANN: *Handbuch der physiologischen Optik, 2. Aufl.*, Hamburg and Leipzig, 1896.

(30a) HELMHOLTZ, HERMANN: *Über den Ursprung und die Bedeutung der geometrischen Axiome* (1870) in: *Vorträge und Reden, 4. Aufl.*, Braunschweig, 1896, Bd. II.

(31) HERTZ, HEINRICH: *Die Prinzipien der Mechanik*, Lpz., 1904.

(32) HÖNIGSWALD, RICHARD: *Über den Unterschied und die Beziehungen der logischen und der erkenntnistheoretischen Elemente in der kritischen Philosophie der Geometrie* (*Verh. des III. internat. Kongr. für Philos.* Heidelb., 1908.

(33) HÖNIGSWALD, RICHARD: *Zum Streit über die Grundlagen der Mathematik.* Heidelb., 1912.

(34) KANT: *Kritik der reinen Vernunft* (cited from Ed. 2, 1787).

(35) KANT: *Werke, herausgegeb. von* Ernst Cassirer, *Bd.* I–X, Berlin, 1911 ff.

(36) KEILL: *Introductio ad veram Physicam*, Oxford, 1702.

(37) KEPLER: *Opera omnia*, ed. Frisch, Vol. I–VIII, Frankf. and Erlangen, 1858 ff.

(38) KNESER, ADOLF: *Mathematik und Natur. Rede.* Breslau, 1911.

(39) KÖNIG, EDMUND: *Kant und die Naturwissenschaft*, Braunschweig, 1907.

(40) LAUE, MAX: *Das Relativitätsprinzip* (*Die Wissenschaft, H.* 38), Braunschweig, 1911.

(41) LAUE, MAX: *Das Relativitätsprinzip. In: Jahrb. f. Philosophie* (s. Nr. 9), Berlin, 1913.

(42) LEIBNIZ: *Mathematische Schriften, Bd.* I–VII, *hg. von* C. J. Gerhardt, Berlin, 1849 ff.

(43) LEIBNIZ: *Philosophische Schriften*, Bd. I–VII, *hg. von* C. J. Gerhardt, Berlin, 1875 ff.

(44) LEIBNIZ: *Hauptschriften zur Grundlegung der Philosophie*, *übersetzt von A. Buchenau, hg. von E. Cassirer*, Bd. I, II, Lpz., 1904/06. (*Philos. Bibliothek*).

(45) LEIBNIZ: *Neue Versuche über den menschlichen Verstand*, *übers. u. herausgeg, von* E. Cassirer, Lpz., 1915 (*Philos. Bibliothek*).

(45a) LENARD, PHILIPP: *Über Äther und Materie*, 2. Aufl., Heidelberg, 1911. *Über Relativitätsprinzip, Äther, Gravitation*, Lpz., 1918.

(46) LORENTZ, H. A.: *Electromagnetic phenomena in a system moving with any velocity smaller than that of light.* *Proceed. Acad. Sc. Amsterd.* 6 (1904), (Cited from No. 47).

(46a) LORENTZ, H. A.: *Das Relativitätsprinzip, drei Vorlesungen, gehalten in Teylers Stiftung zu Haarlem.* Lpz. and Berl., 1914.

(47) LORENTZ-EINSTEIN-MINKOWSKI: *Das Relativitätsprinzip.* *Eine Sammlung von Abhandlungen.* *Mit Anmerk. von A. Sommerfeld und Vorwort von O. Blumenthal.* (*Fortschritte der mathemat. Wissensch., Heft 2, Leipzig and Berlin, 1913.*)

(48) LOBATSCHEFSKY: *Zwei geometrische Abhandlungen*, *übers. v.* Fr. Engel, Lpz., 1898.

(49) MACH, ERNST: *Die Prinzipien der Wärmelehre*, Lpz., 1896.

(50) MACH, ERNST: *Erkenntnis und Irrtum, Skizzen zur Psychologie der Forschung*, Lpz., 1905.

(51) MAXWELL, J. C.: *Substanz und Bewegung (Matter and Motion), dtsch. von* E. v. Fleischel, Braunschw., 1881.

(52) MAYER, ROBERT: *Die Mechanik der Wärme in gesammelten Schriften, hg. von Weyrauch, 3. Aufl.*, Stuttgart, 1893.

(53) MEYERSON, EMILE: *Identité et réalité*, Paris, 1908.

(54) MINKOWSKI, HERMANN: *Raum und Zeit, Vortrag, Cöln*, 1908 (Cited from No. 47).

(55) MÜLLER, ALOYS: *Das Problem des absoluten Raumes und seine Beziehung zum allgemeinen Raumproblem.* Braunschweig, 1911.

(56) NATORP, PAUL: *Die logischen Grundlagen der exakten Wissenschaften.* Lpz. and Berl., 1910.

(57) NEUMANN, CARL: *Über die Prinzipien der Galilei-Newtonschen Theorie*, Lpz., 1870.

(58) NEWTON, ISAAC: *Optice, lat reddid. S. Clarke*, Lausanne and Geneva, 1740.

(59) NEWTON, ISAAC: *Philosophiae naturalis principia mathematica* (1686). *Ausg. von Le Seur und Jacquier*, 4. Bd., Genf, 1739 ff.

(60) OSTWALD, WILHELM: *Vorlesungen über Naturphilosophie*, Leipzig, 1902.

(61) PETZOLDT, JOS.: *Die Relativitätstheorie im erkenntnistheoretischen Zusammenhang des relativistischen Positivismus.* *Verh. der Dtsch. Physik. Gesellschaft XIV.*, Braunschweig, 1912.

(62) PFLÜGER, A.: *Das Einsteinsche Relativitätsprinzip, gemeinverständlich dargestellt, 3. Aufl.*, Bonn, 1920.

(63) PLANCK, MAX: *Das Prinzip der Erhaltung der Energie*, Leipzig, 1887.

(64) PLANCK, MAX: *Zur Dynamik bewegter Systeme.* *Annalen der Physik*, 4. F., XXVI, 1 ff. (1908).

(65) PLANCK, MAX: *Bemerkungen zum Prinzip der Aktion und Reaktion in der allg. Dynamik. Physikal. Zeitschrift* IX (1908).

(66) PLANCK, MAX: *Die Einheit des physikalischen Weltbildes*, Leipzig, 1909.

(67) PLANCK, MAX: *Die Stellung der neuen Physik zur mechanischen Weltanschauung.* (*Verh. der Ges. dtsch. Naturf. u. Arzte in Königsberg*, 1910, Leipzig, 1911, S. 58–75.)

(68) PLANCK, MAX: *Acht Vorlesungen über theoretische Physik*, Leipzig, 1910.

(69) PLANCK, MAX: *Zur Machschen Theorie der physikalischen Erkenntnis. Vierteljahresschr. für wissensch. Philos.*, 1910.

(70) PLANCK, MAX: *Neue Bahnen der physikalischen Erkenntnis* (*Rektoratsrede*), Berlin, 1913.

(71) PLANCK, MAX: *Das Wesen der Lichts, Vortrag*, Berlin, 1920.

(72) POINCARÉ, HENRI: *La Science et l'hypothèse*, Paris.

(73) POINCARÉ, HENRI: *Der Wert der Wissenschaft*, dtsch. v. E. Weber, mit. *Anmerk und Zusätzen von H. Weber*, Berlin and Leipzig, 1906.

(74) POINCARÉ, HENRI: *Science et méthode*, Paris, 1908.

(75) POINCARÉ, LUCIEN: *Die moderne Physik*, dtsch. von Brahn, Leipzig, 1908.

(76) RIEHL, AL: *Der philosophische Kritizismus und seine Bedeutung für die positive Wissenschaft* I/II, Leipzig, 1876–97.

(77) RIEMANN, B: *Über die Hypothesen, welche der Geometrie zugrunde liegen* (1854).

(78) RIGHI: *Die moderne Theorie der physikalischen Erscheinungen*, dtsch. von Dessau, Leipzig, 1908.

(79) SCHLICK, MORTIZ: *Raum und Zeit in der gegenwärtigen Physik*, Berlin, 1917.

(80) SCHLICK, MORITZ: *Allgemeine Erkenntnislehre*, Berlin, 1918.

(81) SELLIEN, EWALD: *Die erkenntnistheoretische Bedeutung der Relativitätstheorie, Kieler Inaug.-Diss.*, Berlin, 1919.

(82) STREINTZ, H.: *Die physikalischen Grundlagen der Mechanik*, Leipzig, 1883.

(83) WEYL, HERMANN: *Raum. Zeit. Materie. Vorlesungen über allgemeine Relativitätstheorie, 3. Aufl.*, Berlin, 1920.

(84) WEYL, HERMANN: *Das Kontinuum. Kritische Untersuchungen über die Grundlagen der Analysis.* Leipzig, 1918.

The work of Dr. Hans Reichenbach on the meaning of the theory of relativity for the concept of physical knowledge (*Die Bedeutung der Relativitätstheorie für den physikalischen Erkenntnisbegriff*) became accessible to me in manuscript while this essay was being printed. I can here only refer to this thorough and penetrating work, which has much in common with the present essay in its way of stating the problem; I cannot, however, completely agree with its results, especially with regard to the relation of the theory of relativity to the Kantian critique of cognition.

INDEX

A CATALOGUE OF
SELECTED DOVER BOOKS
IN ALL FIELDS OF INTEREST

A CATALOGUE OF SELECTED DOVER BOOKS IN ALL FIELDS OF INTEREST

CELESTIAL OBJECTS FOR COMMON TELESCOPES, T. W. Webb. The most used book in amateur astronomy: inestimable aid for locating and identifying nearly 4,000 celestial objects. Edited, updated by Margaret W. Mayall. 77 illustrations. Total of 645pp. 5⅜ x 8½.
20917-2, 20918-0 Pa., Two-vol. set $10.00

HISTORICAL STUDIES IN THE LANGUAGE OF CHEMISTRY, M. P. Crosland. The important part language has played in the development of chemistry from the symbolism of alchemy to the adoption of systematic nomenclature in 1892. ". . . wholeheartedly recommended,"—Science. 15 illustrations. 416pp. of text. 5⅝ x 8¼. 63702-6 Pa. $7.50

BURNHAM'S CELESTIAL HANDBOOK, Robert Burnham, Jr. Thorough, readable guide to the stars beyond our solar system. Exhaustive treatment, fully illustrated. Breakdown is alphabetical by constellation: Andromeda to Cetus in Vol. 1; Chamaeleon to Orion in Vol. 2; and Pavo to Vulpecula in Vol. 3. Hundreds of illustrations. Total of about 2000pp. 6⅛ x 9¼.
23567-X, 23568-8, 23673-0 Pa., Three-vol. set $32.85

THEORY OF WING SECTIONS: INCLUDING A SUMMARY OF AIRFOIL DATA, Ira H. Abbott and A. E. von Doenhoff. Concise compilation of subatomic aerodynamic characteristics of modern NASA wing sections, plus description of theory. 350pp. of tables. 693pp. 5⅜ x 8½.
60586-8 Pa. $9.95

DE RE METALLICA, Georgius Agricola. Translated by Herbert C. Hoover and Lou H. Hoover. The famous Hoover translation of greatest treatise on technological chemistry, engineering, geology, mining of early modern times (1556). All 289 original woodcuts. 638pp. 6¾ x 11.
60006-8 Clothbd. $19.95

THE ORIGIN OF CONTINENTS AND OCEANS, Alfred Wegener. One of the most influential, most controversial books in science, the classic statement for continental drift. Full 1966 translation of Wegener's final (1929) version. 64 illustrations. 246pp. 5⅜ x 8½. (EBE) 61708-4 Pa. $5.00

THOUGHT WITH NO CONSCIOUSNESS

THE PRINCIPLES OF PSYCHOLOGY, William James. Famous long course complete, unabridged. Stream of thought, time perception, memory, experimental methods; great work decades ahead of its time. Still valid, useful; read in many classes. 94 figures. Total of 1391pp. 5⅜ x 8½.
20381-6, 20382-4 Pa., Two-vol. set $17.90

YUCATAN BEFORE AND AFTER THE CONQUEST, Diego de Landa. First English translation of basic book in Maya studies, the only significant account of Yucatan written in the early post-Conquest era. Translated by distinguished Maya scholar William Gates. Appendices, introduction, 4 maps and over 120 illustrations added by translator. 162pp. 5⅜ x 8½.

23622-6 Pa. $3.00

THE MALAY ARCHIPELAGO, Alfred R. Wallace. Spirited travel account by one of founders of modern biology. Touches on zoology, botany, ethnography, geography, and geology. 62 illustrations, maps. 515pp. 5⅜ x 8½.

20187-2 Pa. $6.95

THE DISCOVERY OF THE TOMB OF TUTANKHAMEN, Howard Carter, A. C. Mace. Accompany Carter in the thrill of discovery, as ruined passage suddenly reveals unique, untouched, fabulously rich tomb. Fascinating account, with 106 illustrations. New introduction by J. M. White. Total of 382pp. 5⅜ x 8½. (Available in U.S. only)   23500-9 Pa. $5.50

THE WORLD'S GREATEST SPEECHES, edited by Lewis Copeland and Lawrence W. Lamm. Vast collection of 278 speeches from Greeks up to present. Powerful and effective models; unique look at history. Revised to 1970. Indices. 842pp. 5⅜ x 8½.   20468-5 Pa. $9.95

THE 100 GREATEST ADVERTISEMENTS, Julian Watkins. The priceless ingredient; His master's voice; 99 44/100% pure; over 100 others. How they were written, their impact, etc. Remarkable record. 130 illustrations. 233pp. 7⅞ x 10 3/5.   20540-1 Pa. $6.95

CRUICKSHANK PRINTS FOR HAND COLORING, George Cruickshank. 18 illustrations, one side of a page, on fine-quality paper suitable for watercolors. Caricatures of people in society (c. 1820) full of trenchant wit. Very large format. 32pp. 11 x 16.   23684-6 Pa. $6.00

THIRTY-TWO COLOR POSTCARDS OF TWENTIETH-CENTURY AMERICAN ART, Whitney Museum of American Art. Reproduced in full color in postcard form are 31 art works and one shot of the museum. Calder, Hopper, Rauschenberg, others. Detachable. 16pp. 8¼ x 11.

23629-3 Pa. $3.50

MUSIC OF THE SPHERES: THE MATERIAL UNIVERSE FROM ATOM TO QUASAR SIMPLY EXPLAINED, Guy Murchie. Planets, stars, geology, atoms, radiation, relativity, quantum theory, light, antimatter, similar topics. 319 figures. 664pp. 5⅜ x 8½.

21809-0, 21810-4 Pa., Two-vol. set $11.00

EINSTEIN'S THEORY OF RELATIVITY, Max Born. Finest semi-technical account; covers Einstein, Lorentz, Minkowski, and others, with much detail, much explanation of ideas and math not readily available elsewhere on this level. For student, non-specialist. 376pp. 5⅜ x 8½.

60769-0 Pa. $5.00

THE SENSE OF BEAUTY, George Santayana. Masterfully written discussion of nature of beauty, materials of beauty, form, expression; art, literature, social sciences all involved. 168pp. 5⅜ x 8½.   20238-0 Pa. $3.50

ON THE IMPROVEMENT OF THE UNDERSTANDING, Benedict Spinoza. Also contains *Ethics, Correspondence,* all in excellent R. Elwes translation. Basic works on entry to philosophy, pantheism, exchange of ideas with great contemporaries. 402pp. 5⅜ x 8½.   20250-X Pa. $5.95

THE TRAGIC SENSE OF LIFE, Miguel de Unamuno. Acknowledged masterpiece of existential literature, one of most important books of 20th century. Introduction by Madariaga. 367pp. 5⅜ x 8½.
20257-7 Pa. $6.00

THE GUIDE FOR THE PERPLEXED, Moses Maimonides. Great classic of medieval Judaism attempts to reconcile revealed religion (Pentateuch, commentaries) with Aristotelian philosophy. Important historically, still relevant in problems. Unabridged Friedlander translation. Total of 473pp. 5⅜ x 8½.   20351-4 Pa. $6.95

THE I CHING (THE BOOK OF CHANGES), translated by James Legge. Complete translation of basic text plus appendices by Confucius, and Chinese commentary of most penetrating divination manual ever prepared. Indispensable to study of early Oriental civilizations, to modern inquiring reader. 448pp. 5⅜ x 8½.   21062-6 Pa. $6.00

THE EGYPTIAN BOOK OF THE DEAD, E. A. Wallis Budge. Complete reproduction of Ani's papyrus, finest ever found. Full hieroglyphic text, interlinear transliteration, word for word translation, smooth translation. Basic work, for Egyptology, for modern study of psychic matters. Total of 533pp. 6½ x 9¼. (USCO)   21866-X Pa. $8.50

THE GODS OF THE EGYPTIANS, E. A. Wallis Budge. Never excelled for richness, fullness: all gods, goddesses, demons, mythical figures of Ancient Egypt; their legends, rites, incarnations, variations, powers, etc. Many hieroglyphic texts cited. Over 225 illustrations, plus 6 color plates. Total of 988pp. 6⅛ x 9¼. (EBE)
22055-9, 22056-7 Pa., Two-vol. set $20.00

THE STANDARD BOOK OF QUILT MAKING AND COLLECTING, Marguerite Ickis. Full information, full-sized patterns for making 46 traditional quilts, also 150 other patterns. Quilted cloths, lame, satin quilts, etc. 483 illustrations. 273pp. 6⅞ x 9⅝.   20582-7 Pa. $5.95

CORAL GARDENS AND THEIR MAGIC, Bronsilaw Malinowski. Classic study of the methods of tilling the soil and of agricultural rites in the Trobriand Islands of Melanesia. Author is one of the most important figures in the field of modern social anthropology. 143 illustrations. Indexes. Total of 911pp. of text. 5⅝ x 8¼. (Available in U.S. only)
23597-1 Pa. $12.95

THE PHILOSOPHY OF HISTORY, Georg W. Hegel. Great classic of Western thought develops concept that history is not chance but a rational process, the evolution of freedom. 457pp. 5⅜ x 8½.   20112-0 Pa. $6.00

LANGUAGE, TRUTH AND LOGIC, Alfred J. Ayer. Famous, clear introduction to Vienna, Cambridge schools of Logical Positivism. Role of philosophy, elimination of metaphysics, nature of analysis, etc. 160pp. 5⅜ x 8½. (USCO)   20010-8 Pa. $2.50

A PREFACE TO LOGIC, Morris R. Cohen. Great City College teacher in renowned, easily followed exposition of formal logic, probability, values, logic and world order and similar topics; no previous background needed. 209pp. 5⅜ x 8½.   23517-3 Pa. $4.95

REASON AND NATURE, Morris R. Cohen. Brilliant analysis of reason and its multitudinous ramifications by charismatic teacher. Interdisciplinary, synthesizing work widely praised when it first appeared in 1931. Second (1953) edition. Indexes. 496pp. 5⅜ x 8½.   23633-1 Pa. $7.50

AN ESSAY CONCERNING HUMAN UNDERSTANDING, John Locke. The only complete edition of enormously important classic, with authoritative editorial material by A. C. Fraser. Total of 1176pp. 5⅜ x 8½.
20530-4, 20531-2 Pa., Two-vol. set $16.00

HANDBOOK OF MATHEMATICAL FUNCTIONS WITH FORMULAS, GRAPHS, AND MATHEMATICAL TABLES, edited by Milton Abramowitz and Irene A. Stegun. Vast compendium: 29 sets of tables, some to as high as 20 places. 1,046pp. 8 x 10½.   61272-4 Pa. $17.95

MATHEMATICS FOR THE PHYSICAL SCIENCES, Herbert S. Wilf. Highly acclaimed work offers clear presentations of vector spaces and matrices, orthogonal functions, roots of polynomial equations, conformal mapping, calculus of variations, etc. Knowledge of theory of. functions of real and complex variables is assumed. Exercises and solutions. Index. 284pp. 5⅝ x 8¼.   63635-6 Pa. $5.00

THE PRINCIPLE OF RELATIVITY, Albert Einstein et al. Eleven most important original papers on special and general theories. Seven by Einstein, two by Lorentz, one each by Minkowski and Weyl. All translated, unabridged. 216pp. 5⅜ x 8½.   60081-5 Pa. $3.50

THERMODYNAMICS, Enrico Fermi. A classic of modern science. Clear, organized treatment of systems, first and second laws, entropy, thermodynamic potentials, gaseous reactions, dilute solutions, entropy constant. No math beyond calculus required. Problems. 160pp. 5⅜ x 8½.
60361-X Pa. $4.00

ELEMENTARY MECHANICS OF FLUIDS, Hunter Rouse. Classic undergraduate text widely considered to be far better than many later books. Ranges from fluid velocity and acceleration to role of compressibility in fluid motion. Numerous examples, questions, problems. 224 illustrations. 376pp. 5⅝ x 8¼.   63699-2 Pa. $7.00

THE AMERICAN SENATOR, Anthony Trollope. Little known, long unavailable Trollope novel on a grand scale. Here are humorous comment on American vs. English culture, and stunning portrayal of a heroine/villainess. Superb evocation of Victorian village life. 561pp. 5⅜ x 8½.
23801-6 Pa. $7.95

WAS IT MURDER? James Hilton. The author of *Lost Horizon* and *Goodbye, Mr. Chips* wrote one detective novel (under a pen-name) which was quickly forgotten and virtually lost, even at the height of Hilton's fame. This edition brings it back—a finely crafted public school puzzle resplendent with Hilton's stylish atmosphere. A thoroughly English thriller by the creator of Shangri-la. 252pp. 5⅜ x 8. (Available in U.S. only)
23774-5 Pa. $3.00

CENTRAL PARK: A PHOTOGRAPHIC GUIDE, Victor Laredo and Henry Hope Reed. 121 superb photographs show dramatic views of Central Park: Bethesda Fountain, Cleopatra's Needle, Sheep Meadow, the Blockhouse, plus people engaged in many park activities: ice skating, bike riding, etc. Captions by former Curator of Central Park, Henry Hope Reed, provide historical view, changes, etc. Also photos of N.Y. landmarks on park's periphery. 96pp. 8½ x 11. 23750-8 Pa. $4.50

NANTUCKET IN THE NINETEENTH CENTURY, Clay Lancaster. 180 rare photographs, stereographs, maps, drawings and floor plans recreate unique American island society. Authentic scenes of shipwreck, lighthouses, streets, homes are arranged in geographic sequence to provide walking-tour guide to old Nantucket existing today. Introduction, captions. 160pp. 8⅞ x 11¾. 23747-8 Pa. $7.95

STONE AND MAN: A PHOTOGRAPHIC EXPLORATION, Andreas Feininger. 106 photographs by *Life* photographer Feininger portray man's deep passion for stone through the ages. Stonehenge-like megaliths, fortified towns, sculpted marble and crumbling tenements show textures, beauties, fascination. 128pp. 9¼ x 10¾. 23756-7 Pa. $5.95

CIRCLES, A MATHEMATICAL VIEW, D. Pedoe. Fundamental aspects of college geometry, non-Euclidean geometry, and other branches of mathematics: representing circle by point. Poincare model, isoperimetric property, etc. Stimulating recreational reading. 66 figures. 96pp. 5⅝ x 8¼.
63698-4 Pa. $3.50

THE DISCOVERY OF NEPTUNE, Morton Grosser. Dramatic scientific history of the investigations leading up to the actual discovery of the eighth planet of our solar system. Lucid, well-researched book by well-known historian of science. 172pp. 5⅜ x 8½. 23726-5 Pa. $3.50

THE DEVIL'S DICTIONARY. Ambrose Bierce. Barbed, bitter, brilliant witticisms in the form of a dictionary. Best, most ferocious satire America has produced. 145pp. 5⅜ x 8½. 20487-1 Pa. $2.50

HISTORY OF BACTERIOLOGY, William Bulloch. The only comprehensive history of bacteriology from the beginnings through the 19th century. Special emphasis is given to biography-Leeuwenhoek, etc. Brief accounts of 350 bacteriologists form a separate section. No clearer, fuller study, suitable to scientists and general readers, has yet been written. 52 illustrations. 448pp. 5⅝ x 8¼. 23761-3 Pa. $6.50

THE COMPLETE NONSENSE OF EDWARD LEAR, Edward Lear. All nonsense limericks, zany alphabets, Owl and Pussycat, songs, nonsense botany, etc., illustrated by Lear. Total of 321pp. 5⅜ x 8½. (Available in U.S. only) 20167-8 Pa. $4.50

INGENIOUS MATHEMATICAL PROBLEMS AND METHODS, Louis A. Graham. Sophisticated material from Graham *Dial*, applied and pure; stresses solution methods. Logic, number theory, networks, inversions, etc. 237pp. 5⅜ x 8½. 20545-2 Pa. $4.50

BEST MATHEMATICAL PUZZLES OF SAM LOYD, edited by Martin Gardner. Bizarre, original, whimsical puzzles by America's greatest puzzler. From fabulously rare *Cyclopedia*, including famous 14-15 puzzles, the Horse of a Different Color, 115 more. Elementary math. 150 illustrations. 167pp. 5⅜ x 8½. 20498-7 Pa. $3.50

THE BASIS OF COMBINATION IN CHESS, J. du Mont. Easy-to-follow, instructive book on elements of combination play, with chapters on each piece and every powerful combination team—two knights, bishop and knight, rook and bishop, etc. 250 diagrams. 218pp. 5⅜ x 8½. (Available in U.S. only) 23644-7 Pa. $4.50

MODERN CHESS STRATEGY, Ludek Pachman. The use of the queen, the active king, exchanges, pawn play, the center, weak squares, etc. Section on rook alone worth price of the book. Stress on the moderns. Often considered the most important book on strategy. 314pp. 5⅜ x 8½. 20290-9 Pa. $5.00

LASKER'S MANUAL OF CHESS, Dr. Emanuel Lasker. Great world champion offers very thorough coverage of all aspects of chess. Combinations, position play, openings, end game, aesthetics of chess, philosophy of struggle, much more. Filled with analyzed games. 390pp. 5⅜ x 8½. 20640-8 Pa. $5.95

500 MASTER GAMES OF CHESS, S. Tartakower, J. du Mont. Vast collection of great chess games from 1798-1938, with much material nowhere else readily available. Fully annoted, arranged by opening for easier study. 664pp. 5⅜ x 8½. 23208-5 Pa. $8.50

A GUIDE TO CHESS ENDINGS, Dr. Max Euwe, David Hooper. One of the finest modern works on chess endings. Thorough analysis of the most frequently encountered endings by former world champion. 331 examples, each with diagram. 248pp. 5⅜ x 8½. 23332-4 Pa. $3.95

THE COMPLETE BOOK OF DOLL MAKING AND COLLECTING, Catherine Christopher. Instructions, patterns for dozens of dolls, from rag doll on up to elaborate, historically accurate figures. Mould faces, sew clothing, make doll houses, etc. Also collecting information. Many illustrations. 288pp. 6 x 9. 22066-4 Pa. $4.95

THE DAGUERREOTYPE IN AMERICA, Beaumont Newhall. Wonderful portraits, 1850's townscapes, landscapes; full text plus 104 photographs. The basic book. Enlarged 1976 edition. 272pp. 8¼ x 11¼. 23322-7 Pa. $7.95

CRAFTSMAN HOMES, Gustav Stickley. 296 architectural drawings, floor plans, and photographs illustrate 40 different kinds of "Mission-style" homes from *The Craftsman* (1901-16), voice of American style of simplicity and organic harmony. Thorough coverage of Craftsman idea in text and picture, now collector's item. 224pp. 8⅛ x 11. 23791-5 Pa. $6.50

PEWTER-WORKING: INSTRUCTIONS AND PROJECTS, Burl N. Osborn. & Gordon O. Wilber. Introduction to pewter-working for amateur craftsman. History and characteristics of pewter; tools, materials, step-by-step instructions. Photos, line drawings, diagrams. Total of 160pp. 7⅞ x 10¾. 23786-9 Pa. $3.50

THE GREAT CHICAGO FIRE, edited by David Lowe. 10 dramatic, eyewitness accounts of the 1871 disaster, including one of the aftermath and rebuilding, plus 70 contemporary photographs and illustrations of the ruins—courthouse, Palmer House, Great Central Depot, etc. Introduction by David Lowe. 87pp. 8¼ x 11. 23771-0 Pa. $4.00

SILHOUETTES: A PICTORIAL ARCHIVE OF VARIED ILLUSTRATIONS, edited by Carol Belanger Grafton. Over 600 silhouettes from the 18th to 20th centuries include profiles and full figures of men and women, children, birds and animals, groups and scenes, nature, ships, an alphabet. Dozens of uses for commercial artists and craftspeople. 144pp. 8⅜ x 11¼. 23781-8 Pa. $4.50

ANIMALS: 1,419 COPYRIGHT-FREE ILLUSTRATIONS OF MAMMALS, BIRDS, FISH, INSECTS, ETC., edited by Jim Harter. Clear wood engravings present, in extremely lifelike poses, over 1,000 species of animals. One of the most extensive copyright-free pictorial sourcebooks of its kind. Captions. Index. 284pp. 9 x 12. 23766-4 Pa. $8.95

INDIAN DESIGNS FROM ANCIENT ECUADOR, Frederick W. Shaffer. 282 original designs by pre-Columbian Indians of Ecuador (500-1500 A.D.). Designs include people, mammals, birds, reptiles, fish, plants, heads, geometric designs. Use as is or alter for advertising, textiles, leathercraft, etc. Introduction. 95pp. 8¾ x 11¼. 23764-8 Pa. $4.50

SZIGETI ON THE VIOLIN, Joseph Szigeti. Genial, loosely structured tour by premier violinist, featuring a pleasant mixture of reminiscences, insights into great music and musicians, innumerable tips for practicing violinists. 385 musical passages. 256pp. 5⅝ x 8¼. 23763-X Pa. $4.00

TONE POEMS, SERIES II: TILL EULENSPIEGELS LUSTIGE STREICHE, ALSO SPRACH ZARATHUSTRA, AND EIN HELDEN-LEBEN, Richard Strauss. Three important orchestral works, including very popular *Till Eulenspiegel's Marry Pranks*, reproduced in full score from original editions. Study score. 315pp. 9⅜ x 12¼. (Available in U.S. only)
23755-9 Pa. $8.95

TONE POEMS, SERIES I: DON JUAN, TOD UND VERKLARUNG AND DON QUIXOTE, Richard Strauss. Three of the most often performed and recorded works in entire orchestral repertoire, reproduced in full score from original editions. Study score. 286pp. 9⅜ x 12¼. (Available in U.S. only)
23754-0 Pa. $8.95

11 LATE STRING QUARTETS, Franz Joseph Haydn. The form which Haydn defined and "brought to perfection." (*Grove's*). 11 string quartets in complete score, his last and his best. The first in a projected series of the complete Haydn string quartets. Reliable modern Eulenberg edition, otherwise difficult to obtain. 320pp. 8⅜ x 11¼. (Available in U.S. only)
23753-2 Pa. $8.95

FOURTH, FIFTH AND SIXTH SYMPHONIES IN FULL SCORE, Peter Ilyitch Tchaikovsky. Complete orchestral scores of Symphony No. 4 in F Minor, Op. 36; Symphony No. 5 in E Minor, Op. 64; Symphony No. 6 in B Minor, "Pathetique," Op. 74. Bretikopf & Hartel eds. Study score. 480pp. 9⅜ x 12¼.
23861-X Pa. $10.95

THE MARRIAGE OF FIGARO: COMPLETE SCORE, Wolfgang A. Mozart. Finest comic opera ever written. Full score, not to be confused with piano renderings. Peters edition. Study score. 448pp. 9⅜ x 12¼. (Available in U.S. only)
23751-6 Pa. $12.95

"IMAGE" ON THE ART AND EVOLUTION OF THE FILM, edited by Marshall Deutelbaum. Pioneering book brings together for first time 38 groundbreaking articles on early silent films from *Image* and 263 illustrations newly shot from rare prints in the collection of the International Museum of Photography. A landmark work. Index. 256pp. 8¼ x 11.
23777-X Pa. $8.95

AROUND-THE-WORLD COOKY BOOK, Lois Lintner Sumption and Marguerite Lintner Ashbrook. 373 cooky and frosting recipes from 28 countries (America, Austria, China, Russia, Italy, etc.) include Viennese kisses, rice wafers, London strips, lady fingers, hony, sugar spice, maple cookies, etc. Clear instructions. All tested. 38 drawings. 182pp. 5⅜ x 8.
23802-4 Pa. $2.75

THE ART NOUVEAU STYLE, edited by Roberta Waddell. 579 rare photographs, not available elsewhere, of works in jewelry, metalwork, glass, ceramics, textiles, architecture and furniture by 175 artists—Mucha, Seguy, Lalique, Tiffany, Gaudin, Hohlwein, Saarinen, and many others. 288pp. 8⅜ x 11¼.
23515-7 Pa. $8.95

THE CURVES OF LIFE, Theodore A. Cook. Examination of shells, leaves, horns, human body, art, etc., in *"the* classic reference on how the golden ratio applies to spirals and helices in nature . . . . "—Martin Gardner. 426 illustrations. Total of 512pp. 5⅜ x 8½.   23701-X Pa. $6.95

AN ILLUSTRATED FLORA OF THE NORTHERN UNITED STATES AND CANADA, Nathaniel L. Britton, Addison Brown. Encyclopedic work covers 4666 species, ferns on up. Everything. Full botanical information, illustration for each. This earlier edition is preferred by many to more recent revisions. 1913 edition. Over 4000 illustrations, total of 2087pp. 6⅛ x 9¼.   22642-5, 22643-3, 22644-1 Pa., Three-vol. set $28.50

MANUAL OF THE GRASSES OF THE UNITED STATES, A. S. Hitchcock, U.S. Dept. of Agriculture. The basic study of American grasses, both indigenous and escapes, cultivated and wild. Over 1400 species. Full descriptions, information. Over 1100 maps, illustrations. Total of 1051pp. 5⅜ x 8½.   22717-0, 22718-9 Pa., Two-vol. set $17.00

THE CACTACEAE,, Nathaniel L. Britton, John N. Rose. Exhaustive, definitive. Every cactus in the world. Full botanical descriptions. Thorough statement of nomenclatures, habitat, detailed finding keys. The one book needed by every cactus enthusiast. Over 1275 illustrations. Total of 1080pp. 8 x 10¼.   21191-6, 21192-4 Clothbd., Two-vol. set $50.00

AMERICAN MEDICINAL PLANTS, Charles F. Millspaugh. Full descriptions, 180 plants covered: history; physical description; methods of preparation with all chemical constituents extracted; all claimed curative or adverse effects. 180 full-page plates. Classification table. 804pp. 6½ x 9¼.
                  23034-1 Pa. $13.95

A MODERN HERBAL, Margaret Grieve. Much the fullest, most exact, most useful compilation of herbal material. Gigantic alphabetical encyclopedia, from aconite to zedoary, gives botanical information, medical properties, folklore, economic uses, and much else. Indispensable to serious reader. 161 illustrations. 888pp. 6½ x 9¼. (Available in U.S. only)
                  22798-7, 22799-5 Pa., Two-vol. set $15.00

THE HERBAL or GENERAL HISTORY OF PLANTS, John Gerard. The 1633 edition revised and enlarged by Thomas Johnson. Containing almost 2850 plant descriptions and 2705 superb illustrations, Gerard's *Herbal* is a monumental work, the book all modern English herbals are derived from, the one herbal every serious enthusiast should have in its entirety. Original editions are worth perhaps $750. 1678pp. 8½ x 12¼.
                  23147-X Clothbd. $75.00

MANUAL OF THE TREES OF NORTH AMERICA, Charles S. Sargent. The basic survey of every native tree and tree-like shrub, 717 species in all. Extremely full descriptions, information on habitat, growth, locales, economics, etc. Necessary to every serious tree lover. Over 100 finding keys. 783 illustrations. Total of 986pp. 5⅜ x 8½.
                  20277-1, 20278-X Pa., Two-vol. set $12.00

GREAT NEWS PHOTOS AND THE STORIES BEHIND THEM, John Faber. Dramatic volume of 140 great news photos, 1855 through 1976, and revealing stories behind them, with both historical and technical information. Hindenburg disaster, shooting of Oswald, nomination of Jimmy Carter, etc. 160pp. 8¼ x 11. 23667-6 Pa. $6.00

CRUICKSHANK'S PHOTOGRAPHS OF BIRDS OF AMERICA, Allan D. Cruickshank. Great ornithologist, photographer presents 177 closeups, groupings, panoramas, flightings, etc., of about 150 different birds. Expanded *Wings in the Wilderness*. Introduction by Helen G. Cruickshank. 191pp. 8¼ x 11. 23497-5 Pa. $7.95

AMERICAN WILDLIFE AND PLANTS, A. C. Martin, et al. Describes food habits of more than 1000 species of mammals, birds, fish. Special treatment of important food plants. Over 300 illustrations. 500pp. 5⅜ x 8½. 20793-5 Pa. $6.50

THE PEOPLE CALLED SHAKERS, Edward D. Andrews. Lifetime of research, definitive study of Shakers: origins, beliefs, practices, dances, social organization, furniture and crafts, impact on 19th-century USA, present heritage. Indispensable to student of American history, collector. 33 illustrations. 351pp. 5⅜ x 8½. 21081-2 Pa. $4.50

OLD NEW YORK IN EARLY PHOTOGRAPHS, Mary Black. New York City as it was in 1853-1901, through 196 wonderful photographs from N.-Y. Historical Society. Great Blizzard, Lincoln's funeral procession, great buildings. 228pp. 9 x 12. 22907-6 Pa. $8.95

MR. LINCOLN'S CAMERA MAN: MATHEW BRADY, Roy Meredith. Over 300 Brady photos reproduced directly from original negatives, photos. Jackson, Webster, Grant, Lee, Carnegie, Barnum; Lincoln; Battle Smoke, Death of Rebel Sniper, Atlanta Just After Capture. Lively commentary. 368pp. 8⅜ x 11¼. 23021-X Pa. $11.95

TRAVELS OF WILLIAM BARTRAM, William Bartram. From 1773-8, Bartram explored Northern Florida, Georgia, Carolinas, and reported on wild life, plants, Indians, early settlers. Basic account for period, entertaining reading. Edited by Mark Van Doren. 13 illustrations. 141pp. 5⅜ x 8½. 20013-2 Pa. $6.00

THE GENTLEMAN AND CABINET MAKER'S DIRECTOR, Thomas Chippendale. Full reprint, 1762 style book, most influential of all time; chairs, tables, sofas, mirrors, cabinets, etc. 200 plates, plus 24 photographs of surviving pieces. 249pp. 9⅞ x 12¾. 21601-2 Pa. $8.95

AMERICAN CARRIAGES, SLEIGHS, SULKIES AND CARTS, edited by Don H. Berkebile. 168 Victorian illustrations from catalogues, trade journals, fully captioned. Useful for artists. Author is Assoc. Curator, Div. of Transportation of Smithsonian Institution. 168pp. 8½ x 9½. 23328-6 Pa. $5.00

SECOND PIATIGORSKY CUP, edited by Isaac Kashdan. One of the greatest tournament books ever produced in the English language. All 90 games of the 1966 tournament, annotated by players, most annotated by both players. Features Petrosian, Spassky, Fischer, Larsen, six others. 228pp. 5⅜ x 8½. 23572-6 Pa. $3.50

ENCYCLOPEDIA OF CARD TRICKS, revised and edited by Jean Hugard. How to perform over 600 card tricks, devised by the world's greatest magicians: impromptus, spelling tricks, key cards, using special packs, much, much more. Additional chapter on card technique. 66 illustrations. 402pp. 5⅜ x 8½. (Available in U.S. only) 21252-1 Pa. $5.95

MAGIC: STAGE ILLUSIONS, SPECIAL EFFECTS AND TRICK PHO-TOGRAPHY, Albert A. Hopkins, Henry R. Evans. One of the great classics; fullest, most authoritative explanation of vanishing lady, levitations, scores of other great stage effects. Also small magic, automata, stunts. 446 illustrations. 556pp. 5⅜ x 8½. 23344-8 Pa. $6.95

THE SECRETS OF HOUDINI, J. C. Cannell. Classic study of Houdini's incredible magic, exposing closely-kept professional secrets and revealing, in general terms, the whole art of stage magic. 67 illustrations. 279pp. 5⅜ x 8½. 22913-0 Pa. $4.00

HOFFMANN'S MODERN MAGIC, Professor Hoffmann. One of the best, and best-known, magicians' manuals of the past century. Hundreds of tricks from card tricks and simple sleight of hand to elaborate illusions involving construction of complicated machinery. 332 illustrations. 563pp. 5⅜ x 8½. 23623-4 Pa. $6.95

THOMAS NAST'S CHRISTMAS DRAWINGS, Thomas Nast. Almost all Christmas drawings by creator of image of Santa Claus as we know it, and one of America's foremost illustrators and political cartoonists. 66 illustrations. 3 illustrations in color on covers. 96pp. 8⅜ x 11¼. 23660-9 Pa. $3.50

FRENCH COUNTRY COOKING FOR AMERICANS, Louis Diat. 500 easy-to-make, authentic provincial recipes compiled by former head chef at New York's Fitz-Carlton Hotel: onion soup, lamb stew, potato pie, more. 309pp. 5⅜ x 8½. 23665-X Pa. $3.95

SAUCES, FRENCH AND FAMOUS, Louis Diat. Complete book gives over 200 specific recipes: bechamel, Bordelaise, hollandaise, Cumberland, apricot, etc. Author was one of this century's finest chefs, originator of vichyssoise and many other dishes. Index. 156pp. 5⅜ x 8. 23663-3 Pa. $2.75

TOLL HOUSE TRIED AND TRUE RECIPES, Ruth Graves Wakefield. Authentic recipes from the famous Mass. restaurant: popovers, veal and ham loaf, Toll House baked beans, chocolate cake crumb pudding, much more. Many helpful hints. Nearly 700 recipes. Index. 376pp. 5⅜ x 8½. 23560-2 Pa. $4.95

ILLUSTRATED GUIDE TO SHAKER FURNITURE, Robert Meader. Director, Shaker Museum, Old Chatham, presents up-to-date coverage of all furniture and appurtenances, with much on local styles not available elsewhere. 235 photos. 146pp. 9 x 12. 22819-3 Pa. $6.95

COOKING WITH BEER, Carole Fahy. Beer has as superb an effect on food as wine, and at fraction of cost. Over 250 recipes for appetizers, soups, main dishes, desserts, breads, etc. Index. 144pp. 5⅜ x 8½. (Available in U.S. only) 23661-7 Pa. $3.00

STEWS AND RAGOUTS, Kay Shaw Nelson. This international cookbook offers wide range of 108 recipes perfect for everyday, special occasions, meals-in-themselves, main dishes. Economical, nutritious, easy-to-prepare: goulash, Irish stew, boeuf bourguignon, etc. Index. 134pp. 5⅜ x 8½. 23662-5 Pa. $3.95

DELICIOUS MAIN COURSE DISHES, Marian Tracy. Main courses are the most important part of any meal. These 200 nutritious, economical recipes from around the world make every meal a delight. "I . . . have found it so useful in my own household,"—*N.Y. Times.* Index. 219pp. 5⅜ x 8½. 23664-1 Pa. $3.95

FIVE ACRES AND INDEPENDENCE, Maurice G. Kains. Great back-to-the-land classic explains basics of self-sufficient farming: economics, plants, crops, animals, orchards, soils, land selection, host of other necessary things. Do not confuse with skimpy faddist literature; Kains was one of America's greatest agriculturalists. 95 illustrations. 397pp. 5⅜ x 8½. 20974-1 Pa. **$4.95**

A PRACTICAL GUIDE FOR THE BEGINNING FARMER, Herbert Jacobs. Basic, extremely useful first book for anyone thinking about moving to the country and starting a farm. Simpler than Kains, with greater emphasis on country living in general. 246pp. 5⅜ x 8½. 23675-7 Pa. $3.95

PAPERMAKING, Dard Hunter. Definitive book on the subject by the foremost authority in the field. Chapters dealing with every aspect of history of craft in every part of the world. Over 320 illustrations. 2nd, revised and enlarged (1947) edition. 672pp. 5⅜ x 8½. 23619-6 Pa. $8.95

THE ART DECO STYLE, edited by Theodore Menten. Furniture, jewelry, metalwork, ceramics, fabrics, lighting fixtures, interior decors, exteriors, graphics from pure French sources. Best sampling around. Over 400 photographs. 183pp. 8⅜ x 11¼. 22824-X Pa. $6.95

ACKERMANN'S COSTUME PLATES, Rudolph Ackermann. Selection of 96 plates from the *Repository of Arts,* best published source of costume for English fashion during the early 19th century. 12 plates also in color. Captions, glossary and introduction by editor Stella Blum. Total of 120pp. 8⅜ x 11¼. 23690-0 Pa. $5.00

CATALOGUE OF DOVER BOOKS

THE ANATOMY OF THE HORSE, George Stubbs. Often considered the great masterpiece of animal anatomy. Full reproduction of 1766 edition, plus prospectus; original text and modernized text. 36 plates. Introduction by Eleanor Garvey. 121pp. 11 x 14¾. 23402-9 Pa. $8.95

BRIDGMAN'S LIFE DRAWING, George B. Bridgman. More than 500 illustrative drawings and text teach you to abstract the body into its major masses, use light and shade, proportion; as well as specific areas of anatomy, of which Bridgman is master. 192pp. 6½ x 9¼. (Available in U.S. only) 22710-3 Pa. $4.50

ART NOUVEAU DESIGNS IN COLOR, Alphonse Mucha, Maurice Verneuil, Georges Auriol. Full-color reproduction of Combinaisons ornementales (c. 1900) by Art Nouveau masters. Floral, animal, geometric, interlacings, swashes—borders, frames, spots—all incredibly beautiful. 60 plates, hundreds of designs. 9⅜ x 8-1/16. 22885-1 Pa. $4.50

FULL-COLOR FLORAL DESIGNS IN THE ART NOUVEAU STYLE, E. A. Seguy. 166 motifs, on 40 plates, from Les fleurs et leurs applications decoratives (1902): borders, circular designs, repeats, allovers, "spots." All in authentic Art Nouveau colors. 48pp. 9⅜ x 12¼. 23439-8 Pa. $6.00

A DIDEROT PICTORIAL ENCYCLOPEDIA OF TRADES AND IN-DUSTRY, edited by Charles C. Gillispie. 485 most interesting plates from the great French Encyclopedia of the 18th century show hundreds of working figures, artifacts, process, land and cityscapes; glassmaking, papermaking, metal extraction, construction, weaving, making furniture, clothing, wigs, dozens of other activities. Plates fully explained. 920pp. 9 x 12. 22284-5, 22285-3 Clothbd., Two-vol. set $50.00

HANDBOOK OF EARLY ADVERTISING ART, Clarence P. Hornung. Largest collection of copyright-free early and antique advertising art ever compiled. Over 6,000 illustrations, from Franklin's time to the 1890's for special effects, novelty. Valuable source, almost inexhaustible.
Pictorial Volume. Agriculture, the zodiac, animals, autos, birds, Christmas, fire engines, flowers, trees, musical instruments, ships, games and sports, much more. Arranged by subject matter and use. 237 plates. 288pp. 9 x 12. 20122-8 Clothbd. $15.00

Typographical Volume. Roman and Gothic faces ranging from 10 point to 300 point, "Barnum," German and Old English faces, script, logotypes, scrolls and flourishes, 1115 ornamental initials, 67 complete alphabets, more. 310 plates. 320pp. 9 x 12. 20123-6 Clothbd. $15.00

CALLIGRAPHY (CALLIGRAPHIA LATINA), J. G. Schwandner. High point of 18th-century ornamental calligraphy. Very ornate initials, scrolls, borders, cherubs, birds, lettered examples. 172pp. 9 x 13. 20475-8 Pa. $7.95

GEOMETRY, RELATIVITY AND THE FOURTH DIMENSION, Rudolf Rucker. Exposition of fourth dimension, means of visualization, concepts of relativity as Flatland characters continue adventures. Popular, easily followed yet accurate, profound. 141 illustrations. 133pp. 5⅜ x 8½.

23400-2 Pa. $2.75

THE ORIGIN OF LIFE, A. I. Oparin. Modern classic in biochemistry, the first rigorous examination of possible evolution of life from nitrocarbon compounds. Non-technical, easily followed. Total of 295pp. 5⅜ x 8½.

60213-3 Pa. $5.95

PLANETS, STARS AND GALAXIES, A. E. Fanning. Comprehensive introductory survey: the sun, solar system, stars, galaxies, universe, cosmology; quasars, radio stars, etc. 24pp. of photographs. 189pp. 5⅜ x 8½. (Available in U.S. only)

21680-2 Pa. $3.75

THE THIRTEEN BOOKS OF EUCLID'S ELEMENTS, translated with introduction and commentary by Sir Thomas L. Heath. Definitive edition. Textual and linguistic, notes, mathematical analysis, 2500 years of critical commentary. Do not confuse with abridged school editions. Total of 1414pp. 5⅜ x 8½.   60088-2, 60089-0, 60090-4 Pa., Three-vol. set $19.50

Prices subject to change without notice.

Available at your book dealer or write for free catalogue to Dept. GI, Dover Publications, Inc., 31 East 2nd St. Minieola., N.Y. 11501. Dover publishes more than 175 books each year on science, elementary and advanced mathematics, biology, music, art, literary history, social sciences and other areas.